GAVIN LINGIAH
7, GLENWOOD GARDENS
LENZIE
GLASGOW
G66 4JP

CW01238759

Arabidopsis

The Practical Approach Series

SERIES EDITOR

B. D. HAMES
Department of Biochemistry and Molecular Biology
University of Leeds, Leeds LS2 9JT, UK

See also the Practical Approach web site at **http://www.oup.co.uk/PAS**
★ **indicates new and forthcoming titles**

Affinity Chromatography
Affinity Separations
Anaerobic Microbiology
Animal Cell Culture (2nd edition)
Animal Virus Pathogenesis
Antibodies I and II
Antibody Engineering
Antisense Technology
★ Apoptosis
Applied Microbial Physiology
★ Arabidopsis
Basic Cell Culture
Behavioural Neuroscience
Bioenergetics
Biological Data Analysis
Biomechanics—Materials
Biomechanics—Structures and Systems
Biosensors
★ C. Elegans
Carbohydrate Analysis (2nd edition)
Cell-Cell Interactions

The Cell Cycle
★ Cell Growth, Differentiation and Senescence
★ Cell Separation
Cellular Calcium
Cellular Interactions in Development
Cellular Neurobiology
Chromatin
★ Chromosome Structural Analysis
Clinical Immunology
Complement
★ Crystallization of Nucleic Acids and Proteins (2nd edition)
Cytokines (2nd edition)
Cytoskeleton
Cytoskeleton: signalling and cell regulation
Diagnostic Molecular Pathology I and II
DNA and Protein Sequence Analysis
DNA Cloning 1: Core Techniques (2nd edition)

DNA Cloning 2: Expression Systems (2nd edition)

DNA Cloning 3: Complex Genomes (2nd edition)

DNA Cloning 4: Mammalian Systems (2nd edition)

★ DNA Microarrays

★ DNA Viruses

Drosophila (2nd edition)

Electron Microscopy in Biology

Electron Microscopy in Molecular Biology

Electrophysiology

Enzyme Assays

Epithelial Cell Culture

Essential Developmental Biology

Essential Molecular Biology I and II

★ Eukaryotic DNA Replication

Experimental Neuroanatomy

Extracellular Matrix

Flow Cytometry (2nd edition)

Fmoc solid phase peptide synthesis

Free Radicals

Gel Electrophoresis of Nucleic Acids (2nd edition)

★ Gel Electrophoresis of Proteins (3rd edition)

Gene Probes 1 and 2

★ Gene Targeting (2nd edition)

Gene Transcription

Genome Mapping

Glycobiology

Growth Factors and Receptors

Haemopoiesis

★ High Resolution Chromotography

Histocompatibility Testing

HIV Volumes 1 and 2

HPLC of Macromolecules (2nd edition)

Human Cytogenetics I and II (2nd edition)

Human Genetic Disease Analysis

★ Image Processing and Analysis

★ Immobilized Biomolecules in Analysis

Immunochemistry 1

Immunochemistry 2

Immunocytochemistry

★ Imnundiagnostics

★ *In Situ* Hybridization (2nd edition)

Iodinated Density Gradient Media

Ion Channels

★ Light Microscopy (2nd edition)

Lipid Modification of Proteins

Lipoprotein Analysis

Liposomes

★ Lymphocytes (2nd edition)

Mammalian Cell Biotechnology

Medical Parasitology

Medical Virology

MHC Volumes 1 and 2

Molecular Genetic Analysis of Populations (2nd edition)

Molecular Genetics of Yeast

Molecular Imaging in Neuroscience
Molecular Plant Pathology I and II
Molecular Virology
Monitoring Neuronal Activity
★ Mouse Genetics and Transgenics
Mutagenicity Testing
Mutation Detection
Neural Cell Culture
Neural Transplantation
Neurochemistry (2nd edition)
Neuronal Cell Lines
NMR of Biological Macromolecules
Non-isotopic Methods in Molecular Biology
Nucleic Acid Hybridisation
★ Nuclear Receptors
Oligonucleotides and Analogues
Oligonucleotide Synthesis
PCR 1
PCR 2
★ PCR 3: PCR In Situ Hybridization
Peptide Antigens
Photosynthesis: Energy Transduction
Plant Cell Biology
Plant Cell Culture (2nd edition)
Plant Molecular Biology
Plasmids (2nd edition)
Platelets
Postimplantation Mammalian Embryos
★ Post-translational Processing
Preparative Centrifugation
Protein Blotting
★ Protein Expression
Protein Engineering
Protein Function (2nd edition)
★ Protein Localization by Fluorescence Microscopy
★ Protein Phosphorylation (2nd edition)
Protein Purification Applications
Protein Purification Methods
Protein Sequencing
Protein Structure (2nd edition)
Protein Structure Prediction
Protein Targeting
Proteolytic Enzymes
Pulsed Field Gel Electrophoresis
RNA Processing I and II
RNA-Protein Interactions
Signalling by Inositides
★ Signal Transduction (2nd edition)
Subcellular Fractionation
Signal Transduction
★ Transcription Factors (2nd edition)
Tumour Immunobiology
★ Virus Culture

Arabidopsis
A Practical Approach

Edited by
ZOE A. WILSON
*Plant Science Division,
School of Biological Sciences,
University of Nottingham*

OXFORD
UNIVERSITY PRESS

Great Clarendon Street, Oxford OX2 6DP
Oxford University Press is a department of the University of Oxford
and furthers the University's aim of excellence in research, scholarship,
and education by publishing worldwide in
Oxford New York
Athens Auckland Bangkok Bogotá Buenos Aires Calcutta
Cape Town Chennai Dar es Salaam Delhi Florence Hong Kong Istanbul
Karachi Kuala Lumpur Madrid Melbourne Mexico City Mumbai
Nairobi Paris São Paulo Singapore Taipei Tokyo Toronto Warsaw
and associated companies in Berlin Ibadan

Oxford is a registered trade mark of Oxford University Press

Published in the United States
by Oxford University Press Inc., New York

© Oxford University Press, 2000

All rights reserved. No part of this publication may be reproduced,
stored in a retrieval system, or transmitted, in any form or by any means,
without the prior permission in writing of Oxford University Press.
Within the UK, exceptions are allowed in respect of any fair dealing for the
purpose of research or private study, or criticism or review, as permitted
under the Copyright, Designs and Patents Act, 1988, or in the case
of reprographic reproduction in accordance with the terms of licences
issued by the Copyright Licensing Agency. Enquiries concerning
reproduction outside those terms and in other countries should be
sent to the Rights Department, Oxford University Press,
at the address above.

This book is sold subject to the condition that it shall not, by way
of trade or otherwise, be lent, re-sold, hired out, or otherwise circulated
without the publisher's prior consent in any form of binding or cover
other than that in which it is published and without a similar condition
including this condition being imposed on the subsequent purchaser

Users of books in the Practical Approach Series are advised that prudent
laboratory safety procedures should be followed at all times. Oxford
University Press makes no representation, express or implied, in respect of
the accuracy of the material set forth in books in this series and cannot
accept any legal responsibility or liability for any errors or omissions
that may be made.

A catalogue record for this book is available from the British Library

Library of Congress Cataloging in Publication Data
Arabidopsis : a practical approach / edited by Zoe A. Wilson
(The practical approach series ; PAS/223)
Includes bibliographical references and index.
1. Arabidopsis–Laboratory manuals. 2. Arabidopsis–Molecular aspects–Laboratory
manuals. I. Wilson, Zoe A. II. Practical approach series ; 223.
QK495.C9 A687 2000   583'.64–dc21   99-045851

ISBN 0-19-963565-X (Hbk)
0-19-963564-1 (Pbk)

Typeset by Footnote Graphics,
Warminster, Wilts
Printed in Great Britain by Information Press, Ltd,
Eynsham, Oxon.

Preface

Over the past fifteen years *Arabidopsis thaliana* has changed from being an insignificant weed, to one of the most cultivated laboratory plants around the world. This increase in popularity can be put down to its small genome, low levels of repetitive DNA, small size and fast generation time. As such it is an ideal molecular genetic tool for the analysis of development in higher plants. The acknowledgement of the importance of a model plant species was brought about by the desire of researchers around the world to come together in an International effort which will allow the advancement of Plant Sciences across all plant species.

This book provides an introduction to the key techniques required for the use of *Arabidopsis* as an experimental system. It gives a basic introduction to the optimal growth conditions and the genetic resources available for *Arabidopsis*, how this material should be handled, maintained and used. Individual chapters describe strategies for the identification, mapping and characterization of different mutants by microscopy, molecular cytogenetics and gene expression analysis. Different cloning strategies, using transposons, T-DNA and map position are described in detail and provide a means to generate, identify and characterise genes of developmental interest.

I would like to thank all the contributors for their help and patience in completing their chapters and revealing their innermost laboratory secrets. I would also like to thank Dr. Bernard Mulligan who instigated most of the original ground work for this book.

University of Nottingham Z.A.W
September 1999

Contents

List of Contributors   *xv*
Abbreviations   *xvii*

1. Growth, maintenance, and use of *Arabidopsis* genetic resources   1

Mary Anderson and Fiona Wilson

1. What is *Arabidopsis*?   1
2. What makes *Arabidopsis* such an attractive experimental model?   3
3. *Arabidopsis* genetic resource centres   4
   - *Arabidopsis* genetic resources   4
   - Accessing *Arabidopsis* resources   5
4. Mutants of *Arabidopsis*   5
   - Single gene mutation lines   5
   - Resources for the identification/investigation of novel genes   8
   - Mapping tools   10
5. Considerations of available resources for identifying novel genes   11
   - Forward genetics   12
   - Reverse genetics   14
6. Growing *Arabidopsis*   15
   - How to maintain clean growth conditions   15
   - Growing *Arabidopsis* in the glasshouse   16
   - Chemical control of pests and diseases   19
7. Seed storage   21
8. Growing *Arabidopsis* with specific growth requirements   23
9. Sterile culture of *Arabidopsis*   24

   References   26

2. Preservation and handling of stock centre clones   29

Randy Scholl, Keith Davis, and Doreen Ware

1. Introduction   29

Contents

2. Missions of a plant DNA resource centre — 29

3. Preservation of stocks — 30
  Plasmids with small DNA inserts — 30
  Cosmids — 32
  Phage and phage libraries — 33
  Yeast artificial chromosome (YAC) libraries — 33
  Pools of YAC library cells for PCR screening — 37
  Distribution of YAC libraries arrayed on nylon filters — 38
  Other large-insert libraries — 38
  Yeast expression analysis—'two-hybrid' libraries and complementation testing — 42

4. Verification of stock identity and purity — 42

5. Pooled DNA from T-DNA lines for PCR screening — 44

6. Organization of stock information — 48
  Collecting, maintaining, and disseminating stock data — 48
  Organizing and distributing patron data — 48

7. The future — 48

References — 49

3. Genetic mapping using recombinant inbred lines — 51

Clare Lister, Mary Anderson, and Caroline Dean

1. Introduction — 51

2. Preparation and digestion of *A. thaliana* genomic DNA — 54
  Preparation of genomic DNA — 56
  Identifying an RFLP — 60
  Southern blotting and hybridization — 61

3. Polymorphic markers — 62
  RFLP markers — 62
  PCR-based markers — 66
  Phenotypic and biochemical markers — 68

4. Calculating map positions — 69
  Mapping programs — 69
  NASC mapping service — 70

5. Integration of a mutation into a molecular map — 72

References — 74

Contents

4. *Arabidopsis* mutant characterization; microscopy, mapping, and gene expression analysis   77

Kriton Kalantidis, L. Greg Briarty, and Zoe A. Wilson

1. Generation of mutants and their importance for developmental biology	77
2. Mapping and segregation analysis	78
Mapping of mutations	78
Influence of environment on phenotype	82
3. Microscopy	82
Fresh material characterization	82
Fixed material characterization	84
4. Analysis of plant gene expression	89
RNA isolation	89
Northern analysis	91
In situ hybridizations	92
References	103

5. Classical and molecular cytogenetics of *Arabidopsis*   105

G. H. Jones and J. S. Heslop-Harrison

1. Introduction	105
2. Mitotic chromosome analysis by light microscopy	108
3. *In situ* hybridization to mitotic chromosome preparations	112
Photography of *in situ* hybridizations	117
4. Meiotic chromosome analysis by light microscopy	117
5. Meiotic chromosome analysis by electron microscopy	121
References	123

6. Tissue culture, transformation, and transient gene expression in *Arabidopsis*   125

Keith Lindsey and Wenbin Wei

1. Introduction	125
2. Stable transformation by *Agrobacterium tumefaciens*	128

Contents

   3. Transient gene expression in *Arabidopsis* protoplasts   131
      Reporter gene enzyme assays   134

  Acknowledgements   139

  References   139

7. Transposon and T-DNA mutagenesis   143

Mark G. M. Aarts, Csaba Koncz, and Andy Pereira

  1. Introduction   143

  2. Transposon tagging   143
      Endogenous transposable elements   143
      Transposon tagging systems in *Arabidopsis*   144
      Which system to use?   149
      Genetic and molecular analysis of a putatively transposon tagged mutant   151
      Further applications of transposon tagging   156

  3. T-DNA tagging   158
      The use of T-DNA as insertional mutagen   158
      Random tagging   158
      Available populations of T-DNA transformants   159
      Promoter/enhancer trapping   160
      Analysis of T-DNA mutants and cloning a tagged gene   161
      Further applications of T-DNA tagging   166

  References   166

8. Map-based cloning in *Arabidopsis*   171

Joanna Putterill and George Coupland

  1. Introduction   171

  2. Locating the mutation of interest relative to DNA markers   172
      Determining an approximate map position   172
      Identifying a short genetic interval containing the mutation as a prelude to isolating the gene   175

  3. Placing the gene on the physical map   177
      Chromosome landing   177
      Chromosome walking with YAC clones   178

  4. Identification of the gene   189
      Location of the gene by molecular complementation   189
      Determining the structure of the gene   194

  5. Perspectives   194

  References   195

9. Physical mapping: YACs, BACs, cosmids, and nucleotide sequences   199

Ian Bancroft

1. Introduction   199

2. Genome mapping with YAC clones   199

3. Genome mapping with BAC and P1 clones   202
- Communal resources   202
- Construction of BAC libraries   208
- Genome mapping with BACs   211

4. High resolution mapping with cosmids   216
- Approaches to mapping with cosmids   216
- Construction of cosmid libraries   216

5. Nucleotide sequences—the ultimate mapping tool   221
- The EST sequencing project   221
- The genome sequencing project   221
- Sequence-based mapping   222

References   223

10. Web-based bioinformatic tools for *Arabidopsis* researchers   225

Seung Y. Rhee and David J. Flanders

1. Introduction   225
- What is bioinformatics   225
- Sources of *Arabidopsis* bioinformatic data   226

2. Basic tools for the Internet   227
- Web basics   227
- Getting onto the Web   228
- Using your browser   229
- Browser tips and errors   232
- Privacy issues   235

3. Scenarios of bioinformatic use in *Arabidopsis* research   235

4. Gene information resources   237
- General gene information   237
- *Arabidopsis* gene information   237
- Plant gene information   238

5. Maps   239
- Genetic maps   239
- Physical maps   240

6. Sequencing   242

Contents

The *Arabidopsis* genome initiative (AGI)	243
Annotation of sequences by AGI	243
Caveats in annotation	244
Sequence contigs from AtDB	245

7. Sequence analysis tools — 245
- *BLAST* — 246
- *FASTA* — 250
- EST databases — 252
- Gene identification programs — 253
- Gene family analyses — 255
- Motif analyses — 255
- Protein structures — 258
- Comprehensive sequence analysis tools — 259

8. Current issues and future directions in bioinformatics — 260
- Some important bioinformatic issues — 260
- Bioinformatic tools currently under development for *Arabidopsis* research

9. Conclusion — 261

Acknowledgements — 261

References — 261

Appendix 1 — 263

Glossary — 264

Appendix — 267
Index — 273

Contributors

MARK G. M. AARTS
CPRO-DLO, Postbus 16, 6700 AA Wageningen, The Netherlands.

MARY ANDERSON
John Innes Centre, Colney Lane, Norwich NR4 7UH, UK.

IAN BANCROFT
John Innes Centre, Colney Lane, Norwich NR4 7UH, UK.

L. GREG BRIARTY
Plant Science Division, School of Biological Sciences, The University of Nottingham, University Park, Nottingham NG7 2RD, UK.

GEORGE COUPLAND
John Innes Centre, Colney Lane, Norwich NR4 7UH, UK.

KEITH DAVIS
Arabidopsis Biological Resource Center, The Ohio State University, Columbus, OH, USA.

CAROLINE DEAN
John Innes Centre, Colney Lane, Norwich NR4 7UH, UK.

DAVID J. FLANDERS
PBI Cambridge, Maris Lane, Trumpington, Cambridge CB2 2LQ, UK.

J. S. HESLOP-HARRISON
Karyobiology Group, John Innes Centre, Colney Lane, Norwich NR4 7UH, UK.

G. H. JONES
School of Biological Sciences, University of Birmingham, Birmingham B15 2TT, UK.

KRITON KALANTIDIS
Plant Molecular Biology Unit, Institute of Molecular Biology and Biotechnology, The Forth Institute, Crete.

CSABA KONCZ
MPI für Züchtungsforschung, Carl-von-Linné-Weg 10, D-50829, Köln, Germany.

KEITH LINDSEY
Department of Biological Sciences, University of Durham, South Road, Durham DH1 3LE, UK.

Contributors

CLARE LISTER
John Innes Centre, Colney Lane, Norwich NR4 7UH, UK.

ANDY PEREIRA
CPRO-DLO, Postbus 16, 6700 AA Wageningen, The Netherlands.

JOANNA PUTTERILL
School of Biological Sciences, University of Auckland, Private Bag 92019, Auckland, New Zealand.

SEUNG Y. RHEE
Department of Genetics, School of Medicine, Stanford University, Stanford, CA 94305–5120, USA.

RANDY SCHOLL
Arabidopsis Biological Resource Center, The Ohio State University, 1735 Neil Avenue, 309 Botany and Zoology Building, Columbus, OH, USA.

DOREEN WARE
Arabidopsis Biological Resource Center, The Ohio State University, 1735 Neil Avenue, 309 Botany and Zoology Building, Columbus, OH, USA.

WENBIN WEI
Department of Biological Sciences, University of Durham, South Road, Durham DH1 3LE, UK.

FIONA WILSON
Nottingham Arabidopsis Stock Centre, Plant Science Division, School of Biological Sciences, The University of Nottingham, University Park, Nottingham NG7 2RD, UK.

ZOE A. WILSON
Plant Science Division, School of Biological Sciences, The University of Nottingham, University Park, Nottingham NG7 2RD, UK.

Abbreviations

2,4-D	2,4-dichlorophenoxyacetic acid
2-ip	2-isopentenyladenine
A_{260}	absorbance (260 nm)
A_{280}	absorbance (280 nm)
ABRC	*Arabidopsis* Biological Research Centre
Ac	Activator
AFLP	amplified fragment length polymorphism
AGI	*Arabidopsis* genome initiative
AGR	*Arabidopsis* Genome Resources
AIMS	*Arabidopsis* Information Management System
ARMS	*Arabidopsis* RFLP mapping set
AtDB	*Arabidopsis thaliana* database
ATGC	*Arabidopsis thaliana* Genome Centre
ATP	adenosine triphosphate
BAC	bacterial artificial chromosome
BLAST	basic local alignment search tool
bp	base pair
BSA	bovine serum albumin
CAPS	cleaved amplified polymorphic DNAs
cDNA	complementary DNA
CIAP	calf intestinal alkaline phosphatase
CIM	callus inducing medium
cM	centimorgan
Col	Columbia
CTAB	cetyltrimethylammonium bromide
Cvi	Cape Verde Islands
DAPI	4′,6-diamidino-2-phenylindole
dCTP	deoxycytidine triphosphate
ddW	double distilled water
DMSO	dimethyl sulfoxide
DNA	deoxyribonucleic acid
DNase	deoxyribonuclease
dNTP	deoxynucleoside triphosphate
Ds	Dissociation
dSpm	defective Suppressor-mutator
DTT	dithiothreitol
EDTA	ethylenediaminetetraacetic acid
EM	electron microscopy
EMS	ethylmethane sulfonate
En	Enhancer

Abbreviations

EST	expressed sequence tag
FISH	fluorescence *in situ* hybridization
fm	femtomole
FTP	file transfer protocol
FW	fresh weight
GDE	Genetic Data Environment
GFP	green fluorescent protein
GMI	germination medium I
GUS	β-glucuronidase
HPT	hygromycin phosphotransferase
HSP	high scoring segment pair
hyg^r	hygromycin resistant
I	Inhibitor element
IAA	indole-3-acetic acid
IP	Internet protocol
IPCR	inverse polymerase chain reaction
IPTG	isopropyl-β-D-thiogalactopyranoside
kb	kilobase pair
LA-IPCR	long-range IPCR
L*er*	Landsberg *erecta*
LLR	log likelihood ratio
LM	light microscopy
LMP	low melting point
LUC	luciferase
MAFF	Ministry of Farming and Fisheries
Mb	megabase pair
MES	2-(*N*-morpholino)ethanesulfonic acid
MOPS	morpholinopropanesulfonic acid
mRNA	messenger RNA
MSU	Michigan State University
MU	4-methylumbelliferone
MUG	4-methylumbelliferyl glucuronide
NASC	Nottingham *Arabidopsis* Stock Centre
NBT	nitro blue tetrazolium salt
NCBI	National Centre for Biotechnology Information
NM	nitrosomethyl biuret
NMU	nitrosomethyl urea
NOR	nucleolar organizer region
NPG	*NetPlantGene*
ORF	open reading frame
PBS	phosphate-buffered saline
PCR	polymerase chain reaction
PEG	polyethylene glycol
PFGE	pulsed-field gel electrophoresis

Abbreviations

PIR	International Protein Sequence Database
PMSF	phenylmethylsulfonyl fluoride
RAPD	random amplified polymorphic DNA
r.c.f.	relative centrifugal force
RFLP	restriction fragment length polymorphism
RI	recombinant inbred
RIL	recombinant inbred lines
RNase	ribonuclease
RTF	rich text format
SAM	*S*-adenosylmethionine
SC	synaptonemal complexes
SCOP	Structural Classification of Proteins
SDS	sodium dodecyl sulfate
SEM	shoot elongation medium
SIM	shoot inducing medium
SOM	shoot overlay medium
Spm	Supressor-mutator
SRS	Sequence Retrieval System
SSC	standard saline citrate
SSLP	simple sequence length polymorphism
SSPE	standard saline/phosphate/EDTA
strep^r	streptomycin resistant
TAC	transformable artificial chromosome
TBE	Tris/borate/EDTA buffer
TC	tentative consensus
T-DNA	transfer DNA
TE	Tris/EDTA buffer
TIGR	The Instiue for Genomic Research
TIR	terminal inverted repeat
tRNA	transfer RNA
U.Minn	University of Minnesota
UV	ultraviolet
VAST	Vector Alignment Search Tool
WS	Wassilewskija
WT	wild-type (normal)
X-gluc	5-bromo-4-chloro-3-indoyl glucuronide
X-phosphate	5-bromo-4-chloro-3-indoyl phosphate
YAC	yeast artificial chromosome

1

Growth, maintenance, and use of *Arabidopsis* genetic resources

MARY ANDERSON and FIONA WILSON

1. What is *Arabidopsis*?

The genus *Arabidopsis* belongs to the Sisymbriae tribe, a member of the Brassicaceae (mustard or crucifer) family. Several species belong to the *Arabidopsis* genus, the most well known member and the one most extensively used in research being *A. thaliana* (L.) Heynh 2n = 10 (common name thale cress or mouse eared cress). Others members include *A. griffithiana* (synonymn *A. pumila*) 2n = 32, *A. himialaica* 2n = 32, *A. lasiocarpa* 2n = 32, *A. suecica* 2n = 26, *A. wallichii*, 2n = 16, and *A. korshinskyi* 2n = 48 (1). When reference is made to *Arabidopsis*, this is to *Arabidopsis thaliana*.

Arabidopsis grows vegetatively as a small ground hugging rosette of about 2–5 cm diameter, from which a flowering stem is produced which can extend to a height of 20–70 cm. Flowers which are approximately 3 mm long and 1 mm wide, are produced at intervals as part of an inflorescence or raceme (*Figure 1*). They are typical of the crucifers producing four sepals, four petals, six stamens, and a single ovary consisting of two fused carpels. Many ecotypes produce several flowering stems which themselves can be heavily branched. *Arabidopsis* is autogamous, or self-fertile, producing several hundred seed pods or siliques on fertilization. The siliques contain about 50 seeds and shatter on ripening to allow the distribution of seed. Although *Arabidopsis* does not have any nectaries to attract insect pollination, outcrossing by insects can occur, but only at a very low frequency (2). In the commonly used rapid flowering ecotypes, e.g. Landsberg *erecta*, Columbia, the entire life cycle can be completed in six to eight weeks.

Arabidopsis occurs throughout temperate regions of Europe, Asia, and North Africa and has been widely introduced to other areas including North America and Australia (3). Its natural habitat is open free draining ground, such as sandy or gravely soils.

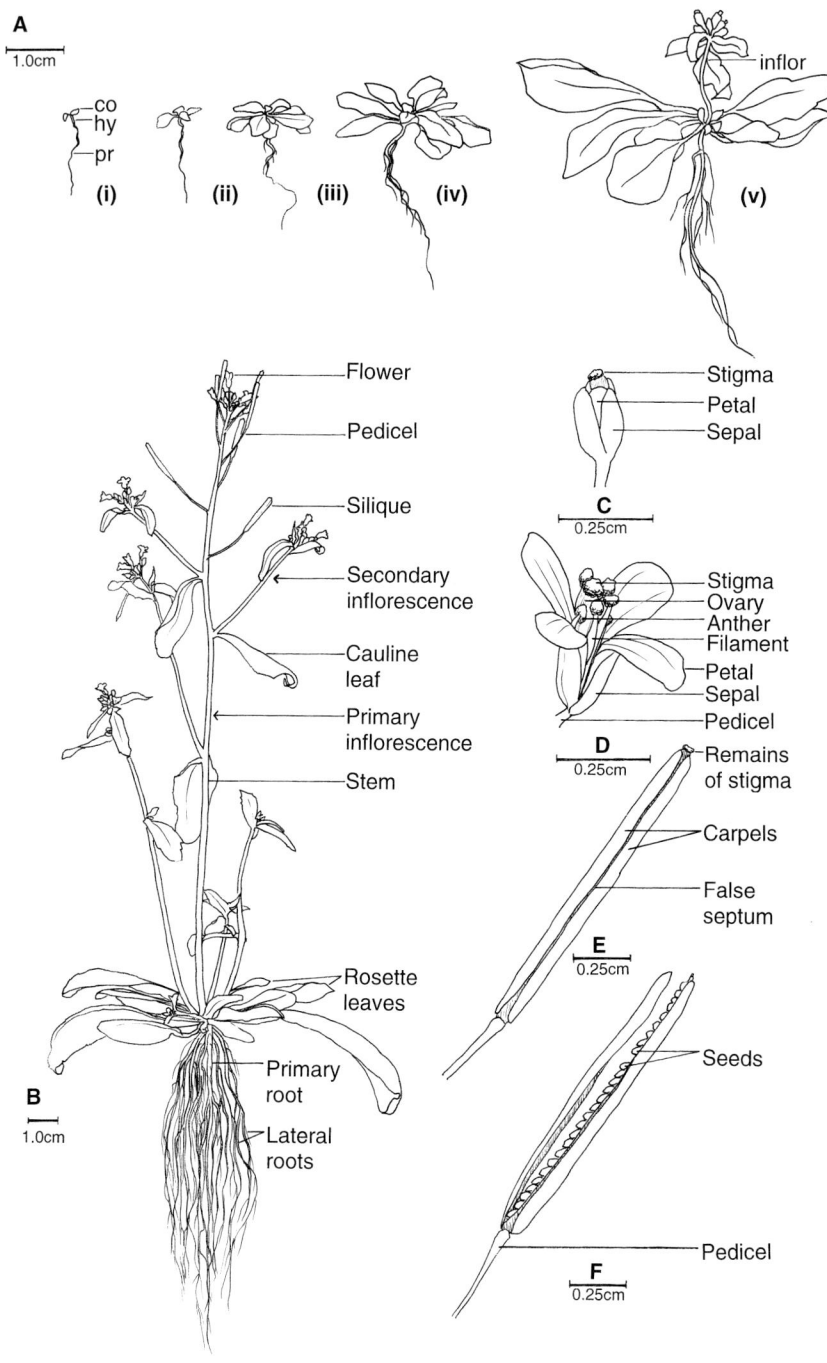

1: Growth, maintenance, and use of Arabidopsis *genetic resources*

Figure 1. (A) Developmental stages of an *Arabidopsis* (Landsberg *erecta*) plant grown at 25°C ± 2°C with a 20 h photoperiod. (i) to (v) represent stages between 2 days and 21 days post-germination. At 2 days post-germination (i) the two cotyledons (co) can be seen subtending a hypocotyl (hy) and a single primary root (pr). Transition from a juvenile vegetative meristem, producing small round leaves, to an adult vegetative meristem, producing larger more spatulate leaves generally occurs between production of leaves four and five (between stages ii–iii). The number of vegetative rosette leaves produced (iv) may vary from five to > 30 depending on the plant genotype and growth conditions. The transition to the reproductive phase of growth (v) is characterized by elongation of the internodes in the inflorescence (inflor). (B) A mature plant of *Arabidopsis* (Landsberg *erecta*) at 35 days post-germination grown at 25°C ± 2°C with an 20 h photoperiod. A primary inflorescence has developed from the apical meristem, secondary inflorescence have arisen from the axils of cauline leaves on the primary inflorescence, and tertiary inflorescence will develop from the equivalent positions on the secondary inflorescence. The structure of the complete inflorescence is termed an open raceme. Individual flowers develop in order from the base to the apice of an inflorescence. The structures of the flowers and fruits of this plant are shown in more detail in C–F. (C) A flower bud showing the sepals enfolding the petals and the gynoecium. At this stage when the stigma can be seen protruding from the centre of the bud self-pollination will already have occurred. (D) The mature flowers of *Arabidopsis* are borne on short stalks (pedicels) and have four sepals, four petals, six stamens (four long and two short), and a superior sessile gynoecium. Each stamen is characterized by the two regions; stamen filament and anther. Outwardly the gynoecium is divided into three regions; the basal ovary is subtended by a very short style at the tip of which is the stigma. The ovary consists of two carpels. (E) On a fully developed silique (pod) of *Arabidopsis* the two carpels (otherwise known as locules or valves) of the ovary can be seen to be separated by a false septum. (F) As the siliques mature and dry they turn from green to yellow/brown and then separate along the false septum releasing the seeds (typically 30–60 per silique).

2. What makes *Arabidopsis* such an attractive experimental model?

(a) In comparison to other angiosperms *Arabidopsis* has a relatively small genome with current estimates ranging from 100–120 × 10^6 bp (4). Tobacco and pea have haploid genome sizes of 1.6 × 10^9 bp and 4.5 × 10^9 bp, respectively. The size of the *Arabidopsis* genome is only 15 times that of the bacterium *Escherichia coli*, and eight times that of the yeast *Saccharomyces cerevisiae*. The main reason that the genome is so small is because there is comparatively little repetitive DNA, with over 60% of the nuclear DNA having a protein coding function (5). This is extremely important when adopting saturation mutagenesis strategies (see Chapter 7), or map-based cloning strategies (see Chapter 8), in order to identify and clone genes of interest.

(b) *Arabidopsis* has a short life cycle. It can cycle from seed to seed in six to eight weeks. Growth of the plant is also non-seasonal. Therefore, unlike other model systems such as maize, several generations can be produced in one year, which facilitates rapid genetic analysis.

(c) It is a diploid (haploid chromosome number n = 5), unlike many other experimental models such as maize and other crop plants, which are polyploid. The identification of recessive traits is therefore more straightforward and complications of gene dosage effects do not occur.

(d) It is easily transformed by *Agrobacterium tumefaciens*.

(e) Many thousands of seeds can be produced per plant which, coupled with its small size, makes it an ideal subject for mutagenesis experiments.

(f) It is self-fertile, which makes the maintenance of homozygous lines straightforward. It is also possible to perform crosses when necessary.

(g) There are a wealth of mutants that have been generated via a range of mutagenesis strategies and natural populations which have been collected in the wild that can be used as tools in the dissection of the biology of this model system.

The focus of this chapter is to outline the types of *Arabidopsis* genetic resources that are available, how they can be used, and how to obtain them. Protocols for the growth, maintenance, and mutagenesis of *Arabidopsis* are given.

3. *Arabidopsis* genetic resource centres

Arabidopsis has been the subject of research for the last fifty years, and in this time a depth of genetic diversity has been collected from the wild and generated through mutagenesis experiments. This diversity is now maintained at two *Arabidopsis* resource centres, the Nottingham *Arabidopsis* Stock Centre (NASC), The University of Nottingham, UK, and the *Arabidopsis* Biological Resource Centre (ABRC), Ohio State University, USA, which share in the task of germplasm conservation and distribution to the research community. Through this collaboration it has been possible to gather and make available over 50 000 accessions of *Arabidopsis*. Seed of some *Arabidopsis* lines can be obtained from Lehle Seeds (for example wild-type and EMS-mutated M1 and M2 seeds, strains for transformation, irradiated seeds). Details on how to contact the resource centres are given in the Appendix. Note: the *Arabidopsis* Biological Resource Centre (ABRC) is a joint seed and DNA resource centre. For information on DNA resources and protocols, please see Chapter 2.

3.1 *Arabidopsis* genetic resources

Mutants are particularly important in the dissection of physiological and developmental pathways. Genes can be identified by causing an aberration in gene function that results in a change in phenotype. Classical methods of mutagenesis have included the use of chemicals such as ethylmethane sulfonate (EMS) (6), nitrosomethyl urea (NMU), nitrosomethyl biuret (NMB), and irradiation with X-rays, fast neutrons, or heavy charged particles (7). Gene aberration can also occur through the insertion of T-DNAs or transposons within the chromosome (see Chapter 7). Novel genes can also be identified by screening wild populations or ecotypes, which are sources of niche-specific genes such as pathogen resistance, flowering time, or salt tolerance. Hence, researchers can now tap a spectrum of diversity from natural populations to populations of highly sophisticated transformed lines. Mutants that are available reflect the current research interests. They carry a wide range of aberrations affecting many developmental, biochemical, and physiological processes. Many mutants are being generated as part of individual investigations or through large mutagenesis programmes and are being shared with the research community by being placed in the resource centres. The various resources can be considered as tools for research as well as demonstration material for teaching. The potential uses of the various categories of resources are outlined in Section 4.

3.2 Accessing *Arabidopsis* resources

Full details of the 50 000 accessions maintained and distributed by NASC and ABRC are carried in their online catalogues. The information can be browsed or searched using criteria such as name, map position, phenotype, type of material. Questions about the stocks can also be sent directly to NASC or ABRC, using the stock centres' online facilities. Stocks can be ordered directly online, by fax, telephone, or letter. Seed stocks carry a handling charge. Refer to the centres' Web sites for the latest charges. Seed orders are dispatched within five working days of receiving the order.

4. Mutants of *Arabidopsis*

4.1 Single gene mutation lines

Several hundred lines that contain a single, mapped mutation are available from the stock centres. This material can be used to:

(a) Study gene function.
(b) Test for allelism with a newly identified mutation.
(c) Fine-map a new mutation.
(d) Investigate interactions between groups of mutations, such as the hormone mutants, through crossing the material and introducing one mutation into the background of another.
(e) Generate demonstration material for teaching.

Single gene mutations can result in a broad range of physiological, developmental, and morphological aberrations and are outlined below. Examples of lines commonly used in mutagenesis experiments and a selection of mutants are shown in *Figure 2*.

Mary Anderson and Fiona Wilson

1: Growth, maintenance, and use of Arabidopsis genetic resources

Figure 2. (A) Four of the *Arabidopsis thaliana* lines commonly used in research; from left to right, Lansberg *erecta* (L*er*), Columbia-4 (Col-4), Wassilewskija-2 (Ws-2), and C-24. Note the variations in whole plant morphology and in flowering time (all plants are at 35 days post-germination). (B) A *brevipedicellus (bp-1)* mutant showing typical short pedicels and downward pointing siliques. (C) A multiple mutant which has *ecer*i*ferum (cer2-2), apetala (ap2-1)*, and *brevipedicellus (bp-1)* mutations. The *cer* mutation results in a reduced, or absent, wax layer on the stems and pods which gives these structures a bright green appearance. (D) Detailed picture of an *agamous (ag-1)* mutant flower in which the whorls of the stamens and gynoecium have been replaced by extra whorls of petals and sepals. (E) Detailed picture of an *apetala (ap1-1)* mutant flower in which the petals are absent or reduced. (F) A typical wild-type *Arabidopsis* flower with four petals in a cruciform arrangement. (G) Siliques from the four lines showed in (A) (same order left to right). Note the distinctive blunt shape of the Lansberg *erecta* siliques. (H) Close-up of the upper portion of flowering stems of the lines shown in (A) and (G) (same order left to right). Note the absence of hairs on the stem of the C-24 plant. (I) Details of a fasciated stem in a Waldmeister (wam1-1) mutant plant. (J) This dwarf mutant has extremely reduced length flowering stems and short multiple leaves. Scale bar in A = 10 mm. Scale bar in B–J = 1 mm.

4.1.1 Biochemical mutants

Mutants are available that are affected in several metabolic pathways including nitrogen (8), fatty acid (9), thiamine (10), and tryptophan metabolism (11), or are deficient in leucine and valine (12) or vitamin C (13).

4.1.2 Colour mutants

Many mutations that cause a change in the colour of plant organs throughout the plant have been isolated. These include albinos (14), chlorina lines whose general phenotype is that of a yellow/green plant with the severity of the yellowness varying according to the mutant locus (15), as well as variegated lines (16).

4.1.3 Flower and flowering time mutants

Many genes play an important role in specifying floral meristem identity. This is reflected in the variety of mutations that have been isolated which are affected in flower morphology. Mutants are available that lack petals (17), or pistils (18). Other mutants are homeotic where one whorl of organs is replaced by another, such as in *agamous* where the inner whorl of stamens and gynoecium are replaced by petals and sepals (19). Other mutants include male steriles that are defective in anther or pollen development (20) and female steriles (21). Other mutants are affected in the timing of flowering, flowering either earlier or later than their wild-type counterpart (22).

4.1.4 Form mutants

This category of mutants covers a vast array of aberrations involved in many aspects of organ formation from the root (23), vegetative meristem (24), rosette (25), or trichomes (26) right through to the embryo (27), seed (28), and silique (29).

4.1.5 Hormone mutants

Mutants have been identified which have altered responses to plant hormones being either insensitive or sensitive to a particular hormone. Lines affected in abscisic acid (30), ethylene (31), gibberellin (32), auxin (33), or brassinosteroid (34) responses are all available.

4.1.6 Photomorphogenic mutants

Mutants in photoreceptors include the phytochrome (35) and blue light receptor (36) lines. Other mutants have been identified whose ability to respond normally to light have been affected (37, 38).

4.2 Resources for the identification/investigation of novel genes

4.2.1 Ecotypes

The term ecotype has been used to identify isolated populations of *Arabidopsis* that have been harvested from the wild. They are an ideal source of niche-specific genes. These lines are used to look for novel genes that result in specific environmental adaptation for instance, that convey salt or drought tolerance, disease resistance, or altered flowering time. Ecotypes have been gathered from throughout the world. The morphologies of these isolated populations are very distinct. Many hundreds of these lines are available.

4.2.2 T-DNA transformed populations

T-DNAs can cause gene aberrations through insertions into exons, promoters, or enhancer regions, and so act as insertional mutagens. Such mutations identify genes in a specific biological process whilst facilitating direct isolation of the gene or its promoter/enhancer, since they are molecularly tagged by the presence of the T-DNA. T-DNA lines have been developed by several groups using different transformation methods including seed imbibition (39), root co-cultivation (40), and vacuum infiltration of whole plants (41) with *Agrobacterium tumefaciens* carrying modified T-DNA. Approximately 30 000 T-DNA lines are currently available. T-DNA insertions have been demonstrated to be well distributed throughout the genome, with on average 1.5 insertions present per transformed line (42). Initial transformation reporter genes included antibiotic resistance to kanamycin, hygromycin, and/or streptomycin (39, 40). More recently a selectable marker that confers resistance to the herbicide Basta (phosphinothricin) has been developed which makes screening for transformants very efficient as seeds can be planted directly in the glasshouse and sub-irrigated with herbicide (41).

In addition to acting as insertional mutagens, modified T-DNAs can facilitate the cloning of genes that are expressed in specific cell types or that are inducible by environmental factors, including pathogens, without displaying

1: Growth, maintenance, and use of Arabidopsis *genetic resources*

an obvious phenotype (see also Chapters 6 and 7). In this instance, the modified T-DNA construct is designed to be used as a promoter trap which contains a promoterless reporter gene such as *gus*A (43). The promoterless *gus*A gene is activated following insertion downstream of a native gene promoter. The native pattern of expression of the 'trapped' gene can be visualized by staining for GUS (β-glucuronidase) activity. One disadvantage of using GUS as a gene expression reporter is that detection requires a destructive assay.

Enhancer trap lines have also been developed to detect enhancers, which are capable of orientation-independent transcriptional activation of a gene (44). Enhancer trap T-DNAs contain a reporter gene, such as GUS, that has a weak or minimal promoter located near the border of the insert. When integrated beside a native enhancer sequence the reporter gene will exhibit the same pattern of expression as the nearby chromosomal gene. The use of the green fluorescent protein (GFP) reporter gene modified from the jellyfish *Aequorea victoria* allows for direct *in vivo* visualization of gene expression with either long wavelength UV (e.g. hand-held lamp) or blue light (e.g. argon laser) (44).

An alternative way of perturbing gene function is to spatially and/or temporally target the mis-expression of genes. A chosen target gene can be cloned under control of a GAL4-responsive promoter, separately transformed into *Arabidopsis* and maintained silently in the absence of GAL4 (44). Crossing of this single line with any of the GAL4-GFP lines will allow specific activation of the target gene in particular tissue and cell types (as indicated by the presence of GFP fluorescence) and the phenotypic consequences of mis-expression, including those lethal to the cells, can be studied.

4.2.3 Transposon tagged lines

Modified maize transposons, Dissociation/Activator (*Ds/Ac*) (45) or Enhancer/Inhibitor (*En/I*) (46) [also known as the Supressor-mutator *Spm/dSpm*] elements have been used to generate large populations of tagged lines (see Chapter 7).

Populations of lines can be generated using an autonomous transposable element where the transposon carries the transposase gene required for its own transposition (47) and is allowed to 'jump' during multiple rounds of replication. This is known as the one-element system. Such a strategy has been used to induce multiple inserts (5–40) within the genome as a means of achieving saturation mutagenesis in a low number of lines (E. Wisman, personal communication). This has benefits for forward and reverse genetics approaches as it reduces the number of lines to be screened. The disadvantage being that once an insert in the gene of interest has been identified, it can be problematic to identify which of the elements is the one responsible for the aberration of the gene of interest.

In the two-element system *Ds* or *I* [*dSpm*], transposons have been modified and cannot move unless transposase is supplied from the *Ac* or *En* [*Spm*] transposon, respectively (45, 46). In nature the *Ac* and *En* [*Spm*] transposon can autonomously jump, but their ability to do so in this system has been removed. Hence, the *Ds* or *I* [*dSpm*] element is stable and will remain at its original location (which can be mapped) until it is crossed to a plant containing the non-mobile *Ac* or *En* [*Spm*] constructs. Following transposition the *Ds* or *I* [*dSpm*] element can be re-stabilized through segregation away from the *Ac* or *En* [*Spm*] activity. Reporter genes are used to monitor transposition and to select plants carrying transposed elements.

Transposon tagged lines offer several advantages over the use of standard T-DNA lines (see Chapter 7). Although transposons can be used to disrupt gene function in the same way as a T-DNA insertion (i.e. direct insertion, promoter/enhancer traps, and gene traps), they can also be used to move around the genome. This means that it is possible to generate large numbers of novel insertion events through rounds of transposition. Remobilization of the element, resulting in reversion of the mutation, provides a rapid means of establishing that a mutation is caused by the insertion. In addition, when excision is imprecise, transposition of an element out of a gene and re-insertion elsewhere can yield further mutations. This strategy can be used to generate an allelic series of mutations. Current research is generating lines that contain mapped elements. These can be used as starting material in targeted tagging of mapped genes.

Several thousand transposon tagged lines are currently available from the stock centres. Importantly, the parental lines of these transposon lines are available and so it is possible to conduct one's own transposition experiments using this material. Lines are available that either induce linked (45) or unlinked (48) transposition events.

4.3 Mapping tools
4.3.1 Multiple marker lines

Multiple mutants or marker lines, as their name suggests, carry several different mutations. These lines have been designed to carry mutations that lie on one or more of the five chromosomes and are a useful resource for mapping novel mutations through classical mapping strategies. Generally NW100 (49), a line that carries ten mutations, two per chromosome, has been used in mapping analysis (see Chapter 4). A second line generated from NW100, NW100F carries nine mutations, but lacks the *ms1* gene mutation, which confers male sterility. This makes mapping with this line more straightforward as it is not necessary to maintain your parental NW100 stock in the heterozygote state. NW100 is difficult to maintain because it carries so many different mutations, so alternative multiple marker lines are also available that have been constructed to include mutations on only two or three chromosomes. Other lines, which carry a number of mutations located on single chromo-

1: Growth, maintenance, and use of Arabidopsis genetic resources

somes are very useful for finer mapping of a mutation once it has been placed on a particular chromosome.

4.3.2 Trisomics

An alternative mapping tool to the multiple mutants, are trisomic lines. These lines were first described by L. M. Steinitz-Sears (50) and are either trisomics or telotrisomics. The phenotype of the trisomic is governed by which chromosome is extra in addition to the normal diploid complement. Trisomics are generally derived from triploids (the result of a 4n female × 2n male cross). They can be maintained through selfing or crossing to the diploid female parent. Trisomics and telotrisomics allow mapping to chromosomes and chromosome arms; this can be carried out by segregation analysis of mutations with the extra chromosome phenotype rather than by cytogenetic analysis (51).

4.3.3 Recombinant inbred (RI) lines

These lines offer a quick way of mapping lines molecularly or through some phenotypic trait such as disease resistance (see Chapter 3). RI lines can be used to map any trait that is polymorphic between the RI parental lines and also offer a vital tool for the integration of the physical and genetic map. Four sets of RI lines are available which have been generated using different parental lines. The canonical mapping population for *Arabidopsis* is the Landsberg *erecta* × Columbia population developed by Lister and Dean (52). The background and practical use of recombinant inbreds is described in Chapter 3.

5. Considerations of available resources for identifying novel genes

Over 40 000 mutagenized lines of *Arabidopsis* that have been generated through chemical means, ionizing radiation, and through transformation with T-DNA and transposons are now available from the stock centres. When screening for a new mutation it is best to use material that has been saturated with mutations. Evidence from the *Arabidopsis* genome initiative (AGI) suggests that *Arabidopsis* has approximately 20 000 genes. Therefore to achieve saturation mutagenesis of the genome, in which each mutant is independent and mutagenesis is random, a population of 20 000 mutagenized lines needs to be generated. However, this is an ideal scenario. In practise, to approach saturation mutagenesis of the *Arabidopsis* genome it will be necessary to produce 100 000 mutated lines to allow for redundancy, whether they be induced by chemical means or through transformation.

The researcher now has the choice of both forward and reverse genetics as a means of identifying genes and consequently studying their function. Forward genetics refers to the classical identification of genes through perturbation of their function through mutagenesis and the subsequent cloning and sequencing of the gene. Reverse genetics exploits the ever-increasing availability of genomic sequence (the genome sequence is scheduled to be complete by the end of the year 2000) (53). Sequence analysis identifies potential coding regions and genes. This information can be used to make primers to amplify out these regions from T-DNA or transposon transformed populations as a means of identifying plants carrying an aberration in the gene of interest.

5.1 Forward genetics

When choosing which resource to exploit for a forward approach to gene cloning it is necessary to consider both the type of gene to be identified, the technical difficulty, time and expense required to identify the appropriate mutant, and the subsequent molecular analysis of this mutant.

5.1.1 T-DNA transformed lines

At this time there are approximately 40 000 T-DNA transformed lines available for screening. They have the bonus that if a new mutation is identified during a screen then it may be tagged, which potentially makes the cloning of the gene very fast. Although large populations of lines are being generated, currently the publicly available populations are not yet of sufficient size to carry insertions in every gene. It is predicted that saturation mutagenesis of the genome with T-DNAs and transposons will be achieved in 1999. The majority of T-DNA or transposon insertions do not result in a visible mutation (54). Nevertheless, there is no doubt that this is already a successful route since many genes have been isolated using this material. Importantly, these lines are also powerful tools for a reverse genetics (see Section 5.2 and Chapter 7).

5.1.2 EMS mutagenized populations

Saturation mutagenesis of the genome can also be achieved using chemicals such as ethylmethane sulfonate (EMS) (see *Protocol 1*). The production of an EMS mutagenized population is technically easier and faster than production of T-DNA or transposon insertion populations. EMS generally causes point mutations. The number of 'hits' per genome (i.e. per germline cell in a treated seed) is greater than for insertional mutagenesis. Consequently the number of M1 individuals required to represent saturation mutagenesis can be significantly less than that required for T-DNA or transposon insertion populations. Therefore when starting to screen for a novel mutation, or to look for new alleles of a mutation, this can be the population of choice. Once a mutant has been identified and mapped, it can be cloned by mobilizing a nearby transposon and obtaining an insertion within the target gene. Alternatively the gene can be cloned using mapped-based cloning strategies (see Chapter 8).

5.1.3 Strategies for screening EMS mutagenized populations

There are several strategies for harvesting the M2 seed after EMS mutagenesis. The most appropriate method will depend on the nature of the mutant screen. The two extremes of the spectrum are:

(a) Keep the seed harvested from each M1 plant separate and generate a collection of M2 families.
(b) Pool all the seed harvested from the M1 plants to give a bulked M2 population (Lehle Seeds supplies a bulked M2).

Although harvesting and screening each M2 family separately is labour-intensive there are two main advantages to this approach. First, it offers the possibility of recovering mutations that are infertile when homozygous via the heterozygote siblings of the mutant plants. Secondly, this strategy guarantees the independence of all mutants isolated. This ensures that mutations isolated from the M2 collection that are subsequently found to be allelic are highly likely to be different alleles at the same locus rather than two isolates of the same mutational event.

If the mutant you are looking for is unlikely to be infertile or if the screen you are carrying out would be much more difficult to do with individual families, then a compromise strategy of screening groups of M2 families together might be preferable.

Theoretical considerations can be given to the size of the M1 population and the frequency at which a new mutant class will be observed in the M2 (55). In practise, such screens are usually limited by the effort involved. If a mutant class is rare, increasing the size of the M1 may increase the probability of recovering it more than simply screening a greater number of M2 plants from the same mutagenesis. For example, if a particular trait cannot be found in a M2 population 20 times the size of the M1 it probably cannot be recovered from that population.

Protocol 1. EMS mutagenesis[a]

Equipment and reagents

- Atmos bags (Sigma)
- 10 ml plastic disposable syringe (Becton Dickinson)
- Glass beakers
- 3MM paper (Whatman)
- Adapted 10 ml syringes[b]
- Ethylmethane sulfonate[c] (EMS) (Sigma)
- 0.1% (w/v) potassium chloride (Sigma)
- 0.1 M sodium phosphate pH 5.0, 5% (v/v) dimethyl sulfoxide, 50–100 mM EMS (Sigma)
- 100 mM sodium thiosulfate (Sigma)

Method

1. Pre-imbibe a known number of seeds overnight in 0.1% (w/v) potassium chloride. (NB: 1000 Landsberg *erecta* seeds weigh approx. 20 mg.)

Protocol 1. *Continued*

2. Soak the seeds for 3–5 h in 0.1 M sodium phosphate pH 5.0, 5% (v/v) dimethyl sulfoxide, and 50–100 mM EMS. Concentrations of EMS in this range give a high frequency of mutation in the M1; the use of 200 mM EMS results in little lethality but M1 plants are sterile. Typically mutagenize 5000 seeds in 20 ml of the above solution.

3. Wash the seeds twice in 100 mM sodium thiosulfate for 15 min. Sodium thiosulfate destroys the EMS. All waste solutions must be discarded into a beaker containing solid sodium thiosulfate.

4. Wash the seeds twice in distilled water for 15 min.

5. Open the 'Atmos' bag, pour the sodium sulfate inactivated waste into the sink, and wash down with plenty of water. Leave all beakers, etc. used in the procedure under running water for half an hour.

6. Transfer the seeds to 3MM paper and allow to dry overnight.

7. Dilute the seeds in dry sand to make distribution easier. Plant at a density of 1 cm^{-2}.

[a] Based on a protocol by O. Leyser and I. Furner with permission.
[b] Steps 2–6 must be carried out in an 'Atmos' bag in a fume hood, which can make manipulating the seed awkward. Use adapted 10 ml syringes to hold the seeds throughout the entire procedure. These are made by cutting off the front end of the syringe and attaching a circle of fine nylon mesh into place. Seeds are placed in the syringe and the plunger can be used to draw the appropriate solutions through the nylon mesh onto the seeds.
[c] EMS is a highly toxic, volatile, colourless liquid and all manipulations with it should be carried out in an 'Atmos' bag in a fume hood.

This generation of plants is called the M1 generation. An indication that the mutagenesis has been successful is the appearance of chlorophyll-deficient sectors at low frequency among the M1 plants (1 in 100–1000 plants).

Plants mature in six to eight weeks at which time the M2 generation can be harvested. Recessive and dominant mutations can be recovered from this generation. In a successful mutagenesis 0.5–3% M1 plants segregate chlorophyll-deficient plants in their M2 progeny.

5.2 Reverse genetics

Reverse genetics is a powerful method of ascertaining the gene function of sequenced genes whose function is previously unknown (56). T-DNA and transposon tagged populations are both essential screening material in a reverse genetics approach using primers based on sequence analysis of potential coding regions. DNA from lines can be pooled for the initial PCR analysis. Further rounds of PCR analysis will identify the plant line containing the insert positioned in the gene of interest which can then be analysed for phenotypic aberrations. To aid in reverse genetics, the stock centres distribute

1: Growth, maintenance, and use of Arabidopsis *genetic resources*

DNA and seed from some of the publicly available T-DNA and transposon tagged populations.

6. Growing *Arabidopsis*

When growing *Arabidopsis* the highest standards of hygiene should be adopted to achieve maximum plant survival which can be important for several reasons.

(a) Achieving uniform plant growth.
(b) Ensuring that potentially useful new mutants are not lost in the early fragile stages of screening.
(c) Maintenance of all members of a segregating population.

Good hygiene is also required to guarantee the minimum chance of cross-pollination amongst plants by insect vectors.

If the plant material that is being grown is transgenic, then it is the responsibility of the researcher to hold any necessary licensing and to abide by the national legislation for the growth and disposal of these lines. In the case of the UK, contact MAFF, Nobel House, London for the latest details regards UK legislation.

6.1 How to maintain clean growth conditions

It is advisable to adopt a strict regime for growing *Arabidopsis*. Glasshouse/growth room local rules should be established and strictly enforced to ensure that even in communal growth areas these rules are observed. Measures should be in place to allow for the removal of material, where experiments are put at risk by practise of others. This strategy works best if someone is appointed to be in charge of the area, such as a glasshouse manager, who can monitor the growth area and act appropriately.

Before starting to grow *Arabidopsis*, purge the glasshouse area or growth room to remove all potential infection.

(a) Use glasshouse-specific laboratory coats that are laundered regularly. Infection can be spread between growth areas on clothing, so do not move between glasshouses and growth rooms in the same laboratory coat.
(b) Keep footwipes containing a general biocide at the entrance to the glasshouse to prevent the movement of seed and potential contaminants on the soles of shoes.
(c) Prevent the entry of flies, aphids, bees, and wasps by installing screens over vents and windows.
(d) Employ fly strips(Gerhardt Pharmaceuticals Ltd.) as insect traps, which also give an indication of the type and level of any infection.

(e) Use fresh clean compost mixes, which may be autoclaved. Avoid leaving bags of compost open to prevent contamination with fly eggs and fungal spores.

(f) Never move infected material between glasshouses.

(g) Avoid having the growth conditions overly wet, or overly dry. Either of these regimes can encourage infections.

(h) Do not let old plant material accumulate in the glasshouse as this can act as a reservoir for infection.

(i) Do not have stagnant pools of water on the benches or on the floor.

(j) When infection is suspected, alert the glasshouse manager or identify the source of infection and act accordingly.

6.2 Growing *Arabidopsis* in the glasshouse

There are many different ways to grow *Arabidopsis* in the glasshouse. *Arabidopsis* can be grown on soil, commercial compost mixes, perlite, and other relatively inert media. However, there are certain criteria that have to be matched.

(a) Always grow material in well drained medium. Drainage can be enhanced by the addition of grit, sand, or perlite to the compost mix.

(b) Avoid over watering since this encourages algal infections and can also provide favourable conditions for scarid fly larvae.

(c) Temperature should be controlled. Low temperatures will slow growth. Mature plants can withstand temperatures above 30°C, but optimal growth will be achieved if a constant temperature of 25°C can be maintained throughout the growth period.

(d) *Arabidopsis* can tolerate low humidity but care should be taken that the material does not dry out. For optimal seed quality humidities of less than 50% are recommended.

(e) *Arabidopsis* requires long days to flower, so the glasshouse should be set to a 12 hour or greater photoperiod. *Arabidopsis* will grow quite satisfactorily in continuous light. Under these conditions it flowers slightly earlier, produces less leaves, and seed production is reduced.

(f) It is not normally necessary to supplement the plants with nutrients, but poor nutrition can cause reduced height, early flowering, and reduced seed set. Proprietary brands of plant fertilizer can be used to feed *Arabidopsis*.

Protocol 2. Growing *Arabidopsis* in the glasshouse[a]

Equipment and reagents

- Plastic Vacupot trays 5 × 5 × 5.5 cm (H. Smith Plastics Ltd.)
- Propagators (Joseph Bentley)
- Capillary matting (Lantor UK Ltd.)
- Perforated polythene (Fargro Ltd.)
- ARACON™ harvesting system (Aracon)
- Cellophane bags (Courtaulds)
- John Innes No. 3 (J. Arthur Bower's)
- Levington M3 (Levington)
- Perlite (medium grade; Silvaperl)
- 2 mm² aperture horticultural riddle (Endecotts Ltd.)
- Bacto agar (Difco) pH 5.0

A. Sowing

1. Prepare a compost mix of six parts Levingtons M3: six parts John Innes No. 3: one part perlite. Each Vacupot tray takes 2 litres of compost mix. Fill each pot and allow to settle by giving the pots a sharp tap and then compress the compost very lightly using finger tips. Place more compost mix on top until the tray is full.

2. Using the riddle, sieve compost onto the top of the pots to give a fine bed on which to sow the seeds.

3. Place the tray in water to soak, until the surface of the compost is damp, i.e. allow the compost to become wetted by capillary action (if applying INTERCEPT or other systemic insecticide, spray onto the surface prior to soaking following the manufacturer's instructions).

4. Sow the seed onto the surface of the compost. Sow seeds individually with a wetted microspatula or the fused end of a glass capillary tube. It is possible to scatter seeds by distributing the seed from a folded card either alone or mixed with sand or agar.[b]

5. Place the trays inside a propagator in a growth room and germinate at 22°C ± 2°C, with a 2°C night temperature drop, under 2000 lux cool fluorescent lighting with an 18 h photoperiod. Growth should be visible after five days. Plants may be grown to maturity under this regime. Faster cycling of the plants can be achieved by growth under continuous illumination. For DNA preparation, plants can be grown under short day conditions, i.e. 8 h or less, to increase the size and number of rosette leaves. Lines can be encouraged to germinate more uniformly if sown and placed at 4°C for about five days before placing in the growth room or glasshouse (stratification).

6. After seven to ten days (i.e. once two to four true leaves have formed), transfer the trays without propagator to the glasshouse. Place the plants on benching covered in capillary matting overlaid with horticultural perforated plastic sheeting[c] and water from underneath. Optimum conditions for growth are 25°C ± 2°C, under 2000 lux cool fluorescent lighting with a 20 h photoperiod. Plants can be grown in

Protocol 2. *Continued*

environments with more variable temperatures from 15–30 °C but at the extremes of this range plant vigour and fertility will be reduced. Alternatively, *Arabidopsis* can be germinated and grown directly in the glasshouse. For standard lines it is not necessary to supplement the growth medium, although higher yields of seed may be obtained if you do so. Standard proprietary plant foods can be used.

B. *Vernalization*

Certain ecotypes or winter annuals will require extended periods of several weeks of 4 °C treatment in order to induce flowering. This is true for many of the ecotypes found in the UK. In nature these populations would germinate in the autumn and then over winter as rosettes and flower in the following April, producing seed in the June or July. The other category of lines, the summer annuals that include the commonly used lines Landsberg *erecta* and Columbia do not require a cold treatment in order to flower.

C. *Methods to prevent crossing*

Arabidopsis is self-fertile, but cross-pollination can occur within a glasshouse environment. In order to minimize cross-pollination, several strategies can be adopted.

1. Good glasshouse hygiene will minimize the number of insects and hence decrease the chance of cross-pollination through insect vectors.
2. Space the material fairly far apart (20 cm) to prevent flowers from different lines coming into contact with one another.
3. Stake the material to prevent lodging. Stakes can be made from commercially available garden wire and pea rings.
4. Use the ARACON™ system, a commercial system that provides a plastic collar to surround the plant as it bolts, so that each plant is isolated. An alternative to this system can be made from cutting down clear plastic drinks bottles.

D. *Harvesting*

After a period of about three to four weeks in optimal conditions, siliques will start to form. The silique takes about two weeks to mature in which time it turns from green to yellow/brown. On maturation the silique dehisces and shatters scattering its contents. Several measures can be taken to collect the seed.

1. Cellophane bags (45 mm × 190 mm) which are custom made by the suppliers are placed over the plants as the first silique ripens and are secured at the bottom of the bag by the use of sticky tape. The bags

1: Growth, maintenance, and use of Arabidopsis *genetic resources*

allow the exchange of air and water and so condensation does not build up on the inside of the bag and the remaining siliques can ripen. Allow the plants to thoroughly dry out. Then cut the plant off at the base, within the harvesting bag, and gently rub it to free the seeds from the siliques. Collect the seeds directly from the bags by cutting a small hole in the corner. This precludes the need to sieve the material and so reduces the risk of mixing or contaminating seed lines.

2. The ARACON™ system allows for the collection of seed and the shattered siliques in the specially designed collector. Some sieving of the material may be required. Careful cleaning of the system is required between use.

3. Plants grown in small pots can be placed on their side within bags.

[a] If transgenic material is being grown, then all appropriate procedures must be followed. This includes autoclaving plants and soil material before disposal and the thorough sterilization of all equipment that will be reused.
[b] Large numbers of seed can also be sown by mixing with cooled 0.1% (w/v) Bacto agar pH 5.0 (the pH stops the agar solidifying) and dispensing from a 50 ml syringe. If mixing the seed with sand, to aid even distribution, then use sufficient sand to give an even covering over the top of the soil surface (e.g. for a 15 cm wide flat use 0.2 ml of sand).
[c] The horticultural perforated plastic prevents algal growth on the capillary matting and can be regularly replaced to maintain hygienic conditions.

6.3 Chemical control of pests and diseases

Arabidopsis can be susceptible to many of the pests that commonly infest glasshouses and growth rooms. In general, the best philosophy to adopt in pest management is that of prevention rather than cure. The best way to minimize infestation of any kind in the glasshouse or growth room is to maintain good hygienic practise as outlined above. Nevertheless, it is almost inevitable that at some time the glasshouse will become infested with some kind of *Arabidopsis* pest. *Table 1* provides some guidelines on how to treat the most common types of infection that affect *Arabidopsis*.

A good systemic insecticide, such as INTERCEPT 70WG, can be used for basic pest control. Prior to sowing the seeds, the insecticide is applied as a compost drench following the manufacturer's instructions. The insecticide gives protection against aphid, scarid fly, and thrip infections. However, to prevent the development of resistant strains, the insecticide is used in a planned programme of pest control, which also includes the use of nicotine shreds. Fly strips are used to act as insect traps that give a useful indication of the kind and extent of a particular infection.

The use of these chemicals should only be carried out by personnel who have received the necessary training to handle these materials and the manufacturer's instructions should be followed at all times. Warning signs indicating the time and type of treatment should be clearly posted. Ideally, all pesticide and fungicide treatments should be applied during the evenings or at

Table 1. Symptoms and treatment of common *Arabidopsis* pests and diseases

Infection	Symptom	Treatment
Aphids (greenfly)	Aphids massed on leaves and flowering stem. Distorted leaves due to aphids feeding on young tissues.	Apply INTERCEPT 70WG (Levington Horticulture Ltd.) following the manufacturer's instructions. For heavy infestation fumigate with 40% nicotine shreds (DowElanco Ltd.).
Thrips	Leaves are mottled with silver or white flecks. Thrips can accumulate in the inflorescence and can cause fertilization to fail.	Apply INTERCEPT 70WG (Levington Horticulture Ltd.) following the manufacturer's instructions. For heavy infestation fumigate with 40% nicotine shreds (DowElanco Ltd.).
Scarid (mushroom fly or fungus gnat)	Plants lack vigour and leaves turn yellow, without visible infection above the soil. Colourless maggots can be observed amongst the root tissue, where they feed. Adult flies gather around the plant and can be caught on fly strips.	Apply INTERCEPT 70WG (Levington Horticulture Ltd.) following the manufacturer's instructions. For heavy infestation fumigate with 40% nicotine shreds (DowElanco Ltd.).
Red spider mite	Infections occur under dry and hot growth conditions. Leaves have a silvery speckled appearance and in severe cases plants will yellow and wilt. Underside of the leaves has the appearance of being dusted with a white powder suspended from thin white silk strands.	In many circumstances it is better to throw all infected material away as infections can be very destructive. Avoid moving infected material between glasshouses/growth rooms as this pest is very easily spread. Fumigate with 40% nicotine shreds (DowElanco Ltd.) while maintaining a temperature of 25°C for at least 8 h.
Whitefly	Found on the underside of leaves, the adults are small white flies and the nymphs are pale green. Infestation results in plants lacking vigour, wilting, and death.	Apply INTERCEPT 70WG (Levington HorticultureLtd.) following the manufacturer's instructions. For heavy infestation fumigate with 40% nicotine shreds (DowElanco Ltd.).
Botrytis	A grey 'fluffy' mould grows on aerial plant parts and may be accompanied by degradation of the plant material.	At the first sign of infection spray plants thoroughly with Supercarb systemic fungicide (Bio), following the manufacturer's instructions. Repeat treatment two weeks later.
Mildew	White powdery patches develop on leaves and flowering stems. Severe infections can result in plant wilt and death.	At the first sign of infection spray plants thoroughly with Supercarb systemic fungicide (Bio), following the manufacturer's instructions. Repeat treatment two weeks later.

1: Growth, maintenance, and use of Arabidopsis *genetic resources*

weekends. Areas that are undergoing treatment should be locked to prevent accidental exposure of personnel. Glasshouses and growth rooms should be thoroughly ventilated following treatment to flush with fresh air before personnel are allowed to re-enter the areas.

7. Seed storage

The most important factors that influence seed longevity are the temperature at which the seed is stored and the moisture content of the seed. Generally, the higher either of these values, the shorter the shelf-life of the seed. In order to maximize seed viability prompt post-harvest treatment is required. How a seed is treated will greatly influence how long it can be stored.

(a) *Short-term.* Seeds can remain viable for two years if stored in a dry atmosphere at room temperature. A better method is to store at 4°C in dry conditions.

(b) *Medium-term.* Store the seeds in negative bags (Kenro Ltd.) or vials at 15°C, 15% relative humidity. Initial seed longevity studies applying the equations for the prediction of seed longevity (57) and following Harrington's 'rule of thumb' (58) suggest that these conditions will increase the longevity of the seeds from two to three years, at ambient conditions, to approximately 30 years.

(c) *Long-term.* Seeds should be dried to 5–6% moisture content then placed in sealed containers (for example 1.5ml Nalgene System100 Cryovials) at –20°C. If plants are well dried in the glasshouse the moisture content of the harvested seed will be about 7%. Seed is dried by placing in controlled environment chambers set at 15°C and 15% relative humidity (Modified Sanyo Gallenkamp environment control chambers—model type PSC061.SSC.MOD). The seed will equilibrate to approximately 5% moisture content within two days to two months depending on the volume of the seed sample, its initial moisture content, and the type of seed container used. Examples of the time taken for different seed lots to reach low seed moisture contents are shown in *Table 2*. Determination of seed moisture content is described in *Protocol 3*.

When removing seed from archival storage at –20°C for use, extreme care must be taken to prevent damaging the seeds. Such freeze–thaw cycles should be kept to a minimum. Containers should be allowed to come to room temperature before opening to prevent the accumulation of condensation on the seeds and in the container. If condensation does accumulate, then the container should be allowed to thoroughly air dry before returning it to the freezer.

Performing a quick germination test on a proportion of the harvested seed can be used to check seed viability. This should be conducted prior to placing

Table 2. Moisture content evaluation of stored seed[a]

Seed sample volume (ml)	Seed container	(%) Initial moisture content	(%) Final moisture content	Time taken to reach moisture content (days)
1–2	Open Petri dish	6.20	5.17	2
1–2	Negative bag[b]	6.20	5.02	2
0.2	1.5 ml cryovial (lid loose)	6.32	5.06	28
15	20 ml click-cap vial (2 mm diameter hole in lid)	5.98–6.56[c]	5.09–5.57[c]	35
20	40 ml screw-cap vial (lid loose)	7.70–7.76[c]	5.55–5.78[c]	143

[a] All % values are based on an average of three replicate samples.
[b] Coated paper bags used to hold photograph negatives (Kenro Ltd.).
[c] The range of values is derived from the results of tests of three different seed lines placed in the environment control chamber at the same time.

seeds in storage in order to identify any poor quality seed. Germination tests can be performed by germinating the seed on perlite (see *Protocol 4*), agar (see *Protocol 5*), or wetted filter paper. Give the seeds a short cold treatment (two to four days) to help break any dormancy prior to placing them in the growth room.

Protocol 3. Determination of moisture content: modified low constant temperature method (59)

Equipment

- Oven (Genlab Ltd.)
- Small light-weight oven proof dish and lid (Sigma)
- Aluminium foil (Sigma)
- Fine balance
- Microspatula
- Desiccator

Method

1. Pre-heat the oven to 103°C ± 1°C.
2. Weigh a light-weight oven proof dish plus cover to four decimal places. Record the weight (W1). Alternatively fold a small piece of aluminium foil to form a 2 × 3 cm packet and weigh this.
3. Add 100–200 mg of seeds to the dish or packet and weigh again (W2).
4. Prepare other replicates or different samples in the same way.
5. Remove the lid and place each dish on its lid in the oven or place the open foil packet in the oven for 17 h ± 1 h.

1: Growth, maintenance, and use of Arabidopsis genetic resources

6. Remove the dish or packet from the oven, replace the lid of the dish or seal the foil packet, and immediately place in a desiccator at room temperature to cool for 30–45 min so the seeds can cool without absorbing water.
7. Weigh each dish or packet (W3). Do not leave the desiccator lid open during each weighing.
8. Calculate the moisture content as a percentage loss in weight using the following formula.

$$\% \text{ moisture content} = \frac{(W2 - W3)}{(W2 - W1)} \times 100$$

8. Growing *Arabidopsis* with specific growth requirements

Certain *Arabidopsis* mutants have specific growth requirements since they lack or have reduced sensitivity to growth regulators, or lack genes in specific metabolic pathways. Such lines can be germinated on Petri dishes (see *Protocol 4*) where their phenotype can be verified and then transferred to soil for growing to maturity. This method can also be used for conducting germination/viability tests on seed stocks.

Protocol 4. Growing *Arabidopsis* with special growth requirements

Equipment and reagents
- Sterile 9 cm diameter plastic Petri dishes (Bibby Sterilin Ltd.)
- Nesco film (Bando Chemical Industry Ltd.)
- Microspatula (Sigma)
- Perlite (Superfine grade; Silvaperl)
- Gibberellin A_3 (Sigma)
- Gamborgs B5 basal medium[a] with minimal organics (Sigma)
- Thiamine (Sigma)
- Abscisic acid (±) *cis, trans* (Sigma)
- 1 mM KNO_3 (Sigma)

A. *Sowing*
1. Sterilize the perlite in a heat proof container by autoclaving at 121°C for 20 min. Half fill a 9 cm diameter Petri dish with sterile perlite.
2. Wet the perlite with 1 mM KNO_3, plus the appropriate growth regulator (see *Table 3*), or with Gamborgs B5, until the perlite has absorbed the liquid (about 10–15 ml) but there is no free liquid when the dish is tipped.
3. Sow the seeds individually onto the surface of the perlite using a wetted microspatula, or the fused end of a glass capillary tube or Pasteur pipette.
4. Seal the Petri dish with Nesco film and place at 4°C for five days.

Protocol 4. *Continued*

B. *Growth*

1. Allow the plants to germinate in the standard growth room conditions outlined in *Protocol 1*.

2. When two to four true leaves have formed transplant the material to compost by gently scooping underneath the seedling with a microspatula, taking care not to damage the roots. Some carry over of the perlite will occur. Place the seedling in the soil and gently firm the soil around the transplanted material.

3. Maintain high humidity by placing the plants in a closed propagator for a few days to prevent excess water loss and to allow the plants to acclimatize.

4. Plants can then be grown onto maturity as described in *Protocol 2* whilst applying the appropriate growth regulator.

[a] Gamborg, O. L., Miller, R. A., and Ojima, K. (1968). Nutrient requirements of suspension cultures of soybean root cells. *Exp. Cell. Res.*, **50**, 151.

Table 3. Storage and use of growth regulators

	Abscisic acid	Gibberellin	Thiamine
Solvent	1 M NaOH	Absolute ethanol	dH_2O
Diluting agent	dH_2O	dH_2O	dH_2O
Powder storage	Below 0°C	Room temperature	Room temperature
Liquid storage	Below 0°C	Below 0°C	4°C
Stock concentration	50 mg/ml	50 mg/ml	10% (w/v)
Working concentration	5–10 µg/ml	10 µg/ml	1% (w/v)

9. Sterile culture of *Arabidopsis*

For some mutants with a very weak growth habit it may be necessary to germinate and establish the plants in tissue culture, then transplant them to soil when the plants have reached the four leaf to small rosette stage. For these plants it is best to also surface sterilize the seed, as described in *Protocol 5*, to remove any bacteria or fungi which could have an inhibitory effect on plant growth. Alternatively, these plants may be started on sterile perlite plates (see *Protocol 4*).

To observe and screen plants with phenotypes expressed at the seedling stage (e.g. absence of the triple response in the presence of ACC [1-aminocylopropane-1-carboxylic acid] in the dark in ethylene-insensitive mutants) it is much easier if the plants are grown on agar media. If the plants are subsequently to be grown to maturity it is better to use the sterile culture

1: Growth, maintenance, and use of Arabidopsis *genetic resources*

procedure described in *Protocol 5*, however if the seedlings will be discarded after scoring then a more basic agar medium can be substituted for the culture medium (e.g. 0.8% (w/v) agar containing 1 mM KNO$_3$). In either case it is necessary to use seed that has been pre-sterilized as described in *Protocol 5*.

Protocol 5. Sterile culture of *Arabidopsis*

Equipment and reagents
- 9 cm filter paper discs (Whatman No. 1)
- 30 ml sterile Universal containers (Bibby Sterilin Ltd.)
- 9 cm diameter sterile plastic Petri dishes (Bibby Sterilin Ltd.)
- Bacto agar (Difco)
- Murashige and Skoog basal medium[a] (Sigma)
- Sucrose (Sigma)
- Myo-inositol (Sigma)
- Thiamine (Sigma)
- Pyridoxine (Sigma)
- Nicotinic acid (Sigma)
- MES (2-(*N*-morpholino)ethanesulfonic acid; Sigma)
- 1 M KOH in distilled water (Sigma)
- Double magenta GA-7 plant cell culture vessels with couplers and GA-7 filters (Sigma)
- Bleach solution: 10% (v/v) commercial bleach (Domestos, Diversey Lever) diluted with distilled water plus Tween 20 (polyoxyethylenesorbitan monolaurate, Sigma) added at 'one drop' per 50 ml of solution
- 70% (v/v) ethanol in distilled water (BDH)
- Sterile distilled water

A. *Preparation of tissue culture medium*

1. Weigh out the culture medium components according to *Table 4* and dissolve in the appropriate volume of distilled water. Correct the pH of this solution to 5.7 using 1 M KOH.

2. Add 0.4 g of agar to each Magenta pot then add to this 50 ml of the solution prepared in step 1. Replace the top section of each Magenta pot.

3. Autoclave the filled Magenta pots at 121°C for 20 min.

4. After autoclaving it may be necessary to place the Magenta pots in a sterile laminar flow cabinet with the two parts open slightly to allow any condensation to evaporate.

5. Alternatively the appropriate amount of agar can be added to the solution prepared in step 1 and autoclaved before pouring into pre-sterilized Petri dishes.

B. *Sterilization and sowing of seed*

1. Wrap the seed in a 9 cm filter paper disc and place it in a 30 ml Universal container.

2. Add 20 ml of 70% (v/v) ethanol to the Universal containing the seeds and shake the container for 1 min.

3. Pour out the 70% (v/v) ethanol and replace it with 20 ml of the bleach solution. Shake the container gently for 10 min.

Protocol 5. *Continued*

4. Pour out the bleach solution and replace it with 20 ml of sterile distilled water. Shake the container for approx. 30 sec then discard the water.
5. Repeat step 4 five more times.
6. Remove the filter paper from the tube and unwrap it using sterile technique (work in a laminar flow hood and use sterile forceps to unwrap the seeds on a sterile Petri dish). Allow the seeds to dry or sow immediately.
7. Use a sterile bacteriological loop, or fused Pasteur pipette, to pick up individual seeds for transfer to the culture medium in Petri dishes or Magenta pots.

C. *Propagation of plants in tissue culture*

1. For short-term propagation (e.g. for seedling screening purposes) seeds can be sown at a density of 100–200 seeds per 9 cm Petri dish.
2. For longer-term propagation seeds can be sown in Magenta pots at a density of five per pot, for growth to the small rosette stage, or one per pot if the plant is to be grown to maturity.
3. Plants can be grown in Petri dishes or Magenta pots under a regime of 16 h light, 2000 lux, and 22 °C ± 2 °C.

[a] Murashige, T. and Skoog, F. (1962). A revised medium for rapid growth and bioassays with tobacco tissue cultures. *Physiol. Plant.*, **15**, 473.

Table 4. Tissue culture medium components

Culture medium component	Concentration in final medium solution
Murashige and Skoog basal medium	4.4 mg/ml
Sucrose	10 mg/ml
Myo-inositol	100 µg/ml
Thiamine	1.0 µg/ml
Pyridoxine	0.5 µg/ml
Nicotinic acid	0.5 µg/ml
MES	0.5 mg/ml

References

1. Redei, G. P. (1970). *Bibliogr. Genet.*, **20**, 1.
2. Snape, J. W. and Lawrence, M. J. (1971). *Heredity*, **27**, 299.
3. Price, R. A., Palmer, J. D., and Al-Shehbaz, I. A. (1994). In *Arabidopsis* (ed. E. M. Meyerowitz and C. R. Somerville), p. 7 Cold Spring Harbor Laboratory Press, New York.

1: Growth, maintenance, and use of Arabidopsis genetic resources

4. Goodman, H., Ecker, J. R., and Dean, C. (1995). *Proc. Natl. Acad. Sci. USA*, **92**, 1083.
5. Meyerowitz, E. (1987). *Annu. Rev. Genet.*, **21**, 93.
6. Lawley, P. D. (1974). *Mutat. Res.*, **23**, 283.
7. Rietz, G., Facius, R., and Bücker, H. (1989). In *Life sciences research in space* (ed. H. Oser and B. Battrick), p. 65. {ESA SP1105} European Space Agency, Paris.
8. Braaksma, F. J. and Feenstra, W. J. (1982). *Theor. Appl. Genet.*, **64**, 83.
9. Browse, J. P., McCourt, P., and Somerville, C. (1986). *Plant Physiol.*, **81**, 859.
10. Koornneef, M. and Hanhart, C. J. (1981). *Arabidopsis Inf. Serv.*, **18**, 52.
11. Last, R. L. and Fink, G. R. (1988). *Science*, **240**, 305.
12. Redei, G. P. and Acedo, G. (1976). In *Cell genetics in higher plants*, p. 39. Akad. Kiado, Budapest.
13. Conklin, P. L., Williams, E. H., and Last, R. L. (1996). *Proc. Natl. Acad. Sci. USA*, **93**, 9970.
14. Relichova, J. (1976). *Arabidopsis Inf. Serv.*, **13**, 25.
15. Hirono, Y. and Redei, G. P. (1963). *Nature*, **197**, 1324.
16. Vizir, I. Y. (1988). *Arabidopsis Inf. Serv.*, **26**, 55.
17. Mandel, M. A., Gustafson-Brown, C., Savidge, B., and Yanofsky, M. F. (1992). *Nature*, **360**, 273.
18. Koornneef, M., van Eden, J., Hanhart, C. J., Stam, P., Braaksma, F. J., and Feenstra, W. J. (1983). *J. Hered.*, **74**, 265.
19. Bowman, J. L., Drews, G. N., and Meyerowitz, E. M. (1991). *Plant Cell*, **3**, 749.
20. van der Veen, J. H. and Wirtz, P. (1968). *Euphytica*, **17**, 371.
21. Robinson-Beers, K., Pruitt, R. E., and Gasser, C. S. (1992). *Plant Cell*, **4**, 1237.
22. Koornneef, M., Hanhart, C. J., and van der Veen, J. H. (1991). *Mol. Gen. Genet.*, **229**, 57.
23. Baskin, T. I., Betzner, A. S., Hoggart, R., Cork, A., and Williamson, R. E. (1992). *Aust. J. Plant Physiol.*, **19**, 427.
24. Barton, M. K. and Poethig, R. S. (1993). *Development*, **119**, 823.
25. Van Lijsebettens, M., Wang, X., Cnops, G., Boerjan, W., Desnos, T., Hofte, H., *et al.* (1996). *Mol. Gen. Genet.*, **251**, 365.
26. Feenstra, W. J. (1978). *Arabidopsis Inf. Serv.*, **15**, 35.
27. Meinke, D. W. and Sussex, I. M. (1979). *Dev. Biol.*, **72**, 62.
28. Leon-Kloosterziel, K. M., Keijzer, C. J., and Koornneef, M. (1994). *Plant Cell*, **6**, 385.
29. Clark, S. E., Running, M. P., and Meyerowitz, E. M. (1995). *Development*, **121**, 2057.
30. Koornneef, M., Jorna, M. L., Brinkhorst-van der Swan, D. L. C., and Karssen, C. M. (1982). *Theor. Appl. Genet.*, **61**, 385.
31. Roman, G., Lubarsky, B., Kieber, J. J., Rothenberg, M., and Ecker, J. R. (1995). *Genetics*, **139**, 1393.
32. Koornneef, M. and van der Veen, J. H. (1980). *Theor. Appl. Genet.*, **58**, 257.
33. King, J. J., Stimart, D. P., Fisher, R. H., and Bleecker, A. B. (1995). *Plant Cell*, **7**, 2023.
34. Fujioka, S., Li, J., Choi, Y. H., Seto, H., Takatsuto, S., Noguchi, T., *et al.* (1997). *Plant Cell*, **9**, 1951.
35. Nagatani, A., Reed, J. W., and Chory, J. (1993). *Plant Physiol.*, **102**, 269.
36. Ahmad, M. and Cashmore, A. R. (1993). *Nature*, **366**, 162.

37. Cabrera, H. L., Poch, C., Peto, C. A., and Chory, J. (1993). *Plant J.*, **4**, 67.
38. Wei, N. and Deng, X. W. (1992). *Plant Cell*, **4**, 1507.
39. Feldmann, K. A. and Marks, M. D. (1987). *Mol. Gen. Genet.*, **208**, 1.
40. Clarke, M., Wei, W., and Lindsey, K. (1992). *Plant Mol. Biol. Rep.*, **10**, 178.
41. Bechtold, N., Ellis, J., and Pelletier, G. (1993). *C. R. Acad. Sci. Paris, Sci. de la Vie/Life Sci.*, **316**, 1194.
42. Azpiroz-Leehan, R. and Feldmann, K. A. (1997). *Trends Genet.*, **13**, 152.
43. Topping, J. F. and Lindsey, K. (1997). *Plant Cell*, **9**, 1713.
44. Haseloff, J. and Siemering, K. In *GFP: green fluorescent protein. Strategies and applications* (ed. M. Chalfie and S. Kain). In press.
45. Bancroft, I. and Dean, C. (1993). *Genetics*, **134**, 1221.
46. Pereira, A. and Aarts, M. G. (1998). *Methods Mol. Biol.*, **82**, 329.
47. Cardon, G. H., Frey, M., Saedler, H., and Grierl, A. (1993). *Plant J.*, **3**, 773.
48. Sundaresan, V., Springer, P., Volpe, T., Haward, S., Jones, J. D., Dean, C., et al. (1995). *Genes Dev.*, **9**, 1797.
49. Koorneef, M., Hanhart, C. J., Van Loenen Martinet, E. P., and van der Veen, J. H. (1987). *Arabidopsis Inf. Serv.*, **23**, 46.
50. Steinitz-Sears, L. M. (1963). *Genetics*, **48**, 483.
51. Koorneef, M. and Stam, P. (1992). In *Methods in Arabidopsis research* (ed. C. Koncz, N.-H. Chua, and J. Schell), p. 81. World Scientific, Singapore, New Jersey, London, Hong Kong.
52. Lister, C. and Dean, C. (1993). *Plant J.*, **4**, 745.
53. Ecker, J. R. (1998). *Nature*, **391**, 438.
54. Bancroft, I., Jones, J. D., and Dean, C. (1993). *Plant Cell*, **5**, 631.
55. Redei, G. P. and Koncz, C. (1992). In *Methods in Arabidopsis research* (ed. C. Koncz, N.-H. Chua, and J. Schell), p. 16. World Scientific, Singapore, NewJersey, London, Hong Kong.
56. Oliver, S. G., Winson, M. K., Kell, D. B., and Baganz, F. (1998). *Trends Biotechnol.*, **16**, 373.
57. Ellis, E. H. and Roberts, R. H. (1980). *Ann. Bot.*, **45**, 13.
58. Harrington, J. F. (1963). *Proc. Int. Seed Test. Assoc.*, **28**, 989.
59. International Seed Testing Association. (1993). *Seed Sci. Technol.*, **21**, Supplement (International Rules for Seeds Testing — Rules 1993), 43.

2

Preservation and handling of stock centre clones

RANDY SCHOLL, KEITH DAVIS, and DOREEN WARE

1. Introduction

The responsibility for preservation and distribution of DNA clones for molecular genetic research and the genome project in *Arabidopsis* is shared by two centres, the *Arabidopsis* Biological Resource Centre (ABRC), Ohio State University, which was established in September 1991, and the European Community DNA Stock Centre at the Max-Planck Institute in Köln, Germany. In addition to DNA resources, the ABRC shares responsibility for seed stock preservation and distribution and dissemination of information for the *Arabidopsis* community. Hence the DNA laboratory is integrated with the centre's other services, allowing these complementary efforts to be fully co-ordinated. While preservation of biological resources including DNA is a primary function of the stock centres, they also strive to serve the rapidly changing needs of plant molecular biology research. This review will address the practical problems associated with all aspects of the ABRC DNA operation. Protocols utilized for clone characterization, preservation, and dissemination will be featured. Other topics which will be discussed are the philosophy of DNA centre operation, organization of information relating to DNA stocks, and the future of DNA centre operations.

2. Missions of a plant DNA resource centre

To provide services in the areas described above, a DNA centre must hold diverse and numerous stocks, maintain all of these without risk of loss, and be able to send samples of any of these items on very short notice. This must be achieved at minimal cost to the supporting research community, and pertinent information about the stocks must be maintained and distributed to the patrons of the centre. The items to be maintained and the relative demand for these evolve rapidly. Hence flexibility of operation and close contact with the research community is imperative.

The types of stocks currently held by ABRC include the following:

(a) Individual genomic clones bearing characterized genes in plasmids.
(b) cDNA clones in plasmids, including expressed sequence tags (ESTs).
(c) Marker clones identifying mapped RFLPs (in vectors such as phage, cosmids, and plasmids).
(d) Genomic libraries maintained as bulk stocks in phage and cosmid vectors.
(e) cDNA libraries isolated from a variety of tissues maintained as bulk bacteriophage or bacterial stocks.
(f) Arrayed large-insert genomic libraries including yeast artificial chromosome (YAC) libraries, bacterial artificial chromosome (BAC) libraries, and a P1 library.
(g) Yeast expression system libraries including two-hybrid libraries.

We will outline the preservation, distribution, and handling issues as well as current techniques associated with each of these stocks.

3. Preservation of stocks

Protocols for long-term storage of DNA stocks are well-tested and reliable for many vector types (1, 2). The stock centre context requires multiple back-ups for all stocks, similar to those devised for germplasm collections (3), assured stable storage conditions, easy sampling and manipulation of the stocks. These criteria are all considered when methods of handling are chosen by ABRC. In some cases, more than one choice is available. We will discuss the alternatives where appropriate.

3.1 Plasmids with small DNA inserts

A stock centre's activities normally are to receive, verify, maintain, and distribute clones, rather than to isolate and characterize them. Hence, the treatment of genomic and cDNA clones is similar. Likewise the ESTs, which are usually received as individual bacterial strains containing plasmids, do not pose unique problems, as they are deposited in standard vectors (4, 5). However, the rate at which ESTs are generated, donated, and ordered requires very careful consideration of all aspects of their handling, and procedures are modified whenever possible to simplify the handling and storage protocols.

3.1.1 Individual clones of genes

Most individual clones are received as plasmids replicating in *E. coli* hosts or as purified DNA. Many preservation methods are available for plasmids (6). Ultra-cold storage of the plasmid within the host bacterial cells is the most convenient method. Lyophilization of cells containing plasmids is also possible, as is cold storage of the dry plasmid DNA. The former method is

2: Preservation and handling of stock centre clones

presently utilized by ABRC. Back-up stocks of dried DNA are employed to ensure that stocks will not be lost. Lyophilization is not currently employed due to the initial start-up costs, effort, and uncertainty regarding long-range storage effects. To ensure that the strains containing specific clones are not lost, multiple samples are stored in ultra-low freezers that can be automatically transferred to back-up electrical power. Well-established techniques are employed in the handling of plasmid clones. Plasmid DNA is prepared using standard procedures such as alkaline lysis and/or commercially available kits (6). Liquid cultures of bacteria are held at –80°C in LB medium containing dimethyl sulfoxide (DMSO) or glycerol (1, 6). It is current centre policy to request a sample of isolated DNA from the donor in addition to a live culture. This greatly facilitates our ability to make the stock available quickly and ensure its safe preservation.

3.1.2 Expressed sequence tags (ESTs)

The same basic methods which are applied to other clones are also utilized for ESTs, except that bulk methods and short cuts are applied wherever possible. The ABRC's policy is to receive the ESTs as frozen glycerol stocks and utilize these stocks as the sole in-house representatives of the clones. Back-up DNA stocks are maintained at Michigan State University (MSU). If an ABRC glycerol stock becomes non-viable, DNA isolation is attempted from the existing glycerol stock, and if this fails DNA is obtained from MSU. This maximizes storage redundancy while avoiding the problems associated with reculturing and establishment of fresh stocks. The effort required to deposit the stocks into replicated storage is minimized, and the errors which might arise during subculturing are avoided. Numerous cultures can be made for shipment from each working stock.

3.1.3 Plasmid multiplication and distribution

The distribution of DNA clones requires consideration of four principles:

- ease and safety of handling
- maintenance of the purity of culture and stability of the insert
- viability of the culture
- costs of replication, storage, and shipping

Governmental regulations regarding shipment of clones must also be considered and can affect the methods employed. Some stocks present few problems in any of these areas, and others pose difficulties in all areas. Most ABRC stocks are maintained as live cultures and shipped as agar stabs from these cultures. The stocks represent starting points for experiments by users, are inexpensive and not overly labour-intensive. Hence, this method has been used extensively as a vehicle for stock distribution. The protocols for re-establishing user storage cultures from these are well defined. Individual stab cultures can be sent to users by regular mail and do not require cold shipping.

3.2 Cosmids
3.2.1 Cosmid storage
In principle, once created, a cosmid stock has the replicative and storage properties of a standard plasmid-carrying strain. However, the larger inserts are more difficult to maintain stably, especially in some of the vector/host combinations used in the pioneering *Arabidopsis* RFLP and physical mapping efforts (7–11).

Cosmid libraries are usually handled in bulk or as subdivided samples (see Chapters 3, 8, and 9) (7). However, the large numbers of cosmid clones associated with physical and genetic map construction are maintained as individual clones. These are stored as glycerol stocks in tubes, with restriction fingerprint data collected for typing purposes. For individual cosmids, live working and back-up cultures are stored as glycerol stocks with an additional dry DNA back-up stock, just as are plasmids. However, very clearly defined modifications of the handling procedures must be adhered to so that full inserts are retained (1, 9).

3.2.2 Cosmid handling and verification by users
For cosmid cultures, many of the vectors used for *Arabidopsis* require special care, and the protocols recommended for user handling of these clones are provided in *Protocol 1*.

Protocol 1. Cosmid maintenance and analysis

Reagents
- LB (Luria-Bertani) agar medium (per litre): 10 g Bacto tryptone, 5 g Bacto yeast extract, 10 g NaCl, 12 g Bacto agar pH 7.5
- Agarose (electrophoresis grade)
- 10 × TBE buffer: 108 g Tris base, 55 g boric acid, 9.3 g Na$_2$EDTA, made up to 1 litre with deionized water

Method
1. Streak cultures on LB medium with the appropriate antibiotic.
2. Incubate at 37 °C for 20 h.
3. Pick five to ten (preferably ten) small individual colonies and isolate cosmid DNA using standard alkaline lysis (6).[a] Make glycerol stocks from cultures (6) and later discard any shown to be deleted/rearranged.
4. Digest isolated DNA with a restriction enzyme. Conduct electrophoresis on digested DNA on 0.8% (w/v) agarose in 0.5 × TBE, and visualize with ethidium bromide (1, 6). The resulting band pattern should be compared to those published in original papers or the AIMS database (http://aims.cps.msu.edu/aims). Typically, the full-sized

2: Preservation and handling of stock centre clones

> Goodman laboratory cosmids result in five to ten bands when digested with restriction enzyme having a six base recognition sequence.

[a] It is important to pick the smallest colonies. Colonies that grow larger usually have deletions and may be unsuitable as probes for the intended genomic regions.

3.3 Phage and phage libraries

Phage can be maintained in several ways. Storage methods include liquid buffer suspensions held at refrigerator temperatures, glycerol stocks held at ultra-low freezer temperatures, or freeze-dried stocks (1, 6). Any of these methods are convenient, with only the latter requiring any specialized equipment. Since the aqueous stocks are relatively stable, easy and inexpensive to ship, and easy to aliquot, they are used for most ABRC libraries. Libraries received as glycerol stocks have also been maintained as such, with much success. Either of these methods results in stable stocks, which are easy to prepare and viable in storage. Care is exercised to assure that libraries are not amplified more times than necessary and that amplification procedures are unbiased. The general policy is that libraries are not accepted if they have been amplified more than twice prior to receipt.

For phage libraries, the buffer/DMSO- or glycerol-stored libraries are sent on dry ice using express carriers. The stocks, stored at the centre at –80°C in 7% (v/v) DMSO/SM buffer (7% (v/v) DMSO, 5.8 g/litre NaCl, 2 g/litre $MgSO_4 \cdot 7H_2O$, 0.05 M Tris–HCl pH 7.5), 0.002% (w/v) gelatine (sterilized) can be refrozen upon arrival. If they are not refrozen, they will survive up to several months at 4°C. Procedures for screening phage libraries, amplifying phage libraries, and isolating phage DNA are well established (1, 6).

3.4 Yeast artificial chromosome (YAC) libraries

3.4.1 YAC preservation

YAC clones present special problems for preservation. They are used for physical mapping and positional cloning, thus maintaining the integrity of these clones is imperative. Several YAC libraries have been received and are distributed (11–17). Fortunately, the genome complexity of *Arabidopsis* is such that a few hundred YAC clones span the genome with high probability (12–14, 17), and it is not necessary to resort to strategies such as pooled storage of clones. Working stocks of *Arabidopsis* YAC libraries are maintained as individual colonies gridded on plates which are stored, replicated, and then distributed in this form (14, 15). The protocols for replenishing and storing such libraries are straightforward, but require care and some specialized equipment. Storage also requires ultra-low temperatures, unless freeze-drying is employed. YACs are stored long-term in liquid glycerol medium at –80°C (see *Protocol 2*). Protocols are outlined below for YAC library replication (*Protocol 2*) as well as the isolation of DNA from an individual YAC clone (*Protocols 3* and *4*) (14, 15).

Protocol 2. Storage and replication of gridded yeast colonies

Reagents

- -URA–glycerol medium: 0.17% (w/v) yeast base without amino acids or ammonium sulfate (Difco, 0335-15-9), 0.07% (w/v) CSM-URA (Bio 101, 4511 222), 0.5% (w/v) ammonium sulfate, 2.0% (w/v) glucose, 20 mg/ml adenine hemisulfate (Sigma, A9126), 200 µl/ml glycerol
- -URA agar medium: as for -URA medium except glycerol is replaced with 1.8% (w/v) Bacto agar

Method

1. Inoculate colonies into microtitre plates containing -URA–glycerol and incubate at 30 °C for three days.
2. Store as permanent stocks in microtitre plates at –80 °C in -URA–glycerol medium.
3. Replicate library copies onto -URA microtitre plates. The plates are used as a working stock to obtain individual YAC clones or to make filter sets for hybridization.

3.4.2 Yeast colony distribution/storage

Individual yeast colonies are replicated onto agar for shipment to users with minimal material cost. If an entire library is sent, shipping costs are charged to the receiver since this requires use of express carriers. The end-user may want to conduct PCR, colony hybridizations, Southern hybridization, and/or recover the insert of an artificial chromosome. *Protocol 3* outlines a procedure for generating colony replicas in agar plugs which can be used for isolation of the artificial chromosome. A procedure for isolation of DNA from the yeast cultures for hybridization is given in *Protocol 4*, and a high density colony hybridization procedure from a library plate is also described (*Protocol 5*).

Protocol 3. Quick preparation of yeast cell plugs for PCR and other analyses

Equipment and reagents

- Parafilm
- Plug mould (Bio-Rad Corp.), or 1 ml syringe
- LET buffer: 0.5 M EDTA, 10 mM Tris–HCl pH 7.5
- -URA medium: 0.17% (w/v) yeast base without amino acids or ammonium sulfate (Difco, 0335-15-9), 0.07% (w/v) CSM-URA (Bio 101, 4511-222), 0.5% (w/v) ammonium sulfate, 2.0% (w/v) glucose, 20 mg/ml adenine hemisulfate (Sigma, A9126)
- YLB buffer: 0.1 M EDTA, 10 mM Tris–HCl pH 7.5, 1% (w/v) SDS
- LMP agarose: 1% (w/v) low melting point agarose in distilled water
- 50 mM EDTA pH 7.5
- 100 mM EDTA pH 7.5
- Zymolase-20T: 20 mg/ml zymolase 20-T (*Arthrobacter luteus*; Seikagaku Corporation)

Method

1. Inoculate 25 ml -URA broth in 50 ml disposable tubes with the yeast clone.
2. Incubate at 30°C shaking for two to three days. A very dense culture is required. The caps must be loose to allow adequate aeration and it is recommended to tape the lids onto the tubes.
3. Harvest cells by centrifuging at 1200 g, for 10 min.
4. Pour off the supernatant and wash cells in 20 ml of 50 mM EDTA pH 7.5. Centrifuge for 10 min at 1200 g.
5. Pour off the supernatant and resuspend in 12.5 ml of 50 mM EDTA pH 7.5.
6. Remove 750 μl to a microcentrifuge tube to be used for a miniprep.[a]
7. Centrifuge for 10 min at 1200 g.
8. Resuspend the pellet in 225 μl of 50 mM EDTA. Transfer the resuspended solution to an 1.5 ml microcentrifuge tube.
9. Add 12 μl of 20 mg/ml zymolase 20-T.
10. Add 375 μl of 1% (w/v) LMP agarose, quickly resuspend by pipetting up and down.
11. The following set of manipulations is conducted using a plug mould (step 12). If a plug mould is unavailable, a 1 ml syringe can be substituted, as outlined in step 13.
12. Plug mould procedure:
    (a) Pipette the agarose solution into the plug moulds.
    (b) After the plugs solidify carefully remove the plugs using a spatula into a Petri dish with 7 ml of LET buffer.
    (c) Incubate at 37°C overnight.
    (d) Pour off the LET buffer and add 7 ml of YLB buffer.
    (e) Incubate overnight at 40–50°C. At this point you should see release of pigment into buffer. This indicates efficient lysis.
    (f) Equilibrate plugs in 100 mM EDTA at room temperature.
13. Syringe procedure:
    (a) Draw the sample up past the neck of the syringe. Wrap Parafilm over the opening, and place on ice.[b]
    (b) After the agarose solidifies, cut off the tip of the syringe with scissors and transfer the material to a 15 ml centrifuge tube containing 7 ml of LET buffer.
    (c) Incubate at 37°C overnight.
    (d) Pour off the LET buffer and add 7 ml of YLB buffer.

Protocol 3. *Continued*

    (e) Incubate overnight at 40–50 °C. At this point you should see release of pigment into buffer. This indicates that the lysis has worked.

    (f) Equilibrate the 'plug' in 100 mM EDTA at room temperature.

    (g) Return the plug to the syringe. Pour off the buffer, allow the plug to come to the edge of the tube. Place the opening of the syringe on the top of the plug, pull back the plunger, and the plug will flow back into the syringe.

    (h) Seal the syringe with Parafilm.

14. Store plugs at 4 °C.

[a] The cell suspension can be frozen for miniprep isolation at a later date.
[b] Avoid air bubbles in the syringe, since they will cause a break in the plug.

Protocol 4. Yeast DNA miniprep employing liquid culture

Reagents

- Zymolase 20-T (see *Protocol 3*)
- TE buffer: 10 mM Tris–HCl, 1 mM EDTA pH 8.0
- 10% (w/v) SDS
- 5 M potassium acetate
- Absolute ethanol
- 80% (v/v) ethanol
- Phenol:chloroform (1:1): 1 vol. molecular biology grade phenol (saturated with 1 M TE pH 8) :1 vol. chloroform
- 20 mg/ml DNase-free ribonuclease

Method

1. Use the cells saved from the yeast plug preparation, or use fresh two to three day yeast culture.

2. Centrifuge cells in a microcentrifuge at 5000 r.p.m. for 5 min, and aspirate off the supernatant.

3. Resuspend the cells in 300 μl of 0.9 M sorbitol, 0.1 M EDTA pH 7.5 and 1% (v/v) 2-mercaptoethanol.

4. Add 5 μl of 20 mg/ml zymolase 20-T and incubate at 37 °C for 60 min.

5. Centrifuge spheroplasts for 10 sec in a microcentrifuge and resuspend in 300 μl TE pH 8.0.

6. Disperse cells by gentle pipetting.

7. Add 30 μl of 10% (w/v) SDS, and incubate at 65 °C for 30 min.

8. Add 100 μl of 5 M potassium acetate and incubate for at least 15 min at 4 °C or overnight.

9. Centrifuge for 15 min to pellet precipitate and transfer supernatant to a new tube. If the yeast cells had pigmentation the supernatant

2: Preservation and handling of stock centre clones

should be coloured. If the supernatant is not coloured the proteolytic lysis of the cell walls was efficient and the yield of genomic DNA will be low.

10. Extract with phenol:chloroform (1:1).
11. Add an equal volume of ice-cold ethanol.
12. Centrifuge at 4°C for 15 min (may be stored at −20°C prior to centrifugation).
13. Rinse the pellet with 80% (v/v) ethanol.
14. Centrifuge for 15 min, take off supernatant, and dry down in Speed Vac or air dry.
15. Resuspend in 200 µl of TE pH 8.0.
16. Add 1 µl of 20 mg/ml DNase-free RNase and incubate at 37°C for 30 min.
17. Check the DNA by electrophoresis on an 0.8% (w/v) gel with 30 µl of DNA stock. (Suggested loading, 30 µl for one genomic Southern.)

3.5 Pools of YAC library cells for PCR screening

Probing of the CIC YAC library (17) can be conducted via PCR in addition to the somewhat more traditional colony hybridization approach. In fact, when cultures of the library members are pooled in an appropriate combinatorial fashion, PCR amplification on a finite set of isolated DNAs can implicate the single clone carrying an insertion of interest. The CIC library consists of twelve 96-well microtitre plates, representing approximately five genome equivalents. Hence, the probability that a clone containing a specific gene exists in the library is high, and its identification can be achieved with minimal effort via PCR.

The CIC, three-dimensional pooling is as follows. First, the library is divided into 'superpools' within which the pooling is conducted. A superpool represents a set of clones small enough that the probability of finding at least one clone is less than 0.5, and the likelihood of a sequence being represented more than once within the superpool is low (i.e. less than one genome equivalent). This is necessary in that a single positive clone is identified uniquely by the screening process, but multiple clones result in exponentially increasing numbers of non-positive clones being implicated by the screening process. Consequently, there are four superpools (i.e. three plates) of clones, within each of which the three-dimensional pooling is conducted. Within the three plates, the dimensions are 'rows' (the actual rows of the plates), 'columns' (also columns of the plates), and 'half-plates'. Hence, within the three plates of a superpool, there are six half-plates, eight rows, and six columns (the first and seventh, second and ninth columns, etc. of each plate are considered a single 'column' for pooling purposes). This pooling strategy generates 20 total PCR samples within each superpool, for which a single positive in each of one row, column, and half-plate uniquely identifies a clone. Using this approach, a specific clone can be identified utilizing 24 PCR reactions.

The centre has received the cells of this library, pooled as described above, from R. Buell and C. Somerville. The cells are maintained in liquid culture and are sent as a total of 80 samples. DNA can be isolated from these cells directly, or the cultures can be grown to obtain samples from which DNA can be isolated to allow large numbers of primer pairs to be screened against the population (23). DNA can be prepared from the cultures using a scaled up method from Matallana *et al.* (23). PCR of the pools can be conducted using standard PCR enzymes and reagents, with annealing temperatures appropriate to the primers sequences/sizes being employed.

3.6 Distribution of YAC libraries arrayed on nylon filters

The distribution of filters ready for probing represents an attractive alternative to whole-library distribution for users wishing to perform a preliminary screen of a YAC library. Some added effort is required of the centre to produce filters suitable for hybridization, but this is offset by the opportunity to have this critical step conducted under controlled conditions and to generate multiple filters from a few plates (see *Figure 1*). These filters are randomly tested to ensure quality. Hence, users unacquainted with yeast culture technology can simply hybridize three nylon filters to locate colonies of interest and thus do not have to establish a myriad of procedures. For procedures for the preparation and hybridization of YAC filters, see Chapter 8.

3.7 Other large-insert libraries

BAC (20, 21) and P1 (18, 19) libraries may replace, or assist, YAC libraries for positional cloning in *Arabidopsis*. Both P1 and BAC libraries accommodate large inserts (approximately 100 000 bp, and potentially larger) and are maintained in *E. coli* strains that have inserts recoverable by standard miniprep procedures. These libraries are maintained as gridded sets of individual colonies in glycerol-containing medium on microtitre plates. They are normally distributed as filters ready for hybridization. Similar hybridization protocols can be used for these as for YAC filters (see Chapter 8 or ref. 6). When necessary, the entire library can be provided.

Individual clones of these libraries are distributed in the form of culture stabs, as a follow-up to filter probing. Hence, the procedures for handling these cultures are similar to those for plasmid and cosmid clones. The simplicity of DNA isolation and insert recovery for both P1 and BAC clones makes these highly attractive as vehicles for contig construction for positional cloning. The P1 lytic replicon (multicopy) is under the control of the *lacZ* promoter/operator to enable amplification of plasmid copy number for DNA isolation. However, Liu (personal communication) recently found that the copy number

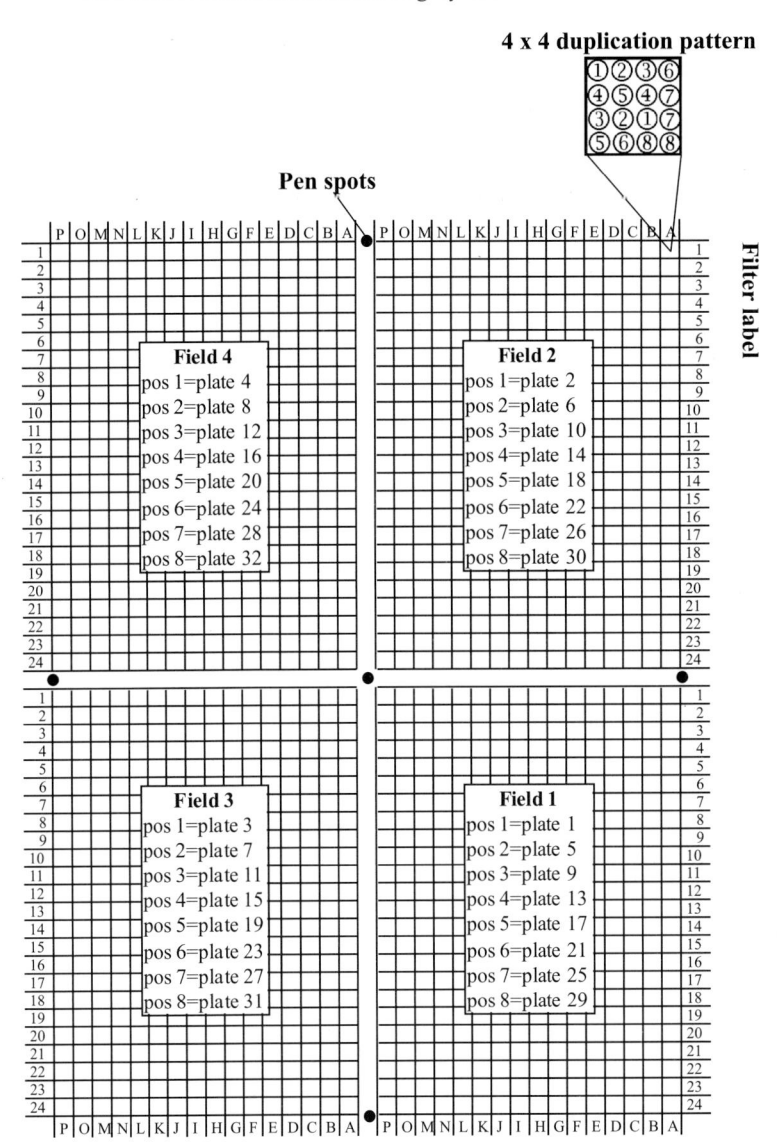

2: Preservation and handling of stock centre clones

4 x 4 duplication pattern

①	②	③	⑥
④	⑤	④	⑦
③	②	①	⑦
⑤	⑥	⑧	⑧

Pen spots

Field 4
pos 1=plate 4
pos 2=plate 8
pos 3=plate 12
pos 4=plate 16
pos 5=plate 20
pos 6=plate 24
pos 7=plate 28
pos 8=plate 32

Field 2
pos 1=plate 2
pos 2=plate 6
pos 3=plate 10
pos 4=plate 14
pos 5=plate 18
pos 6=plate 22
pos 7=plate 26
pos 8=plate 30

Field 3
pos 1=plate 3
pos 2=plate 7
pos 3=plate 11
pos 4=plate 15
pos 5=plate 19
pos 6=plate 23
pos 7=plate 27
pos 8=plate 31

Field 1
pos 1=plate 1
pos 2=plate 5
pos 3=plate 9
pos 4=plate 13
pos 5=plate 17
pos 6=plate 21
pos 7=plate 25
pos 8=plate 29

Filter label

Figure 1. Layout of BAC filters distributed for hybridization by users. Note that the filter is the size of four plates but that colonies of eight plates are superimposed on each of these four areas. Further, each plate position is spotted twice in a specific design noted as '4 × 4 pattern' on the figure. Hence each colony from a total of 32 plates can be unambiguously identified on the single filter hybridization.

in an *Arabidopsis* P1 library does not respond to IPTG, even at concentrations of 1–10 mM. In fact, IPTG was found to be counter-productive, lowering DNA yields by about 20%, due to the poor growth of cells when it is present. Even though the plasmid copy number is not amplified, a moderate amount (0.5–1 μg DNA/ml culture) of clone DNA can be obtained from a mini- or midi-preparation. The modified alkaline procedure outlined in *Protocol 5* was developed for P1 clones (18, 19). It was optimized for P1 DNA isolation by Dr Yao-Guang Liu over the course of his research at the Mitsui Plant Biotechnology Research Institute. This procedure should be applicable for BAC clones as well. It is designed for mini-preparation from 3 ml of culture, although DNA isolations can be scaled up proportionately if necessary. For larger scale preparations commercially available isolation kits such as the one from Qiagen may be employed.

Protocol 5. Preparation of P1 DNA, by a modified alkaline lysis method

Reagents

- LB agar medium (see *Protocol 1*)
- LB, 25 μg/ml kanamycin medium (per litre): 10 g Bacto tryptone, 5 g Bacto yeast extract, 10 g NaCl, 25 mg kanamycin pH 7.5
- 10 mg/ml lysozyme in 10 mM Tris–HCl pH 8.0
- TE, 20 μg/ml RNase A: 20 μg/ml RNase A, 10 mM Tris–HCl pH 8.0 1mM EDTA pH 8.0
- Solution I: 50 mM glucose, 25 mM Tris–HCl, 10 mM EDTA pH 8.0
- Solution II: 0.2 μl NaOH (freshly diluted from 10 N stock), 1% (w/v) SDS
- Solution III: 60 ml of 5 M potassium acetate, 11.5 ml glacial acetic acid, 28.5 ml water; chill to –20°C before use
- Phenol:chloroform (see *Protocol 4*)
- 70% (v/v) ethanol
- Absolute ethanol
- TE buffer: 10 mM Tris–HCl pH 8.0 1mM EDTA pH 8.0
- PEG solution: 1 M $MgCl_2$, 0.5 vol. of 40% polyethylene glycol 6000 or 8000 (note that no NaCl is included)
- 0.3 M sodium acetate

Method

1. Day one. Streak out cells containing P1 clone onto an LB agar plate containing 25 μg/ml kanamycin, culture at 37°C for 20–24 h.

2. Day two. Inoculate several colonies, each separately, into 4 ml (or more) of LB containing 25 μg/ml kanamycin (without IPTG). Cultures can be inoculated directly from storage stocks, but this often results in lower DNA yields or potential amplification of contaminating cells.

3. Culture at 37°C for 12–15 h, allowing the culture to reach saturation. Longer culturing is not recommended. However, the cells may be stored at 4°C for up to several days at this point without noticeable reduction in DNA yield or quality.

4. Day three. Set a refrigerated microcentrifuge to 4°C. Collect cells from 3 ml of culture into one 1.5 ml tube by pelleting cells twice at

2: Preservation and handling of stock centre clones

    8000 r.p.m. for 1 min each time. Remove any medium with a micropipette.

5. Resuspend the cells in 100 µl solution I (about 30–35 µl solution I for every 1 ml of culture) by vortexing. Add 10 µl of lysozyme solution and mix.

6. Add 200 µl of solution II. Mix by GENTLY inverting the tubes 10–15 times. Store the tubes on ice or at room temperature for 5–6 min, but no longer.

Note: in all the following steps, VORTEXING MUST BE AVOIDED. Vortexing will shear high molecular weight DNA, not only fragmenting the clone molecules, but increasing the levels of host chromosomal DNA contamination, as well.

7. Add 150 µl of solution III. Mix by gently inverting the tubes ten times. Store the tubes on ice for 3–5 min.

8. Invert the tubes three times, then centrifuge at maximum speed for 5 min at 4°C. At the end of this spin, set the microcentrifuge to 15°C.

9. Transfer the supernatant to fresh tubes. Add 0.8–1 vol. phenol:chloroform and mix by gently inverting. Centrifuge at maximum speed for 5 min. This phenol:chloroform extraction step should not be omitted; it helps reduce contamination by the host chromosomal DNA along with other impurities.

10. Add 2 vol. of ethanol at room temperature, mix by inverting, let stand for 2 min, and centrifuge at 12 000–16 000 r.p.m. for 5–10 min. Wash the pellet extensively with 70% (v/v) ethanol.

11. Air dry, avoiding over-drying. Redissolve the DNA pellet in TE containing 20 µg/ml RNase A (use 10 µl of TE per ml of culture) by heating at ~ 50°C and gently mixing.[a]

12. Store the DNA at 4°C. Although the DNA can be used directly for various enzymatic manipulations, the following steps are recommended to increase DNA purity and improve restriction by more finicky enzymes.

13. To each sample prepared from 3 ml of culture, add TE to a final volume of 110 µl or more. Insufficient dilution will lead to co-precipitation of impurities in the following steps. If either DNA yield or impurity levels are unusually high, more dilution is called for. Extract again with phenol:chloroform.

14. Add 0.01 vol. of PEG solution. Mix well, then pellet by centrifugation at maximum speed for 10 min at 20°C.

15. Wash the DNA pellet with 70% (v/v) ethanol and redissolve in 30 µl of 0.3 M sodium acetate. Reprecipitate with 2.5 vol. of cold ethanol.

Protocol 5. *Continued*

16. Final expected DNA concentration is 50–100 ng/ml.[b] DNA can be additionally purified by phenol extraction. Store it at 4°C. Shearing of DNA is recommended before use in random prime probe preparation.

[a] If the same DNA was prepared in more than one tube, samples can be pooled at this point.
[b] Correspondence regarding the method can be addressed to Dr Daisuke Shibata (tsu01136@koryu.statci.go.jp).

3.8 Yeast expression analysis — 'two-hybrid' libraries and complementation testing

The large number of selectable markers, expression systems, and transformation procedures developed for yeast make it very attractive for study of expression of *Arabidopsis* genes. This includes the identification of protein–protein interactions of *Arabidopsis*. The cloning system which identifies the interaction of the translation products of members of a eukaryotic cDNA library with the product of a clone of interest by the expression of both in yeast in conjunction with an interaction assay was perfected only in 1993 (22). Its use has been immediately adopted in all model eukaryotic organisms, including *Arabidopsis*. Since the basic library components are theoretically useful for identifications of any protein–protein interactions, one such library can be shared by a large number of users of a species. Hence, the ABRC has received and is distributing several two-hybrid libraries.

For the yeast two-hybrid system, the basic cDNA expression library is cloned into a lambda or plasmid expression vector; in the former case the insert is excised as a plasmid prior to transformation into yeast. These libraries are distributed either as phage suspensions from which the users excise plasmid in bacterial colonies from or as previously excised plasmid DNA to be used for direct transformation of yeast.

The system for expressing cDNA clones in yeast and identifying pairs of interacting proteins utilizes several yeast strains including one carrying the transformed cDNA library. This technique was initially developed by Durfee and co-workers (22). The currently distributed two-hybrid expression libraries (26) were supplied by the Walker and Theologis laboratories. Various modifications of techniques are in use for two-hybrid screening. The literature for specific libraries should be consulted for handling and strategies employed with a particular resource.

4. Verification of stock identity and purity

Clone identity and purity must be rigidly maintained and documented. Adherence to established microbial genetic practice largely accomplishes the

2: Preservation and handling of stock centre clones

required degree of control. Verification and checking by restriction analysis and occasional hybridization experiments constitute the main checks. Documentation of the results by photography represents an effective means of record keeping.

Individual donated plasmid clones are always streaked on appropriate selective medium, single resistant colonies picked, and separate minipreps grown and analysed by restriction analysis. A new culture, which will represent the ABRC stock, is established from a single colony having the correct restriction pattern and antibiotic phenotype.

Parallel procedures are employed for bacteriophage clones and cosmids. Failure will be assured if cosmid clones are handled improperly; particularly since the vectors for many of the established *Arabidopsis* cosmid vectors delete at high rates, and the colonies having the full insert grow very slowly. In this case, the very smallest colonies must be selected in the first step, and checked carefully prior to use of the stock as a probe.

Whenever a clone is received from the centre, it is important that the investigator carefully check the clone prior to use to avoid wasted effort (1, 6). The number of single clones picked for analysis varies for the type of stock. Two plasmid clones or phage plaques are usually sufficient. This applies to genomic clones and cDNAs, including ESTs. Adherence of the end-user to microbial practice and verification is important for ESTs in that these are not checked by the centre, due to the very large size of the current collection (over 25 000 ESTs), it is highly recommended that EST clones be sequenced to confirm the identity prior to use. Picking of up to ten colonies is recommended for cosmid clones.

In most cases quality control checks of phage libraries do not require large efforts. With modern technology, phage analysis is reasonably convenient. Enough phage DNA for analysis (500 µg) can be recovered from 30 ml liquid lysates by polyethylene glycol precipitation. Other procedures are carried out according to standard manuals (6).

Yeast artificial chromosome libraries are maintained as sets of individual clones, but these clones cannot be checked individually for correctness. However, some quality control is possible by preparation of filters blotted to the culture plates and hybridization of these to several known *Arabidopsis* probes. The location of hybridizing colonies is checked for each library, and this constitutes a reasonable statistical check. The integrity of the libraries is evaluated using hybridization against plastid, repetitive DNA, and ribosomal repeat sequences. In addition, all YAC libraries are maintained under appropriate biosynthetic selection, and released from selection only in the growth cycle immediately preceding a hybridization experiment. Genomic and cDNA libraries, which are maintained as bulk stocks, are not characterized extensively by the centre. Such libraries will typically be plated and hybridized with a few clones as a minimal check for representation.

5. Pooled DNA from T-DNA lines for PCR screening

There is great interest in conducting reverse genetics in all eukaryotic model organisms. This can done in *Arabidopsis* by one of several approaches. The currently most successful approach, and one for which the centre distributes stocks is PCR screening of T-DNA pools (see also Chapters 1 and 7). T-DNA lines are generated in larger numbers, DNA is isolated from these lines, and PCR is conducted using primers from a gene of interest paired with primers from the inserted T-DNA. These primer pairs theoretically only produce a PCR-generated band when a T-DNA insertion has occurred within or near the gene of interest, so that the two primer sites are opposed and a unique amplification product arises from PCR utilizing the primer pair. On this basis, the insertion line can be identified by simple PCR and hybridization techniques. The associated pools are obtained. Plants can be isolated from the pools and grown to evaluate the phenotypic effect of the insertion. This approach has been utilized to isolate a number of gene-specific, T-DNA insertion mutant lines.

Clearly, the task of locating an insertion occurring in a short genomic segment from a population of randomly generated T-DNA lines could be daunting. It would require 150000+ insertions to have a reasonable probability of knocking-out any single *Arabidopsis* gene at least once in the population. While screening the lines individually for insertion in a gene of interest is not feasible, PCR is sensitive enough that a single insertion line having opposed genomic T-DNA priming sites can be identified even if its DNA is pooled with as many as 1000 additional non-positive lines. Hence, it can be determined whether an insertion exists in a large population of T-DNA lines with a minimum number of PCR reactions.

If DNA isolation and pooling is conducted in a strategic fashion, the process of sorting through a population to locate the line of interest can also be conducted efficiently and lends itself well to the distribution facilities already existing at the resource centre. The ABRC is currently distributing pooled DNA from 38000 T-DNA lines and expects to receive additional lines so that the total number of insertions in-house will be more than 100000. The pooling strategy associated with these lines is as follows:

(a) Lines are organized into sets of 1000 (called 'superpools').

(b) Within each set of 1000, pools of 100 are established by combining lines into these pools in a two-dimensional matrix, combinatorial scheme. Hence, each line is represented in one of ten row pools of 100 and one of ten column pools of 100. The screening of these 20 DNA samples, following identification of a positive PCR in a superpool, locates the positive insertion to a single set of ten lines.

(c) Following a screen of the row and column pools of 100, the centre can then send the ten individual lines or a pool of these lines as seed for the experimenter to identify and isolate the desired line. Based on this strategy, when users request this stock, pools of 1000 are sent initially, the row–column DNAs are sent as follow-up for appropriate superpools, and then seeds of a single set of ten lines is sent following each successful analysis of the pools of 100. The entire distribution process then becomes a simple three-step procedure.

The protocol used for isolating plant genomic DNA from pooled lines (25) is described in *Protocol 6*. This protocol is for 10 g of frozen tissue using 250 ml Nalgene bottles. The DNA is isolated by a modified CTAB method and is diluted to 5 ng/μl for distribution, and 20 μl of each pool of 1000 are sent for initial screening. DNA quantities of 20 ng or less can be successfully employed for a single PCR reaction (see *Protocol 7*). Hence, it should be possible to run 20–40 reactions with each DNA pool included in the CD5-7 stock.

Protocol 6. Isolation of DNA from large quantities of plant tissue from pooled lines

Equipment and reagents

- Nalgene (250 ml) bottles
- Sterile Corex tubes
- Liquid nitrogen, mortar and pestle
- Buffer B: 2% (w/v) cetyltrimethyl-ammonium bromide (CTAB) pH 8.0 (Sigma)
- 100 mM Tris–HCl pH 8.0
- 20 mM EDTA pH 8.0
- 1.4 M NaCl
- CTAB: 10% in distilled water, pH to 8.0 with NaOH; 65°C
- Chloroform:isoamyl alcohol (24:1)
- TE (see *Protocol 4*)

Method

1. Add 10 ml of buffer B to liquid nitrogen ground tissue and place in a water-bath at 65°C for a minimum of 60 min (or until thoroughly heated). Swirl a few times during incubation and open slightly to release pressure.

2. Remove bottles from the water-bath and let cool. This is very important—care is needed to add chloroform slowly to hot solution.

3. Add 1 vol. (10 ml) of chloroform:isoamyl alcohol to the tissue/buffer B. Take the precaution to wear safety glasses and gloves whenever working with chloroform. Screw the lid tightly on the bottle and vortex for ~ 5 sec, release pressure by opening the lid, and vortex again for 15 sec.

4. Centrifuge for 10 min at 10 000 r.p.m. (GSA rotor/16G). Remove the bottles from the centrifuge, taking care not to mix the phases.

5. Immediately remove the aqueous (top) phase into new Nalgene bottles, using a pipette. This step is to be done as quickly as possible.

6. Repeat steps 3–5, EXCEPT at step 3, vortex just enough to mix; and at

Protocol 6. *Continued*

    step 5, pipette the top layer into a Corex tube. When pipetting the aqueous phase, be sure not to carry over any debris or chloroform.

7. If the solution is still greenish or not clear, repeat the chloroform extraction step.
8. Add 0.1 vol. (1 ml) of 10% (w/v) CTAB Mix by swirling. The solution will become cloudy after adding the CTAB, but will clear after several minutes. Place the bottles into the 65°C water-bath to help speed-up the clearing process.
9. Add 1.5 vol. of double distilled water (ddW), mix by swirling. Let the DNA precipitate for a minimum of 30 min at 4°C, or preferably overnight. Add additional ddW if no loose sediment is seen.
10. Centrifuge for 10 min at 10 000 r.p.m. in a GSA rotor; a pellet should be visible. If not, centrifuge for another 10 min.
11. Decant the supernatant and add 4 ml (8 ml total for 50–100 g tissue samples) of 1 M NaCl. Dissolve the pellet by placing the bottle into a 65°C water-bath for at least 5 min.
12. Transfer to smaller tubes (30 ml Corex, or 50 ml disposable centrifuge tubes). Add 10 ml (20 ml total for 50–100 g tissue samples) of 100% (v/v) ethanol, mix, and let the DNA precipitate for minimum of 30 min at 4°C.
13. Centrifuge for 6 min at 3000 *g*.
14. Be sure the pellet sticks firmly to the tube before decanting. Decant the supernatant.
15. Wash the pellet by adding 25 ml of 70% (v/v) ethanol, vortex gently until pellet loosens and breaks into pieces.
16. Centrifuge for 6 min at 3000 *g*.
17. Decant the supernatant and let the pellet partially air dry or vacuum oven dry. The pellet should not dry completely otherwise it will not dissolve.
18. Dissolve the pellet in 200–800 µl (3 ml for 50–100g of tissue) of TE, depending on the size of the pellet, overnight at room temperature.
19. Transfer the solution to an appropriate tube and check the DNA concentration by 260/280 nm UV spectrophotometry or fluorimetry.

PCR can be somewhat problematic for untested primers and substrates. All unique primers should be checked on genomic DNA prior to use of the DNA from T-DNA pools. If difficulties arise the control primers can be applied to the appropriate control DNA pool of the kit. Otherwise, it is suggested that the pool DNA can be conserved by omitting the second step. The control primers for the CD5-7 pools are for *act2* (2) and this mutant is included in pool 1 of CD5-7. The associated 0.9 kb DNA can be amplified using the included control right border and *ACT2* primers. The positive *act2-1* line of the control T-DNA pool possesses right borders on each side of the insert (2), so primers homologous to the right border are necessary. This product can also be identified in PR600 and PC1000 of the Feldmann pools.

The amplification procedure is described in *Protocol 7* (24, 25). It should be strongly emphasized that PCR amplification be followed-up with Southern hybridization to authenticate potential positive products, since border–border bands are extremely common. Note that since pBR vector sequences are included in the T-DNA vector, the probes employed in these hybridizations should therefore be pBR-free. It is also very useful to employ negative PCR controls such as PCR reactions with each single primer employed in the experiments. A small sample of a putative positive band can also be re-amplified, subcloned, and sequenced for verification.

Protocol 7. PCR of T-DNA pools to identify specific insertions

Equipment and reagents

- 1 × PCR buffer: 20 mM Tris-HCl pH 8.4, 50 mM KCl, 1.5 mM MgCl$_2$, 0.1 mM dNTPs (the 10 × buffer supplied with the *Taq* DNA polymerase can be used)
- Control primer stocks: 10 μM of *ACT2-1* and RB primer, respectively, in TE
- 50 μl reactions with 25 picomoles of each primer per tube are convenient
- 'Hot start' *Taq* DNA polymerase: enzyme provided with commercial PCR kits. The high-fidelity versions of the enzyme are preferable. Less can be used, although it is advisable to utilize higher quantities for the border primer.

Method

1. Program PCR machine. Use 'hot starts' for all PCR reactions employing the border/gene specific primer pairs and the T-DNA pools.[a] Initial denaturation at 95 °C for 1–15 min is needed, followed by 30–45 cycles of 94 °C 30 sec; 60 °C[b] 30 sec; 72 °C 2 min; and then a 10 min 72 °C extension cycle. Note that for pools of 100 lines, 30 cycles are usually sufficient, but 45 cycles may be required for pools of 1000.
2. Make 1 × PCR reaction mixture with all components except DNA using 49 μl per reaction, and store on ice. It is sometimes necessary to vary the Mg^{2+} concentration to achieve optimal amplification.
3. Add 1 μl of pooled DNA to PCR tubes on ice.
4. Transfer tubes to PCR block set at 4 °C.
5. Add PCR reaction mixture to PCR tubes containing DNA.
6. Add mineral oil if necessary. Start cycles.

[a] Hot starts are essential for optimal results.
[b] PCR annealing step temperatures should be set to optimize conditions for the specific primer of each PCR primer pair; lower temperatures result in amplification of complicating non-specific bands.

6. Organization of stock information

The data associated with DNA stocks are varied, complex, rapidly evolving, and interrelated. Hence, data management is a complex issue. Stock centres do not serve as the primary organizational point for all such data, but must have access to these data, and must provide detailed information about clones to researchers.

6.1 Collecting, maintaining, and disseminating stock data

The solutions to the data collection problems of a stock centre are twofold. First the stock collection process is designed so that stock donors provide detailed data at the time of donation, and secondly, creating electronic access to the summarized data so that users have maximum use of the information. Donation forms asking for all pertinent information in an organized format are vital as is the identification of publications relating to the stocks. Users are asked to provide all donation information in electronic form, whenever possible. Nucleotide sequence and related information such as homology search results should be accessible. Some data must be maintained at the stock centre while others, such as sequence data, may best be organized elsewhere —specifically at the large sequence databases dedicated to providing this service. Even with ready access to databases, a local database maintained on either a microcomputer or a workstation is required for the organization of stock centre data and provides additional back-ups to ensure data is not lost. Allocation of significant fractions of the centre's personnel time is required for data management.

6.2 Organizing and distributing patron data

A stock centre must carefully organize and update all information about its users. The necessity to communicate with the researchers allied to the centre, and in some cases with specific sub-groups of the users, requires the careful organization and rapid mobilization. A number of microcomputer programs are very efficient for these tasks. The extent to which these lists are made public are at the discretion of each centre. It is ABRC policy that all stock request records are public and can be accessed through AIMS (http://aims.cps.msu.edu/aims).

7. The future

It is clear that demand for DNA stocks will continue to increase rapidly in the near future and that the number and complexity of stocks with which stock centres will have to deal will also increase. In some cases, adding new types of stocks will mean greater need for financial support, but the labour and technical demands may actually be reduced by others. The recent development of various types of large-insert libraries which are easier to use is a good example. The centre has received an *Agrobacterium*-transformable Kazusa DNA BAC library ('TAC' library) from Research Institute (29), which can been used to rapidly clone genes using positional/complementation principles. Also, robotics is allowing the centre to provide additional services. For example, a set of over 11 000 non-redundant ESTs has been developed in conjunction with the cDNA analyses conducted by Dr Steven Rounsley, formerly of The Institute for Genomic Research (TIGR), and these are now distributed as a set.

A physical map of *Arabidopsis* is now essentially complete (27, 28). The majority of the associated markers including BACs, YACs, and mapped ESTs are available from the resource centre (18, 20). ABRC possesses the BAC libraries utilized by the *Arabidopsis* genome initiative. BAC and P1 clones are easier than are YACs for the user to manipulate, although the insert sizes are currently smaller. Hence, distribution of these as either filter sets, pooled DNAs, or grids of clones require automated handling. The primary solution to the distribution problems posed by these resources may be to distribute the full library only as filters for hybridization, with individual clones from the library available for direct shipping.

In conclusion, the ABRC has established an extensive collection of DNA and seed stock items and has sent over 250 000 samples to researchers throughout the world during the first six years of stock distribution. The centre has already moved several steps beyond its initially envisioned scope by acquiring the large collection of ESTs, large-insert libraries, and isolated DNA for PCR screening. The next foreseeable steps will involve organized sets of mutants (e.g. a systematic insertion mutant collections representing all known *Arabidopsis* genes, a minimum-tiling-path clone set, and a complete set of cDNAs which may consist of full-length clones). It is obvious that similar sudden, unanticipated reorientations will be necessary in the future. The challenge will be to serve researchers with divergent needs while maintaining reasonable economy of operation.

References

1. Ausubel, F. M., Brent, R., Kingston, R. E., Moore, D. D., Seidman, J. G., Smith, J. A., *et al.* (1986). *Current protocols in molecular biology*. Wiley Interscience, New York.
2. Maniatis, T., Fritsch, E. F., and Sambrook, J. (ed.) (1982). *Molecular cloning: a laboratory manual*. Cold Spring Harbor Press, Cold Spring Harbor, NY.
3. Stanwood, P. C. (1985). In *Cryopreservation of plant cells and organs* (ed. K. K. Kartha). CRC Press, Boca Raton, FL.
4. Newman, T. C., de Bruijn, F. J., Green, P., Keegstra, K., Kende, H., McIntosh, L., *et al.* (1994). *Plant Physiol.*, **106**, 1241.
5. Höfte, H., Amselem, J., Chiapello, H., Caboche, M., Moisan, A., Jourjaon, M. F., *et al.* (1993). *Plant J.*, **4**, 1051.

6. Sambrook, J., Fritsch, E. F., and Maniatis, T. (ed.) (1989). *Molecular cloning: a laboratory manual*, 2nd edn. Cold Spring Harbor Press, Cold Spring Harbor, NY.
7. Nam, H. G., Giraudat, J., den Boer, B., Moonan, F., Loos, W. D. B., Hauge, B. M., *et al.* (1989). *Plant Cell*, **1**, 699.
8. Giraudat, J., Hanley, S., Goodman, H. M., and Hauge, B. M. (1991). *Symp. Soc. Exp. Biol.*, **45**, 45.
9. Hauge, B. M., Hanley, S. M., Cartinhour, S., Cherry, J. M., Goodman, H. M., Koornneef, M., *et al.* (1993). *Plant J.*, **3**, 745.
10. Hauge, B. M., Giraudat, J., Hanley, S., Huang, I., Kohchi, T., and Goodman, H. M. (1990). *J. Cell. Biochem.*, (suppl.) **14**, 259.
11. Huang, I., Kohchi, T., Hauge, B. M., Goodman, H. M., Schmidt, R., Cnops, G., *et al.* (1991). *Plant J.*, **1**, 367.
12. Grill, E. and Somerville, C. (1991). *Mol. Gen. Genet.*, **226**, 484.
13. Ward, E. R. and Jen, G. (1990). *Plant Mol. Biol.*, **14**, 561.
14. Guzman, P. and Ecker, J. (1990). *Plant Biol.*, **11**, 95.
15. Guzman, P. and Ecker, J. (1988). *Nucleic Acids Res.*, **16**, 11091.
16. Ecker, J. (1990). *Methods*, **1**, 186.
17. Creusot, F., Fouilloux, E., Dron, M., Lafleuriel, J., Picard, G., Billault, A., *et al.* (1995). *Plant J.*, **8**, 763.
18. Liu, Y.-G., Mitsukawa, N., Vezquez-Tello, A., and Whittier, R. F. (1995). *Plant J.*, **7**, 351.
19. Liu, Y.-G. and Whittier, R. F. (1994). *Nucleic Acids Res.*, **22**, 2168.
20. Zhang, H.-B., Zhao, Z., Ding, X., Paterson, A. H., and Wing, R. A. (1995). *Plant J.*, **7**, 175.
21. Woo, S.-S., Jiang, J., Gill, B. S., Paterson, A. H., and Wing, R. A. (1994). *Nucleic Acids Res.*, **22**, 4922.
22. Durfee, T., Becherer, K., Chen, P.-L., Yeh, S.-H., Yang, Y., Kilburn, A. E., *et al.* (1993). *Genes Dev.*, **7**, 555.
23. Matallana, E., Bell, C. J., Dunn, P. J., Lu, M., and Ecker, J. R. (1992). In *Methods in Arabidopsis research* (ed. C. Koncz, N. Chua, and J. Schell), pp. 144–69. World Scientific Publishing.
24. Winkler, R. G. and Feldmann, K. A. (1998). In *Methods in molecular biology: Arabidopsis protocols* (ed. J. M. Martinez-Zapater and J. Salinas) V. 82 p. 129, Humana Press, Totowa, N.J.
25. McKinney, E. C., Ali, N., Traut, A., Feldmann, K. A., Belostotsky, D. A., McDowell, J. M., *et al.* (1995). *Plant J.*, **8**, 613.
26. Kim, J., Harter, K., and Theologis, A. (1997). *Proc. Natl. Acad. Sci. USA*, **94**, 11786.
27. Meinke, D. W., Cherry, J. M., Dean, C., Rounsley, S. D., and Koornneef, M. (1998). *Science*, **282**, 662.
28. Rhee, S. Y., Weng, S., Flanders, D., Cherry, J. M., Dean, C., Lister, C., *et al.* (1998). *Science*, **282**, 663.
29. Kaneko, T., Kotani, H., Nakamura, Y., Sato, S. Asamizu, E., Miyajima, N., *et al.* (1998). *DNA Research*, **5**, 131.

3

Genetic mapping using recombinant inbred lines

CLARE LISTER, MARY ANDERSON, and CAROLINE DEAN

1. Introduction

New markers can be positioned on a genetic or molecular map by comparison of the segregation of parental alleles in a population containing recombinant chromosomes, with the segregation data of other markers in the same population. Marker order is determined by identifying which markers (e.g. i–vii) co-segregate in a number of recombinant chromosomes (schematically illustrated in *Figure 1*). The fewer the recombination events that separate two markers the closer together they map on a chromosome. If one recombination event separates two markers in 100 chromosomes, then the two markers are considered to map 1 centimorgan (cM) apart.

The central requirement for genetic mapping is therefore the ability to distinguish the parental alleles at each locus. Phenotypic markers were the basis of the original genetic maps. The first molecular markers were restriction fragment length polymorphism (RFLP) markers (1) (see Section 3.1). These are genomic DNA fragments that show a difference in size between restriction digests of parental DNA, as assayed using Southern blot hybridization, with a particular probe. RFLP markers are generally low-copy sequences and have the advantage over phenotypic markers in that they are co-dominant, i.e. heterozygotes can be scored by virtue of containing both alleles. More recently, a number of polymerase chain reaction (PCR)-based markers have been developed that significantly speed-up the mapping procedure (see Section 3.2).

The first two *Arabidopsis thaliana* RFLP maps were generated using F2/F3 populations, derived from crosses between different ecotypes (2, 3). 15 or more individual F3 plants, derived from segregating F2 plants (see *Figure 2a*) were pooled and the extracted DNA used for mapping. The major disadvantage of using F3 families for mapping is that when the F3 seed are exhausted new populations must be generated and the segregation of markers re-analysed to produce a new map.

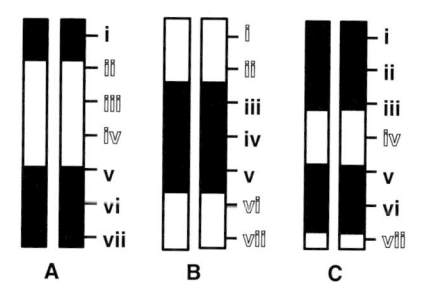

Figure 1. Schematic illustration of recombinant chromosomes. Three pairs of recombinant chromosomes (A, B, and C) are homozygous for all loci shown. Black represents one genotype, white the other genotype. The composition of each recombinant chromosome is determined by analysis of the segregation of the RFLP markers at the seven loci (i–vii). The order of the markers is determined by the position of recombination events.

An alternative mapping population, used successfully in maize (4), wheat (5), and pea (6) mapping is a recombinant inbred (RI) population. A large number of F2 plants are selfed and then one progeny F3 plant, chosen at random from each F2, is allowed to self and set seed. This process is repeated for sufficient generations until the lines are highly homozygous (see *Figure 2b*). Theoretically, at the F8 generation only 0.78% heterozygosity should remain in the population.

RI lines have many advantages for mapping markers compared with F3 families:

(a) They constitute a stable and near-homozygous population.

(b) There is an inexhaustible supply of seed which can be distributed to many laboratories.

(c) All mapping can be carried out on the same population, providing an increasingly dense and useful molecular map.

Since mapping still requires the identification of a polymorphism between the parental lines it is therefore advantageous to have multiple mapping populations, derived from different parents. If a polymorphism is not detected between one set of parental ecotypes, it can often be detected between another. A number of *A. thaliana* RI populations are available (7–10). These lines have been generated from different pairs of parental ecotypes. Reiter et al. (7) used Wassileskija (WS) and W100 (a multi-marker line based on the Landsberg *erecta* phenotype; see Chapter 1) as the parents. Lister and Dean (8) used Landsberg *erecta* (L*er*) and Columbia (Col) and these RI lines have become the canonical lines for mapping in *A. thaliana*. Two sets of lines, from crosses between the ecotypes Col-0 with Nd-1 and Wei with Ksk-1 were generated by Holub and Beynon (9). These RI lines were specifically developed to map fungal resistance genes (where one parent is sensitive to a particular pathogen and the other is resistant), but may also be used for mapping with

3: *Genetic mapping using recombinant inbred lines*

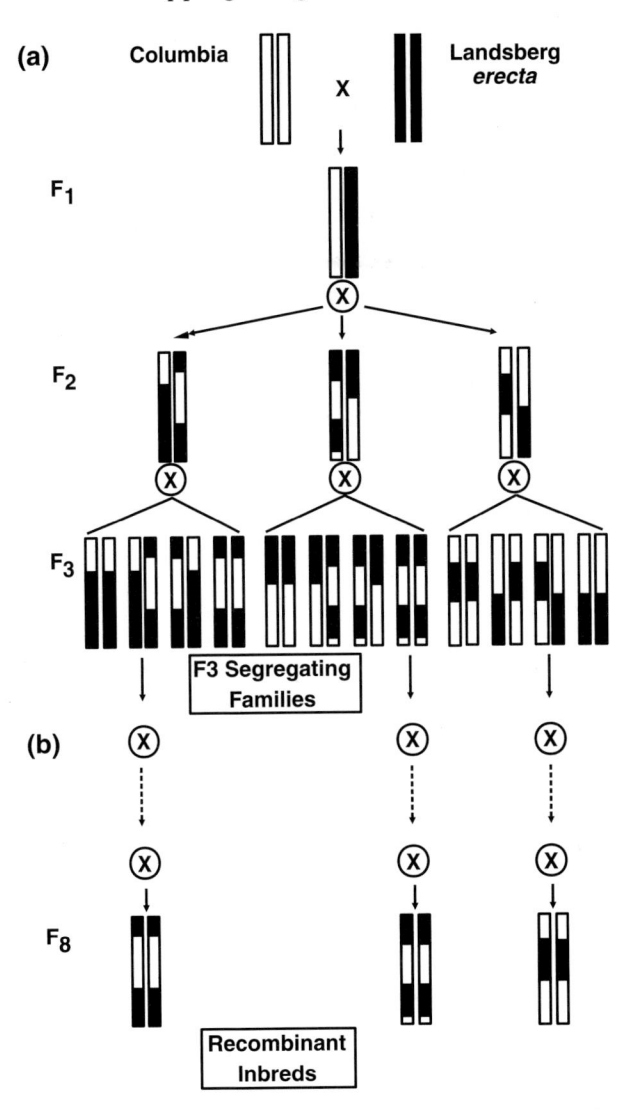

Figure 2. Production of recombinant inbred lines. Recombination, occurring at meiosis in the F1, results in recombinant chromosomes which assort randomly in the F2. Additional rounds of recombination can also occur in the F2 or subsequent generations. For simplicity these additional recombination events have been omitted from the figure. (a) F2 plants are selfed to produce F3 segregating families. (b) A single plant from each F3 family is selfed and taken by single-seed descent through at least five generations until it becomes virtually homozygous.

other markers or traits. The most recent RI population was generated by Alonso-Blanco et al. (10) from a cross between the ecotypes L*er* and Cape Verde Islands (Cvi). The parents of the L*er* × Col RI lines are both common laboratory stocks, whereas the ecotype Cvi comes from a distant geographical location and was expected to show a greater degree of polymophism with Col and L*er* than the level of variation between Col and L*er*; this was confirmed by AFLP examination of the rates of polymorphism between these three ecotypes (see Section 3) (10).

All these RI lines are available from the Nottingham *Arabidopsis* Stock Centre (NASC) (see Section 4 and Chapter 1).

At least 312 RFLP markers, mainly RAPDs (random amplified polymorphic DNA) (11), have been positioned on the RI lines of Reiter et al. (7). Approximately 800 markers have been positioned on the RI lines of Lister and Dean (8) (NASC Internet page; see Section 4) and an additional 395 AFLP markers have been mapped on these lines (10). A total of 321 AFLP markers were mapped on the L*er* × Cvi RI lines and the map of these RI lines has been integrated with the map of the Col × L*er* RI lines using 49 common markers (10).

The mapping process consists of the following:

(a) Grow the parental ecotypes and the RI lines to generate plant material from which to prepare genomic DNA (see *Protocols 1* and *2*).

(b) Prepare the genomic DNA (see *Protocols 3* and *4*).

(c) Identify a polymorphism between the parental ecotypes.

(d) Score the polymorphism on genomic DNA from each of the RI lines, or the phenotype.

(e) Analyse the data obtained using mapping programs.

2. Preparation and digestion of *A. thaliana* genomic DNA

It is important that the DNA used in mapping is of sufficient quality to digest completely with restriction enzymes or pure enough for PCR reactions. Partially restricted DNA may lead to the incorrect identification or missing of a polymorphism. Genomic DNA containing a high concentration of salts or oxidized phenolic compounds (which makes the DNA pellet brown) may not be suitable for restriction endonuclease digestion or PCR reactions. Therefore the conditions of growth of plant material for DNA preparation and the preparation method itself are important; both leaf material or *in vitro* grown root material can be used for DNA isolation. *Protocols 1* and *2* describe optimal conditions for the generation of leaf and root material, respectively.

3: Genetic mapping using recombinant inbred lines

Protocol 1. Growth of *Arabidopsis* leaf material for DNA preparation

Equipment and reagents

- −80 °C freezer
- Small plant pot[a]
- Clingfilm
- Propagator trays with lids[b]
- Glasshouse or controlled environment cabinet[c]
- Aluminium foil
- *A. thaliana* seeds (parental ecotypes and RI lines)
- *Arabidopsis* compost mix (1:1:1, John Innes Compost No. 1: grit: peat)
- Liquid nitrogen

Method

1. Fill a small pot with *Arabidopsis* compost mix, water well, and sprinkle the seed onto the surface.
2. Cover with a propagator lid or Clingfilm and place in a glasshouse or growth cabinet.[c]
3. Allow the seeds to germinate and grow to about the four leaf stage.
4. Prick out the seedlings into a propagator tray containing moist *Arabidopsis* compost mix and cover.
5. Grow the plants until they start to bolt.
6. Harvest 3–5 g of leaf material,[d] wrap it in aluminium foil, freeze immediately in liquid nitrogen, and store at −80 °C.

[a] In which to germinate the seed.
[b] Trays may contain individual cells for each plant, if not then seedlings should be pricked out with 3–4 cm between each plant.
[c] To maximize the amount of leaf material generated plants can be grown in short day (SD) conditions (10 h light, 20–25 °C). Plants flower later in SD conditions, therefore producing more leaves before flowering.
[d] Where possible avoid using leaf material containing high levels of anthocyanin or which has started to senesce.

Protocol 2. Generation of *Arabidopsis* root material for DNA preparation

Equipment and reagents

- Sterile 250 ml flasks
- Sterile 1.5 ml plastic tubes
- Sterile plastic pipette tips
- Sterile plastic Petri dishes
- Sterile 7 cm filter paper discs (Whatman No. 1)
- Laminar flow hood
- Rotary shaker, in controlled environment room (20 °C ± 2 °C, 16 h photoperiod)
- Paper towels
- Aluminium foil
- −80 °C freezer
- *A. thaliana* seeds (parental ecotypes and RI lines)
- Sterile IM media: 1 × Gamborg's B5 medium[a] (without sucrose and agar), 2% (w/v) glucose, 0.5 g/litre MES, pH 5.7 with 1 M KOH

Protocol 2. *Continued*

- Liquid nitrogen
- 70% (v/v) ethanol in distilled water
- Sterile distilled water
- Sterilization solution: 5% (v/v) sodium hypochlorite, 0.1% (v/v) Tween 20 (Sigma) in distilled water

Method

1. Place the seed to be sterilized into a sterile 1.5 ml tube and add 1 ml 70% (v/v) ethanol.
2. Shake the tube gently for 2 min, allow the seeds to settle, and remove the ethanol.
3. Add 1 ml of sterilization solution and shake gently occasionally, for 15 min. Allow the seeds to settle and remove the sterilization solution.
4. Add 1 ml of sterile water and mix. Allow the seeds to settle and remove the water.
5. Repeat step 4 four more times.
6. Remove the seeds from the tube and dry on sterile filter paper in a sterile Petri dish, in a laminar flow hood.
7. Add 50 ml of IM to a sterile 250 ml flask and add 10–15 seeds.[b,c]
8. Shake the seeds gently, at approx. 150 r.p.m., on a rotary shaker for four to six weeks until the flask is full of roots.[d]
9. Remove root mass from flask and dry on paper towels. Wrap in aluminium foil and freeze immediately in liquid nitrogen. Store at −80 °C.

[a] Imperial Laboratories (Europe) Ltd.
[b] This number of seeds are required in case of any residual heterozygosity in the RI lines.
[c] The remaining sterilized dry seed in the Petri dish can be stored at room temperature, after sealing the Petri dish.
[d] Discard any root cultures that become contaminated with fungi or bacteria or become brown in colour.

2.1 Preparation of genomic DNA

The first method described (see *Protocol 3*) has been shown to reliably produce a good yield of relatively pure genomic DNA from leaf or root material, without the requirement for CsCl gradients, and is suitable for both restriction endonuclease digestion and PCR reactions. The second method (*Protocol 4*) is a small scale, rapid method which yields enough DNA for one or two restriction digests, and is suitable for use with most PCR markers.

3: Genetic mapping using recombinant inbred lines

Protocol 3. Large scale preparation of *A. thaliana* genomic DNA[a]

Equipment and reagents
- Pestle and mortar
- Sterile plastic tubes (50 ml, 15 ml, 1.5 ml)
- Sterile plastic pipette tips
- Sterile glass pipettes (10 ml, 25 ml)
- Rotator/shaker
- Benchtop centrifuge/low speed centrifuge
- Microcentrifuge
- Tissue paper
- 'Parafilm' (American National Can, supplied by Merck Ltd.)
- Metal spatula
- 65°C water-bath
- Vacuum desiccator
- Liquid nitrogen
- Chloroform
- Spatula
- CTAB buffer: 140 mM sorbitol, 220 mM Tris–HCl pH 8.0, 22 mM EDTA pH 8.0, 800 mM NaCl, 1% (w/v) Sarkosyl, 0.8% (w/v) cetyltrimethylammonium bromide (CTAB) pH 8.0, autoclaved at 121°C for 20 min
- Phenol (saturated with 10 mM Tris–HCl pH 8.0, 1 mM EDTA)
- Phenol (saturated):chloroform (1:1)
- Isopropanol
- 100% ethanol
- TE: 10 mM Tris–HCl, 1 mM EDTA pH 8.0
- 3 M sodium acetate pH 5.5
- 4 M lithium acetate
- Undigested lambda genomic DNA (Gibco BRL)
- 0.7 % (w/v) agarose gel

Method
1. Grind 3–5 g of the frozen leaf or root material to a very fine powder in a liquid nitrogen-cooled pestle and mortar. Do not allow the plant material to thaw at any point.
2. Transfer the powder to a 50 ml tube containing 20 ml of CTAB buffer at room temperature.[b]
3. Incubate the tube at 65°C for 20 min, with occasional mixing by inverting.
4. Add 10 ml of chloroform, mix well, and place it on a rotator/shaker for 20 min at room temperature.
5. Centrifuge the tube for 10 min at 2500 g in a benchtop/low speed centrifuge.
6. Transfer the aqueous phase to a fresh 50 ml tube,[c] add 17 ml of isopropanol, mix, and place on ice for 10 min.
7. Centrifuge the tube for 5 min at 2500 g to collect the precipitate.
8. Drain off the liquid and dry the sides of tube using tissue paper; do not dry the pellet.
9. Add 4 ml of TE and dissolve the pellet by gentle shaking.
10. Add 4 ml of 4 M lithium acetate and incubate on ice for 20 min.
11. Centrifuge for 10 min at 2500 g.
12. Transfer the supernatant to fresh 50 ml tube, add 16 ml of 100% ethanol, and incubate on ice for 20 min.
13. Centrifuge for 5 min at 2500 g to collect the precipitate.

Protocol 3. *Continued*

14. Drain off the liquid and dry the sides of the tube using tissue paper; do not dry the pellet.
15. Add 4 ml of TE and dissolve the pellet by gentle shaking, then add 400 µl of 3 M sodium acetate pH 5.5.
16. Transfer to a 15 ml tube containing 4.5 ml phenol.
17. Mix thoroughly by inverting and centrifuge for 10 min at 2500 g.
18. Remove the aqueous layer to a fresh 15 ml tube containing 4.5 ml phenol:chloroform (1:1).[d]
19. Mix thoroughly by inverting and centrifuge for 10 min at 2500 g.
20. Remove the aqueous layer to a fresh 15 ml tube containing 4.5 ml chloroform.
21. Mix thoroughly by inverting and centrifuge for 10 min at 2500 g.
22. Remove the supernatant to a fresh 50 ml tube, add 2 vol. of 100% ethanol, and incubate on ice for 20 min.
23. Centrifuge for 15 min at 2500 g to collect the DNA.
24. Drain off the liquid and invert the tube on a piece of Parafilm, tap the DNA pellet onto the film, and then tip it into a 1.5 ml tube.
25. Centrifuge for 5 min at 15 000 g in a microcentrifuge and remove the liquid with a pipette tip. Repeat until no more liquid comes off.
26. Dry the pellet in vacuum desiccator for about 10 min and dissolve the pellet in 200 µl TE.[e]
27. Determine DNA concentration on a 0.7% (w/v) agarose gel, using undigested lambda DNA as a concentration standard.[f,g]

[a] Adapted from ref. 12.
[b] Use a liquid nitrogen-cooled spatula to transfer the powdered material.
[c] Use the 'wrong' end of a 10 ml glass pipette to remove the aqueous layer; this minimizes disturbance of the interface.
[d] Cut the end off a blue plastic pipette tip to transfer the aqueous layer; this minimizes disturbance of the interface.
[e] Do not let the pellet get too dry or it is very difficult to redissolve.
[f] There is often polysaccharide which co-purifies with the DNA; this does not affect the DNA digestion, but can make the solution very viscous.
[g] Yields of DNA can vary considerably but an average yield would be approx. 15 µg/g leaf material.

Protocol 4. Small scale preparation of *A. thaliana* genomic DNA[a]

Equipment and reagents

- Micro-pestle and grinder (Anachem)
- Sterile plastic 1.5 ml tubes
- Sterile plastic pipette tips
- CTAB buffer see (*Protocol 3*)
- Chloroform
- Phenol (saturated):chloroform (1:1)

3: Genetic mapping using recombinant inbred lines

- 65°C water-bath
- Microcentrifuge
- Vacuum desiccator
- −20°C freezer
- Liquid nitrogen
- Glass powder (Sigma)
- Spatula

- Isopropanol (−20°C)
- 100% ethanol
- 70% (v/v) ethanol in distilled water
- TE: 10 mM Tris–HCl, 1 mM EDTA pH 8.0
- 3 M sodium acetate pH 5.5
- Undigested lambda genomic DNA (Gibco BRL)
- 0.7 % (w/v) agarose gel

Method

1. Harvest a small, single plant or an excised leaf (approx. 0.05 g) into a 1.5 ml tube and place immediately into liquid nitrogen.[b]
2. Remove the tube from the liquid nitrogen and add a small spatula full of glass power to the tube. Chill the tube by immersing in liquid nitrogen and allow a small amount to enter the tube.
3. Use a chilled micro-pestle in the grinder to grind the plant material into a fine powder. Do not allow the material to thaw.
4. Add 1 ml of CTAB buffer and mix. Incubate the tube at 65°C for at least 15 min, mixing occasionally.[c]
5. Add 0.4 ml of chloroform and mix well by inverting the tube.
6. Centrifuge for 10 min, at 15 000 g, in a microcentrifuge and transfer the aqueous phase to a fresh 1.5 ml tube.
7. Add 0.7 ml cold isopropanol and incubate at −20°C for 30 min.
8. Centrifuge for 10 min, at 15 000 g, in a microcentrifuge to pellet the DNA. Remove the isopropanol and allow the pellet to briefly air dry.
9. Redissolve the pellet in 200–400 µl of TE.
10. Add an equal volume of phenol:chloroform (1:1) and mix well by inverting the tube.
11. Centrifuge for 10 min, at 15 000 g, in a microcentrifuge and transfer the aqueous phase to a fresh 1.5 ml tube.
12. Add 0.1 vol. of 3 M NaAc pH 5.5 and 2 vol. of 100% ethanol and place at −20°C for 1 h.
13. Centrifuge for 10 min, at 15 000 g, in a microcentrifuge to pellet the DNA. Add 1 ml of 70% ethanol and centrifuge for 5 min.
14. Dry the pellet in a vacuum desiccator and redissolve in 50 µl of TE.
15. Determine the DNA concentration[d] on a minigel (0.7% (w/v) agarose), using undigested lambda DNA as a concentration standard.

[a] Adapted from ref. 12.
[b] Use 1.5 ml tubes which are safe for immersion into liquid nitrogen.
[c] It is convenient to do these preps in batches of 12. Leave the tubes at 65°C until the last preparation has had a 15 min incubation.
[d] Yields of DNA can vary considerably but an average yield would be approx. 2 µg/0.05 g leaf material.

2.2 Identifying an RFLP

The first step in hybridization-based RFLP mapping is to identify a marker-specific polymorphism between the parental ecotypes. Enzymes that have proved useful for finding an RFLP between Landsberg *erecta* and Columbia are *Bcl*I, *Bgl*II, *Bst*EII, *Cfo*I, *Dra*I, *Eco*RI, *Eco*RV, *Hinc*II, *Hin*dIII, *Sac*I, and *Xba*I. Other restriction enzymes which can also be tested are those with an AT-rich recognition site; the *A. thaliana* genome is AT-rich (13), or those with 4 bp or 5 bp recognition sites. Try to avoid those enzymes which are sensitive to methyl-cytosine as they may digest only partially. The digested parental DNA then needs to be hybridized with the marker-specific probe. Once an RFLP has been identified the genomic DNA of the RI lines can be digested with the same enzyme and probed with the specific probe. *Protocol 5* describes standard conditions for the restriction digestion of *Arabidopsis* DNA.

Protocol 5. Restriction endonuclease digestion of *Arabidopsis* genomic DNA

Equipment and reagents

- Sterile 1.5 ml plastic tubes
- 37 °C water-bath or incubator
- Microcentrifuge
- 0.7% (w/v) agarose minigel
- Loading dye: 50% (v/v) glycerol, 5 mM EDTA pH 8.0, 10 mM NaCl, 0.04% (w/v) bromophenol blue, 0.04% (w/v) xylene cyanol FF, 250 µg/ml RNase A
- 0.7–1.5% (w/v) agarose gel (approx. 20 × 24 cm)
- Concentrated restriction enzyme buffer[a]
- 20 mg/ml BSA (Boehringer Mannheim UK)
- 100 mM spermidine (Sigma)
- Sterile distilled water
- Restriction enzymes
- Ethidium bromide (Sigma)

Method

1. Set up a reaction mix in a 1.5 ml tube consisting of 1–2 µg of genomic DNA, 1 × restriction enzyme buffer,[a] 1 µl of 100 mM spermidine, 10 µg of BSA, to a volume of 30 µl with distilled water.[b]

2. Add 10 U of restriction enzyme for each µg of genomic DNA, mix gently, and spin briefly in a microcentrifuge.

3. Incubate at 37 °C[c] for a minimum of 5 h to overnight.[d]

4. Check that the digestion is complete by running a 2–3 µl aliquot of the digestion reaction, with 2 µl of loading dye and 5 µl of TE added, on a 0.7% (w/v) agarose minigel, with an appropriate DNA size marker and an aliquot of undigested genomic DNA.[e]

5. If digestion is complete add 5 µl of loading dye to the rest of the digest. Load onto a 0.7–1.5% (w/v) agarose gel[f] with an appropriate DNA size marker.

6. Electrophorese the digested DNA overnight at 1–5 V/cm.[g]

3: Genetic mapping using recombinant inbred lines

7. Stain the gel in 1 μg/ml ethidium bromide for 1 h and photograph.

[a] The restriction enzyme buffer may either be that which comes with the enzyme from the manufacturer (this is usually 10 × concentrated) or one can make a 5 × 'Cuts All' buffer (100 mM Tris–HCl pH 7.5, 35 mM MgCl$_2$, 500 mM KCl) which is suitable for most enzymes tested.
[b] Larger volumes can be used, but this usually requires precipitation and resuspending the DNA before loading onto a gel.
[c] Most enzymes require 37 °C but not all, therefore check manufacturers instructions before use.
[d] If incubating overnight use a 37 °C plate incubator rather than a water-bath as this reduces condensation.
[e] Digestion can usually be confirmed by the appearance of visible bands (due to repeated sequences) in the digested DNA.
[f] The concentration of agarose required in the gel depends on the frequency at which the restriction enzyme cuts the DNA. 4 bp cutters and those 6 bp cutters which are AT-rich will require a higher % gel than those which are CG-rich.
[g] Electrophorese until the bromophenol blue is 2–3 cm from the bottom of the gel; the actual voltage will depend on the type of gel apparatus used.

2.3 Southern blotting and hybridization

After electrophoresis the DNA is transferred from the agarose gel to a nitrocellulose or nylon membrane, in a process known as 'Southern blotting' (14, 15). The use of nylon membranes is preferable as the probe can be removed and the filter rehybridized many times without damage. Each manufacturer of membranes usually include their own specific protocols and buffer recipes for both Southern blotting and hybridization and it is recommended that these are followed. However, brief outlines of the Southern blotting (capillary transfer method) and hybridization procedures are described in Section 2.3.1.

2.3.1 Southern blotting

After electrophoresis the agarose gel undergoes a series of treatments prior to DNA transfer. These are:

(a) Depurination of the DNA, facilitating the transfer of DNA over 10 kb in size, using 0.25 M HCl or UV light.

(b) Denaturation of the DNA with 0.5 M NaOH, 1.5 M NaCl.

(c) Neutralization with 1.5 M NaCl, 1.0 M Tris–HCl pH 8.0.

After these treatments the membrane (i.e. Hybond N+, Amersham International plc) is placed on top of the gel, over a reservoir of transfer buffer (20 × SSC; 3 M NaCl, 0.3 M Na$_3$ citrate). The membrane is overlaid with paper towels which draw up the buffer and transfer the DNA to the membrane by capillary action. The blotting process usually requires overnight for efficient transfer of the DNA from the gel to the filter. After transfer the DNA must be fixed to the filter by baking at 80 °C (under vacuum if using nitrocellulose) and/or by UV treatment.

2.3.2 Hybridization

There are five steps to the hybridization process:

(a) Labelling of the denatured DNA probe with radioactivity, [α-^{32}P]dCTP is commonly used. The usual methods of labelling DNA are:

   (i) nick translation (16, 17)
   (ii) random priming (18, 19)

   and a number of commercial kits are available for these (i.e. Gibco BRL, Boehringer Mannheim UK).

(b) Pre-hybridization of the filters in a large volume of hybridization buffer.
(c) Hybridization of the probe to the filter in a small volume of hybridization buffer.
(d) Washing the filters to remove the hybridization solution.
(e) Exposure of the filters to X-ray film.

3. Polymorphic markers

Many different types of polymorphic marker have been mapped onto the *A. thaliana* genetic maps. Those that can be scored as RFLP markers include:

- cosmid or plasmid clones (3, 20)
- lambda clones (2)
- cDNAs and expressed sequence tags (ESTs) (21) (see Chapters 2 and 9)
- sequences flanking T-DNA or transposon insertions (see Chapter 7)
- *A. thaliana* RFLP mapping set (ARMS) markers (22)

Those which can be scored using PCR-based techniques include:

- random amplified polymorphic DNA (RAPD) (11)
- cleaved amplified polymorphic sequence (CAPS) markers (23)
- microsatellites (24)
- amplified fragment length polymorphisms (AFLPs) (25, 26)

Phenotypic and biochemical markers can also be scored on the RI lines (9, 27–29).

3.1 RFLP markers

RFLP markers are analysed using Southern blot hybridization. Any cloned or isolated DNA fragment can be used as a RFLP marker. The larger the fragment used, such as a cosmid or lambda clone, the greater the chance of finding a polymorphism, however if the probe hybridizes to many restriction frag-

3: Genetic mapping using recombinant inbred lines

ments scoring the polymorphism may be difficult. In addition the polymorphic bands may be cross-hybridizing sequences that map to a different position in the genome. For small fragments (e.g. cDNAs, ESTs, and sequences flanking transposon or T-DNA inserts), many different restriction enzymes may need to be tested before a polymorphism is identified.

The RFLP markers mapped using the segregating F2 populations (see Section 1) were lambda (2) and cosmid (3) clones carrying random genomic fragments. Some of these clones have been mapped on the RI populations but there are many that have not been mapped therefore *Protocols 6* and *7* describe basic methods for the preparation of these clones.

Protocol 6. Preparation of lambda RFLP markers[a]

Equipment and reagents

- Sterile Pasteur pipettes
- Sterile 50 ml and 1.5 ml plastic tubes
- Sterile 250 ml flask
- 37°C orbital incubator
- 37°C water-bath
- Microcentrifuge
- Vortex
- Low speed centrifuge and rotor for 50 ml tubes
- –20°C freezer
- 0.5% (w/v) agarose minigel
- Vacuum desiccator
- Phenol (saturated):chloroform (1:1)
- Chloroform
- Isopropanol
- 70% (v/v) ethanol in distilled water

- LB + MCM: yeast extract (Difco) (5 g/litre), tryptone (Difco) (10 g/litre), NaCl (10 g/litre), 0.4% (w/v) maltose, 10 mM $MgCl_2$, 10 mM $CaCl_2$
- 1 mg/ml DNase (Boehringer Mannheim UK)
- 10 mg/ml RNase A (Boehringer Mannheim UK)
- PEG 8000 solution: 20% (w/v) PEG 8000 (Sigma), 2 M NaCl, 10 mM $MgCl_2$, 10 mM Tris–HCl pH 8.0
- LD buffer: 10 mM $MgSO_4$, 10 mM Tris–HCl pH 7.5
- 0.5 M EDTA pH 8.0
- TE: 10 mM Tris–HCl pH 8.0, 1 mM EDTA
- 3 M sodium acetate pH 5.5
- Undigested lambda DNA (Gibco BRL)

A. *Preparation of phage stock*

1. Pick a single plaque using a sterile Pasteur pipette.[b]
2. Add the plug to 25 ml of LB + MCM in a sterile 250 ml flask.
3. Incubate at 37°C with fast shaking (300 r.p.m.) for 8–12 h until lysis occurs.[c]
4. Add 100 μl of chloroform and shake for 1 min, then leave at 4°C overnight.
5. Remove the cell debris by centrifugation for 10 min at 11 000 g in a low speed centrifuge and transfer the supernatant to a fresh sterile tube.[d]

B. *DNA preparation*

1. Transfer 20 ml of the supernatant to a 50 ml tube, add 20 μl of 1 mg/ml DNase, 20 μl of 10 mg/ml RNase A, and incubate at 37°C for 45 min.
2. Add 20 ml of PEG 8000 solution and incubate on ice for 1–3 h.

Protocol 6. *Continued*

3. Centrifuge the phage precipitate for 10 min at 11 000 g, 4°C. Pour off the supernatant and drain the tube upside down for 5 min. Resuspend the phage pellet in 400 μl of LD buffer by vortexing and transfer to a 1.5 ml tube.

4. Add 400 μl of phenol:chloroform and mix thoroughly by gently inverting the tube. Centrifuge for 5 min at 15 000 g and remove the aqueous layer to a fresh 1.5 ml tube.

5. Repeat the extraction twice, the second time using chloroform only.

6. Add 25 μl of 0.5 M EDTA, 40 μl of 3 M sodium acetate, and 330 μl of isopropanol.

7. Incubate at room temperature for 5 min.

8. Centrifuge for 5 min at 15 000 g, wash pellet with 70% (w/v) ethanol, and dry briefly in a vacuum desiccator.

9. Resuspend the pellet in 50 μl TE and electrophorese 2 μl, plus 2 μl of loading dye and 6 μl TE, on a 0.5% (w/v) agarose gel using uncut lambda DNA as a concentration standard.

10. Digest the rest of the preparation with a common restriction enzyme (i.e. *Hind*III).[e] Store at −20°C.

[a] According to the method of Davies *et al.* (30) as modified by G. Mardon (personal communication from B. Staskawicz).
[b] From a fresh plate that is less than one week old. The plug includes both phage and bacteria.
[c] The culture becomes turbid, then clears on lysis when cell debris is evident.
[d] The supernatant should have titre of 10^9 p.f.u./ml.
[e] This will increase the labelling efficiency.

Protocol 7. Preparation of cosmid RFLP markers[a]

Equipment and reagents

- Sterile MacCartney bottles
- Sterile plastic 1.5 ml tubes
- Sterile pipette tips
- Sterile 1.5 ml plastic screw-cap tubes (Sarstedt) containing 200 μl sterile glycerol
- 37°C plate incubator
- 37°C orbital incubator
- −80°C freezer
- −20°C freezer
- Microcentrifuge
- Vacuum desiccator
- Vortex
- LB agar plus selective antibiotic plate

- LB liquid media plus selective antibiotic
- Buffer I: 50 mM glucose, 10 mM EDTA, 25 mM Tris–HCl pH 8.0
- Buffer II: 0.2 M NaOH, 1% SDS (w/v)
- Buffer III: 3 M potassium, 5 M acetate pH 4.8
- Phenol (saturated):chloroform:isoamyl alcohol (25:24:1)
- 100% ethanol
- 70% (v/v) ethanol in distilled water
- TE: 10 mM Tris–HCl pH 8.0, 1 mM EDTA pH 8.0
- 0.7% (w/v) agarose minigel

3: Genetic mapping using recombinant inbred lines

Method

1. Streak out the bacteria containing the cosmid clone onto selective medium and grow overnight at 37 °C.
2. Pick the smallest single colonies[b] and inoculate into 3 × 10 ml LB liquid media with the selective antibiotic, in MacCartney bottles. Grow in a shaking incubator overnight at 37 °C.
3. Add 800 μl of the overnight culture to 200 μl of sterile glycerol in a 1.5 ml screw cap tube, vortex briefly, and store at –80 °C.[c]
4. Centrifuge the remaining culture in the MacCartney bottles in a low speed centrifuge for 10 min at 2500 g, discard the supernatant, and allow the bottles to drain for 5 min upside down.[d]
5. Resuspend the pellet in 100 μl of ice-cold buffer I and transfer to a 1.5 ml tube. Incubate on ice for 5 min.
6. Add 200 μl of buffer II, mix by inverting, and incubate on ice for 5 min.
7. Add 150 μl of buffer III, vortex the tube in an inverted position, and incubate on ice for 5 min.
8. Centrifuge for 3 min at 15 000 g and transfer to a fresh tube containing 500 μl phenol:chloroform:isoamyl alcohol (25:24:1).
9. Vortex the tube for 10 sec and centrifuge for 3 min at 15 000 g. Transfer the supernatant to a fresh 1.5 ml tube containing 1 ml of 100% ethanol.
10. Incubate at room temperature for 10 min and centrifuge for 10 min at 15 000 g.
11. Wash pellet with 70% (w/v) ethanol and dry in a vacuum desiccator.
12. Resuspend the pellet in 20 μl of TE. Digest 5 μl of the cosmid DNA with a common restriction enzyme (i.e. *Hin*dIII), and run on a 0.7% (w/v) agarose gel to check that all three preparations are identical and have the correct sized bands.
13. Pool the remainder of all the verified preparations and digest with *Hin*dIII.[e] Store at –20 °C.

[a] According to the methods of Birnboim and Doly (31) and Ish-Horowicz and Burke (32), with minor modifications.
[b] Large cosmids may often be unstable, therefore the smallest, slowest growing colony should be picked. Several colonies should be tested and checked against each other for deletions (at step 13).
[c] This is for long term storage.
[d] As the cosmids are large the yield of DNA is often rather low.
[e] This will increase the labelling efficiency.

Often fragments from larger clones have been subcloned into smaller plasmid vectors, such as cDNA clones. The polymorphic fragments within some of the lambda RFLP markers have also been subcloned into the pBS plasmid and termed ARMS markers (22). All the ARMS markers can be mapped using genomic DNA digested with *Eco*RI using the Columbia × Landsberg *erecta* RI lines. ARMS markers can also be used to integrate a mutation into the RFLP map (see Section 5).

The sequences flanking transposon and T-DNA inserts are typically isolated by inverse PCR (IPCR) and usually subcloned into a plasmid vector (see Chapter 7). All these RFLP markers, cloned into small multicopy plasmids, can be prepared by standard methods (33).

Once the RFLP markers have been prepared they can be labelled radioactively (16–19) and then hybridized to the genomic DNA that has been transferred onto the membranes. Initially this would be to the parental ecotypes, to identify an RFLP, and subsequently to the RI lines, digested with the same restriction enzyme. The RFLP is then simply scored as one parental genotype or the other. *Figure 3a* shows the hybridization of a lambda marker, m220 (2) to a number of RI lines. 'C' indicates those lines which have the Columbia band and 'L' those lines with the Landsberg *erecta* band. As mentioned in Section 1 at the F8 generation there remains approximately 0.8% heterozygosity, therefore occasionally an RFLP marker may detect both alleles and the marker is then scored as heterozygous, or 'H'. To analyse the data with the mapping programs (see Section 4) an entry must be made for every RI line with every marker; therefore a marker that is unscored or unscorable on a particular RI line must also be recorded, as 'U'.

3.2 PCR-based markers

The use of PCR-based markers considerably speeds-up the process of mapping. In many cases the procedure involves only a PCR reaction followed by agarose electrophoresis, so a result can be obtained within a day. A further advantage is that it does not require the use of radioactivity. There are however a few problems with some PCR-based markers. PCR can be sensitive to impurities in the DNA, which prevent the polymerase reaction working, and even trace contamination with other DNA could lead to false results.

3.2.1 RAPDs

The simplest PCR-based markers are RAPDs (11). RAPDs are DNA bands amplified from genomic DNA, using short oligonucleotide primers (usually 9 bp or 10 bp long) and may be polymorphic between different ecotypes. There are usually multiple bands amplified by a primer so, practically, RAPDs are scored as dominant markers. This is partly because it may not be possible to identify both parental alleles, if present, from a multiplicity of bands. Alternatively a band may be amplified in only one ecotype, either due to the absence of annealing sites or because the fragment would be too large to amplify. Many different primers need to be screened to identify ones that are poly-

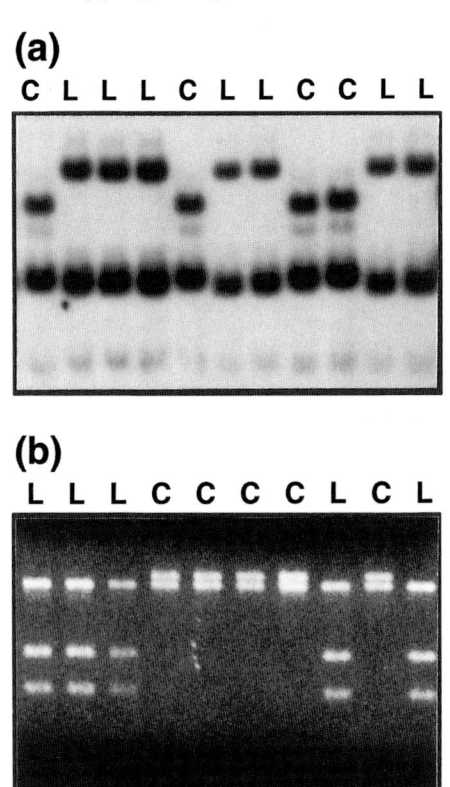

3: Genetic mapping using recombinant inbred lines

(a)
C L L L C L L C C L L

(b)
L L L C C C C L C L

Figure 3. Mapping different markers on the RI lines. (a) The pattern of the Columbia (C) and Landsberg *erecta* (L) alleles identified by the RFLP marker m220 (2). (b) The pattern of the C and L alleles identified by the CAPS marker *GAPC* (8, 23, 34).

morphic between the specific ecotypes. The majority of markers on the map of Reiter *et al.* (7) are RAPD markers.

3.2.2 CAPS markers

The 18 original CAPS markers were mainly generated from cloned and sequenced genes (23). The sequence data allowed the design of primers across small regions of these genes and the identification of potentially polymorphic restriction endonuclease sites between the Landsberg *erecta* and Columbia ecotypes. Using the available sequence data it is possible to design primers which could reveal a polymorphism between two ecotypes and because of this the use of CAPS is a common method used for mapping.

All the original CAPS markers have been analysed on the Columbia × Landsberg *erecta* RI lines (34). *Figure 3b* shows the segregation of the CAPS marker *GAPC* on a number of the RI lines.

3.2.3 Microsatellite sequences

Microsatellites are short genomic regions consisting of simple nucleotide repeats of variable length (also known as simple sequence length polymorphisms; SSLPs). The variability in length of the microsatellites is being exploited to generate markers for mapping in both mammals and plants.

A number of dinucleotide microsatellites have been isolated from *Arabidopsis* and sequenced. Primers have been designed which anneal to the sequences flanking the repeats and are used to amplify the variable repeat region (24). The variability in the length of the microsatellite region between different *Arabidopsis* ecotypes means that each pair of primers for a microsatellite marker can be used to identify multiple alleles at the same locus, i.e. they are very likely to be polymorphic in crosses between different ecotypes. At least 30 microsatellite sequences have been mapped on the Columbia × Landsberg *erecta* RI lines (24) (see Section 4). The latest SSLP information can be obtained at `http://genome.bio.upenn.edu/SSLP×info/SSLP.html`.

3.2.4 AFLPs

AFLPs represent the latest PCR-based markers available (25, 26). The technique is easy to perform and 50 or more different fragments may be examined for polymorphisms, from one primer pair. The technique involves restriction digestion of the genomic DNA with two different enzymes; one with a 4 bp recognition site and the other with a 6 bp recognition site and the attachment of adapters. The majority of fragments will be generated by the frequently-cutting enzyme only, but a proportion will be generated by both and these are the fragments that are utilized. Using primers homologous to the adapter sequences the fragment can be amplified. A radioactive nucleotide is included in the amplification reaction so that after polyacrylamide gel electrophoresis all the amplified bands can be visualized by autoradiography. Alternatively fluorescently labelled primers can be used and the bands analysed by an automated sequencer.

Alonso-Blanco *et al.* (10) have mapped 395 AFLP markers on the L*er* × Col RI lines and 321 AFLPs on the L*er* × Cvi lines. They showed that AFLP markers are distributed over most of the chromosomes but are particularly clustered around the centromeres. The clustering confirms the position of the centromeres on chromosomes 1, 2, 4, and 5 that had previously been mapped either genetically and/or physically. It also allowed positioning of the centromere on chromosome 3.

3.3 Phenotypic and biochemical markers

Several phenotypic and biochemical polymorphic markers have been mapped on the Columbia × Landsberg *erecta* RI lines. This is possible if there is natural variation between the two parental ecotypes. Each RI line was scored as being

3: Genetic mapping using recombinant inbred lines

like the Columbia phenotype (C) or the Landsberg *erecta* phenotype (L) for each trait and the segregation data compared with that of 67 RFLP loci (8).

(a) The *erecta* mutation could be mapped on these lines because *erecta* was segregating in the population. Plants were scored as wild-type (C) or *erecta* (L) and the *ERECTA* locus was mapped between RFLP markers g6842 and m220 on chromosome 2 (8).

(b) Columbia and Landsberg *erecta* have a differential response to the No-Co isolate of the fungal pathogen *Peronospora parasitica*. Columbia shows the sensitive phenotype while Landsberg *erecta* shows the resistant phenotype. A single locus conferring this resistance phenotype (*RPP5*) segregated in the RI lines. *RPP5* was mapped to chromosome 4 after scoring the lines as either resistant (L) or sensitive (C) and comparing the segregation data to the 67 loci data set (27). As a result of using the RI lines in this manner Holub and Beynon (9) generated RI lines from the crosses Col-0 × Nd-1 and Wei × Ksk-1 for use in experiments with downey mildew and white blister. Deslandes *et al.* (28) have used the Col-0 × Nd-1 RI lines to map a locus (*RRS1*) determining resistance to the bacterial pathogen *Ralstonia solanacearum*, which causes bacterial wilt.

(c) Aliphatic glucosinolates are thioglucosides which are synthesized by *Arabidopsis*, *Brassica*, and related genera. These secondary metabolites meditate pest–pathogen interactions and the structure of the side chain determines the products produced on tissue damage. The length and structure of the side chains of several glucosinolates vary between Columbia and Landsberg *erecta*. Glucosinolate profiles of the RI lines were determined using gas chromoatography and classified as Landsberg *erecta* type (L) or Columbia type (C), and single loci, determining specific alterations, were mapped (29).

4. Calculating map positions

Once the segregation of a polymorphic marker has been scored on the RI lines, initially as either C (Columbia), L (Landsberg *erecta*), H (heterozygote), or U (unscored/unscorable) the map position can be calculated using one of several mapping programs, or can be submitted to the mapping service at the Nottingham *Arabidopsis* Stock Centre (NASC) where the canonical Lister and Dean RI map is maintained and the map position will be calculated on your behalf.

Both the accuracy of scoring and the number of RI lines scored are very important in determining a reliable map position for a marker.

4.1 Mapping programs

Marker position can be obtained by using one of several mapping programs; the two which will be described are *MapMaker* (35) and *JoinMap*® (36,

CPRO-DLO). Each program has versions which can be used with either PC, Macintosh, or UNIX operating systems. The version required will depend mainly on the size of the data set to be analysed. The descriptions of these programs have been greatly simplified here, therefore it is essential that the program manuals should be read and the 'tutorials' carried out before using the programs. It should also be noted that while the segregation data was recorded as C, L, H, and U these are not usually the genotype symbols accepted by these programs.

One particular feature to consider is the mapping function. This is the relationship between the map distance and the recombination frequency. The Kosambi mapping function (37), as opposed to the Haldane mapping function (38), appears to give the best fit for recombination data generated in *A. thaliana*, Koornneef and Stam (39). This takes into account the fact that a second recombination event is unlikely to occur close to another, a phenomenon termed interference.

4.1.1 *MapMaker*

Initially two-point analysis of the data classifies the markers into linkage groups and for *Arabidopsis* the markers should fall into five linkage groups. Three-point and multipoint analyses are then performed for each linkage group. This produces a number of possible marker orders and an indication of their likelihood. This marker order is then translated into a map which indicates the centimorgan distances between the markers, as well as the recombination fraction (R = the proportion of recombinant individuals).

MapMaker (Version 3.0) can be obtained as freeware from: http://www.genome.wi.mit.edu/ftp/distribution/software/mapmaker3.

4.1.2 *JoinMap®*

JoinMap® (Version 2) (36, CPRO-DLO) is designed to handle F2, backcross, and RI data. Like *MapMaker* the initial mapping step consists of a pairwise analysis of the recombination frequencies between markers to establish the linkage groups. Once the linkage groups have been established *JoinMap®* sequentially builds up a map for each linkage group, starting from the marker pair which has the most linkage data.

A further feature of *JoinMap®* is that it can be used to statistically integrate segregation data originating from different populations to produce an integrated map. These can be both RI and F2 populations and consist of different ecotypes. An integration of the two RFLP maps generated on the F3 segregating families (2, 3) with the genetic map (40) has been carried out (41).

4.2 NASC mapping service

In July 1996 NASC took over the provision of the *Arabidopsis* mapping service which had been previously maintained by Dr Clare Lister at the John Innes Centre, Norwich.

4.2.1 Using the NASC mapping service

Each week mapping data that has been submitted is analysed using *MapMaker* (30) and a special subset of 94 markers that have been chosen to give good coverage over the five linkage groups (chromosomes). These scores are available for private use from NASC (http://nasc.nott.ac.uk/ RI_data/94_markers.html).

Segregation data should be submitted to NASC using the data submission forms (http://nasc.nott.ac.uk/RI_data/RI_submit.html). The first form allows the marker names to be validated before any data entry has taken place. It is important to check the existing locus names in the RI mapping database (http://nasc.nott.ac.uk/RI_data/new_ri_ map.html) before deciding on what to name a new locus. For use within *MapMaker* (30) the locus name must take the form: *LOCUS_NAME. The locus name itself must NOT exceed eight characters, including the asterix. A second form is used for the segregation data. Mapping data is processed once a week and the results are either e-mailed or faxed back to the investigator. This small scale analysis of the mapping data with the subset of 94 markers gives only an approximate map position.

Three times a year the canonical map is recalculated; typically in April, August, and December. This includes every marker that has been submitted to the mapping service and which has given a reliable map position using the subset of 94 reference markers (i.e. November 1998, 810 markers).

The canonical map is calculated using 'framework' markers that act as a consistent backbone for the map. Each new map is constructed from this framework thus maintaining conformity through each generation of the RI maps. Markers that appear on both the physical map and the recombinant inbred map for chromosomes II, IV, and V are included in the framework with the order set by that found on the physical map. For chromosomes I and III, markers that have been scored on more than 90 lines are used as 'framework' markers. Once the framework is set for each chromosome, the remaining markers are assigned to the map by linkage analysis. Two-point data is calculated for the entire collection of markers and the results of this calculation are used to assign markers to chromosomes based on their linkage to the framework markers. The most probable position for each marker is then determined using three-point analysis. The map is published from the NASC mapping database at (http://nasc.nott.ac.uk/RI_data/ new_ri_map.html) and mirrored at the AtDB project (http://www-genome.stanford.edu/Arabidopsis/) and is included within the *Arabidopsis* Genome Resource (AGR) (http://synteny.nott.ac.uk/ agr/agr.html) (see Chapter 10). All marker scores are published with each map release (http://nasc.nott.ac.uk/RI_data/complete_marker _set.txt).

4.2.2 How to interpret the RI maps

When viewing a map with its order of markers and map distances between the markers there is a general reassurance that the order and to some extent the distances between them are 'right'. However, it cannot be emphasized enough that the accuracy of a map is entirely influenced by the quality of the data that is used in its construction. Consequently it is best to score the marker on as many of the RI lines as possible; scoring on only 20 or 30 lines can only yield a very approximate placement and affects the quality of the map overall.

The maps that are calculated are not 'carved in stone'; they are the most probable representation of the data. As new data is added then maps can change with both the order of markers as well as the distances between markers differing from the previous map. In addition all the markers on a map are not placed there with the same certainty; some may have a unique map position, but others may have multiple possible map positions and the one that is shown is the most likely. Maps are displayed as images and text.

At NASC the maps are displayed as interactive images (see *Figure 4*). The framework markers are drawn to the left of the central line (which represents the chromosome) and are coloured blue. The markers that are assigned to the chromosome by linkage are displayed to the right of this line. Unique placements are coloured black and markers that potentially have more than one position are coloured red. All the marker information is linked to more detailed information, including the details of which chromosome the marker maps to, how many possible positions the marker has, the date when the marker was last mapped onto the large map, and contact information for that marker. Full details of the scores for the marker on each of the lines and the two-point data can also be obtained. Locus names (including synonyms) are searchable from the mapping database. Maps are also published as text files which include information on the map position and the number of possible positions a marker may have. Selecting the marker name from this file accesses the same information that is available through the image files.

5. Integration of a mutation into a molecular map

Chapter 7 describes how mutations can be induced and isolated in *A. thaliana* using T-DNA or transposon insertions. These mutations can be mapped and subsequently analysed by isolating the sequences flanking the insertion using IPCR and following the procedure for molecular mapping as described in Sections 2 and 3. If however a mutation was induced using a chemical or physical mutagen, or is in an ecotype that is not one of the parents of the RI lines, it will not segregate in the RI population and therefore cannot be mapped onto the RI genetic map. If the mutation is mapped relative to a number of markers which are also present on the RI map it is possible to

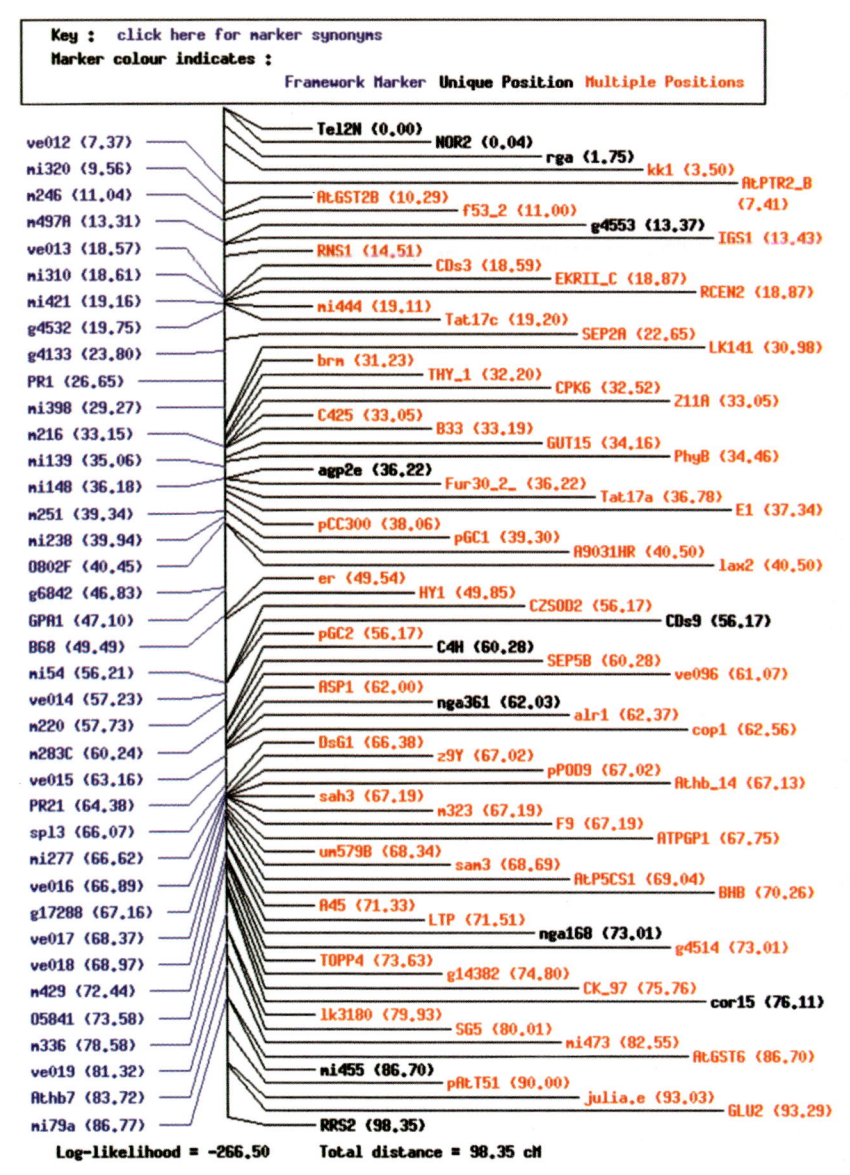

3: Genetic mapping using recombinant inbred lines

Figure 4. Recombinant inbred map of *Arabidopsis thaliana* chromosome II (24 November 1998). The image is from the WWW site at Nottingham *Arabidopsis* Stock Centre (NASC, http://nasc.nott.ac.uk/). Markers are coloured according to the confidence of their allocated position. Framework markers are coloured blue, markers mapping to a single location are black, and markers mapping to more than one location are red. In the latter case, the most probable position is displayed. On the WWW, clicking a marker name queries an underlying database system revealing two-point data, position (cM), and a contact name with e-mail address. The database system may also be queried with a marker name.

statistically integrate the position of the mutation with the RI map using the program *JoinMap®* (31). This integration involves:

(a) Generation of an F2 population between the line containing the mutation and another ecotype.
(b) Scoring the segregation of the mutation in the F2 population.
(c) Identification of polymorphic markers between the parental ecotypes of the F2 population for 10–15 markers.
   (i) These markers must be on the RI map.
   (ii) There should be two or three markers per chromosome.
   (iii) CAPS (18, 29), ARMS (22), or microsatelite (24) markers are suitable.
(d) Scoring the segregation of the markers on the F2 population.
(e) Entering the segregation data of these markers with data from the RI lines into *JoinMap®* (36) for analysis.

JoinMap® gives an approximate position for the mutation relative to the markers on the RI map. This should predict other markers that are more closely linked to the mutation. These additional markers then need to be tested on the F2 population to determine in which interval the mutation maps.

References

1. Botstein, D., White, R. L., Skolnick, M., and Davies, R. W. (1980). *Am. J. Hum. Genet.*, **32**, 314.
2. Chang, C., Bowman, J. L., DeJohn, A. W., Lander, E. S., and Meyerowitz, E. M. (1988). *Proc. Natl. Acad. Sci. USA*, **85**, 6856.
3. Nam, H.-G., Giraudat, J., den Boer, B., Moonan, F., Loos, W. D. B., Hauge, B. M., et al. (1989). *Plant Cell*, **1**, 699.
4. Burr, B., Burr, F. A., Thompson, K. H., Albertson, M. C., and Stuber, C. W. (1988). *Genetics*, **118**, 519.
5. Snape, J. W., Flavell, R. B., O'Dell, M., Hughes, W. G., and Payne, P. P. (1985). *Theor. Appl. Genet.*, **69**, 263.
6. Ellis, T. H. N., Turner, L., Hellens, R. P., Lee, D., Harker, C. L., Enard, C., et al. (1992). *Genetics*, **130**, 649.
7. Reiter, S. R., Williams, J. G. K., Feldmann, K. A., Rafalski, J. A., Tingey, S. V., and Scolnik, P. A. (1992). *Proc. Natl. Acad. Sci. USA*, **89**, 1477.
8. Lister, C. and Dean, C. (1993). *Plant J.*, **4**, 745.
9. Holub, E. and Beynon, J. (1997). *Adv. Bot. Res.*, **24**, 228.
10. Alonso-Blanco, C., Peeters, A. J. M., Koornneef, M., Lister, C., Dean, C., van den Bosch, N., et al. (1998). *Plant J.*, **14**, 259.
11. Williams, J. G. K., Kubelik, A. R., Livak, K. J., Rafalski, J. A., and Tingey, S. V. (1990). *Nucleic Acids Res.*, **18**, 6531
12. Dean, C., Sjodin, C., Page, T., Jones, J., and Lister, C. (1992). *Plant J.*, **2**, 69.
13. Leutwiler, L. S., Hough-Evans, B. R., and Meyerowitz, E. M. (1984). *Mol. Gen. Genet.*, **194**, 15.

14. Southern, E. (1975). *J. Mol. Biol.*, **98**, 503.
15. Wahl, G. M., Stern, M., and Stark, G. R. (1979). *Proc. Natl. Acad. Sci.USA*, **76**, 3683.
16. Rigby, P. W. J., Dieckmann, M., Rhodes, C., and Berg, P. (1977). *J. Mol. Biol.*, **113**, 237.
17. Koch, J., Kolvraa, S., and Bolund, L. (1986). *Nucleic Acids Res.*, **14**, 7132.
18. Feinberg, A. P. and Vogelstein, B. (1983). *Anal. Biochem.*, **132**, 6.
19. Feinberg, A. P. and Vogelstein, B. (1988). *Anal. Biochem.*, **137**, 266.
20. Liu, Y.-G., Mitsukawa, N., Lister, C., Dean, C., and Whittier, R. F. (1996). *Plant J.*, **10**, 733.
21. Newman, T., de Bruijn, F. J., Green, P., Keegstra, K., Kende, H., McIntosh, L., *et al.* (1994). *Plant Physiol.*, **106**, 1241.
22. Fabri, C. O. and Schaeffner, A. R. (1994). *Plant J.*, **5**, 149.
23. Konieczny, A. and Ausubel, F. M. (1993). *Plant J.*, **4**, 403.
24. Bell, C. J. and Ecker, J. R. (1994). *Genomics*, **19**, 137.
25. Vos, P., Hogers, R., Bleeker, M., Reijans, M., van de Lee, M., Hornes, M., *et al.* (1995). *Nucleic Acids Res.*, **23**, 4407.
26. Zabeau, M. amd Vos, P. (1993). *European Patent Application*. Publication No: EP 0 534 858.
27. Parker, J., Szabo, V., Staskawicz, B., Lister, C., Dean, C., Daniels, M., *et al.* (1993). *Plant J.*, **4**, 821.
28. Deslandes, L., Pileur, F., Liaubet, L., Camut, S., Can, C., Williams, K., *et al.* (1998). *Mol. Plant-Microbe Interact.*, **11**, 659.
29. Magrath, R., Morgner, M., Bano, F., Parkin, I., Sharpe, A., Lister, C., *et al.* (1994). *Heredity*, **72**, 290.
30. Davies, R. W., Botstein, D., and Roth, J. R. (1980). In *Advanced bacterial genetics (a manual for genetic engineering)* (ed. R. W. Davies, D. Botstein, and J. R. Roth), p. 109. Cold Spring Harbor Laboratory Press, New York.
31. Birnboim, H. C. and Doly, J. (1979). *Nucleic Acids Res.*, **7**, 1513
32. Ish-Horowicz, D. and Burke, J. F. (1981). *Nucleic Acids Res.*, **9**, 2989.
33. Sambrook, J., Fritsch, E. F., and Maniatis, T. (ed.) (1989). *Molecular cloning, a laboratory manual* (2nd edn), pp. 1.21–1.45. Cold Spring Harbor Laboratory Press, New York.
34. Jarvis, P., Lister, C., Szabo, V., and Dean, C. (1994). *Plant Mol. Biol.*, **24**, 685
35. Lander, E. S., Green, P., Abrahamson, J., Barlow, A., Daly, M. J., Lincoln, S. E., *et al.* (1987). *Genomics*, **1**, 174.
36. Stam, P. (1993). *Plant J.*, **3**, 739.
37. Kosambi, D. D. (1944). *Ann. Eugen.*, **12**, 172.
38. Haldane, J. B. S. (1919). *J. Genet.*, **8**, 299.
39. Koornneef, M. and Stam, P. (1992). In *Methods in Arabidopsis research* (ed. C. Koncz, N.-H. Chua, and J. Schell), pp. 83–99. World Scientific Publishing Co. Pte. Ltd, Singapore.
40. Koornneef, M. (1990). In *Genetic maps* (ed. S. J. O'Brian), pp. 6.95–6.97. Cold Spring Harbor Laboratory Press, New York.
41. Hauge, B. M., Hanley, S. M., Cartinhour, S., Cherry, J. M., Goodman, H. M., Koornneef, M., *et al.* (1993). *Plant J.*, **3**, 745.

4

Arabidopsis mutant characterization; microscopy, mapping, and gene expression analysis

KRITON KALANTIDIS, L. GREG BRIARTY, and ZOE A. WILSON

1. Generation of mutants and their importance for developmental biology

Mutants provide a valuable resource for the study of developmental biology, since the effects of genes can be identified and their role in development assessed in homozygous and heterozygous states. One problem with gene identification by mutagenesis can be the cloning of the specific gene (see Chapters 7 and 8). This chapter will cover a range of approaches which can be used to characterize mutants of *Arabidopsis thaliana*. These include the initial mapping of the mutation and determination of allelism between other mutations with similar phenotypes, both by classical linkage analysis and by molecular approaches. Different approaches are presented to allow the detailed analysis of phenotype by microscopy. The final part of the chapter gives protocols to permit the analysis of gene expression by Northern and *in situ* hybridization.

Mutations can be induced by a number of methods including chemical (e.g. EMS; see Chapter 1) and irradiation (e.g. gamma, X-rays), or by insertional mutagenesis (e.g. T-DNA tagging and transposon tagging; see Chapter 7). In all cases it is necessary to ensure that the observed phenotype is due to a single mutation. A number of backcrosses to the wild-type line, combined with continual selection for the mutant phenotype is therefore necessary. Mutant lines should be taken through to at least generation F6, however in practice little variation in phenotype is seen after four generations.

2. Mapping and segregation analysis
2.1 Mapping of mutations
2.1.1 Genetic mapping using visible markers

Mapping information can be determined by linkage analysis using multi-marker lines and the segregation of mapped visible markers. Preliminary mapping data can be determined by a single cross to the multiple marker line NW100 (NASC; see also Chapter 1). This line carries ten recessive mutations, two on each of the five chromosomes (*Table 1*).

Protocol 1 describes a procedure for making crosses in *Arabidopsis*. The ease of conducting crosses depends on whether both partners contain viable pollen. If both lines are fertile, then emasculation of the 'female line' is necessary to ensure that self-pollination does not occur. If only one line has viable pollen (e.g. one is male sterile) then the cross can be conducted without the need for emasculation, using the male sterile plant as the 'female line'.

Table 1. Homozygous, recessive markers in the *Arabidopsis thaliana* linkage tester line NW100

Gene symbol	Name (Chr – cM)[a]	Phenotype	Map position
an-1	angustifolia	Narrow leaves and slightly crinkled siliques	1–55.2
ap1-1	apetela	No or rudimentary petals	1–99.3
py	pyrimidine requiring	Leaves except cotyledons white, thiamine restores to normal	2–49.1
er-1	erecta	Compact inflorescence, fruits more blunt	2–43.5
hy2-1	long hypocotyl	Elongated hypocotyl, more slender plant	3–11.5
gl1-1	glabra	Trichomes absent on leaf surface and stems	3–46.2
bp-1	brevipedicellus	Short pedicels, siliques bend downwards, plant height reduced	4–15.0
cer2-2	eceriferum	Bright green stems, siliques bend downwards, plant height reduced	4–51.9
ms1-1	male sterile	No functional pollen, therefore no outgrowth of siliques	5–22.5
tt3-1	transparent testa	Yellow seeds and no anthocyanin in any part of the plant	5–57.4

[a] Chr = chromosome; cM = centimorgan.

4: Arabidopsis *mutant characterization*

Protocol 1. Making crosses using *Arabidopsis thaliana*

Equipment and reagents
- Watchmakers' forceps
- Dissecting microscope
- Cotton
- Clingfilm
- ARACON™ harvesting system (Aracon) or cellophane bags (Courtaulds) — optional
- 100% ethanol

A. *Emasculation of Arabidopsis flowers*

1. Look for buds that are still closed, but the stigma is starting to protrude from the end of the bud. Select buds for emasculation that are approximately two buds younger than this stage.
2. Using sterilized (100% (v/v) ethanol; air dried) watchmakers' forceps, remove any buds from the cluster that are not going to be emasculated.
3. Using sterile watchmakers' forceps and a dissecting microscope carefully remove all parts of the flower bud except the pistil.[a]
4. Once all the anthers and most/all of the surrounding tissues have been removed, the buds/pistils should be left for three to five days to mature and to check that emasculation has been successful. After two days keep a daily check on the buds to ensure that the pistils are at the correct stage, they should appear very hairy at the correct time for pollination.
5. At this point they can be used for pollination.

B. *Crossing Arabidopsis flowers*

1. Plants should be taken when the flowers are fully open, which is optimal on sunny mornings.
2. Unless using emasculated lines, choose fully opened flowers that are to be pollinated and mark them with loosely tied cotton. Using forceps, that have been previously sterilized by immersion in ethanol and air drying, remove any flowers from the bud cluster that have not opened or are starting to form pods.
3. Select a fully open flower from the 'pollen donor' and using sterile forceps detach the flower.
4. Lightly brush the stigma of the flower to be pollinated with the anthers/flower of the pollen donor.[b]
5. Repeat this process with one or two more flowers. Sterilize the forceps after each pollination.
6. The 'pollinated' flowers can then be lightly wrapped in 'Clingfilm' for one to three days until pod elongation commences.[c]

Protocol 1. *Continued*

7. Once pod elongation has started, remove the Clingfilm, and allow the pod to mature and dry before harvesting. Plants can be grown in ARACON™ harvesting systems or cellophane bags to avoid seed loss/cross-contamination during pod shatter. Harvest into seed bags when the pods are turning brown.

[a] This is best done by removing a few tissues at a time and taking care to minimize the shearing force on the base of the bud/pistil.
[b] Pollen donor flowers can be used successfully on more than one flower.
[c] This is not essential, but ensures that cross-fertilization by any greenhouse insects or personnel does not occur.

i. Preliminary mapping

A homozygous NW100 plant should be crossed with the unmapped homozygous mutant. The F1 seed should be grown and the plants allowed to self. Approximately 100–200 F2 plants (depending on the level of mapping accuracy required) should then be grown and analysed for the segregation of markers with the mutant phenotype. Plants should be scored for the presence of markers progressively during development. Note: homozygous NW100 plants require a weekly feed of 1% (w/v) aqueous thiamine due to the *py* mutation.

Linkage can be calculated in the F2 by the product ratios of each class in repulsion. Each marker must be scored individually with the unmapped mutation.

The phenotypic classes are designated as:

- XY (a) = wt both loci
- Xy (b) = wt for marker; mutant for unmapped locus
- xY (c) = mutant for marker; wt for unmapped locus
- xy (d) = mutant both loci

The repulsion products are: (a × d)/(b × c).

The ratio of products can then be compared using F2 ratio tables to determine the cross-over value, this value multiplied by 100 gives the frequency of recombination. The standard error can also be calculated by taking the error value from the table and dividing it by the square root of the population size. To convert the recombination frequency into a map distance in cM it is necessary to use a mapping function. There are a number of different methods that can be used, however the Kosambi function (1) has been frequently used and has tended to give good fits for *Arabidopsis* (2).

Kosambi mapping function:

$$D = 25 \ln \frac{(100 + 2r)}{(100 - 2r)}$$

where D = distance cM; r = recombination frequency.

4: Arabidopsis mutant characterization

The standard error of map distance can also be converted into cM (Kosambi):

$$SD = \frac{2500}{2500 - r^2} Sr$$

where SD = standard error cM; Sr = standard error of recombination frequency; r = recombination frequency.

This analysis can also be conducted by various computer programs, e.g. RECF2, GenMap, JoinMap (2).

If the mutation to be mapped influences fertility (e.g. male sterility) then the marker line NW100F, which does not contain the *ms1-1* gene and is fully fertile, should be used.

ii. Complementation analysis

The NW100 segregation analysis will provide an approximate map position which can then be used to identify possible allelic mutants. These can then be crossed to confirm the numbers of loci involved; for recessive mutations cross homozygous mutants and analyse the phenotypes of the F1 population. If the mutations are non-allelic complementation should occur and the phenotype should resemble wild-type, if they are allelic all the resultant progeny should carry the mutant phenotype. If the mutations influence fertility, e.g. male sterile, then the allelism cross must be conducted between a homozygous and a heterozygous mutant. The F1 offspring will either be wild-type if the mutations are non-allelic or will be 50:50 wild-type:mutant if a single locus is involved.

iii. Fine-scale mapping

The NW100 linkage analysis provides a chromosome-specific approximate map location. However, to accurately map a mutation it is then necessary to repeat the segregation analysis using chromosome-specific, multiple marker lines depending on the gene location (these are available from the NASC; see Chapter 1).

2.1.2 Genetic mapping using molecular markers

Mapping mutations by the use of molecular markers can provide a very detailed assessment of map position, however it can involve a large number of crosses and hybridizations. An alternative approach to rapidly map a mutation has been described by Fabri and Scäffner (3) which is based upon RFLPs. This approach has the advantage that the mutant phenotype does not interfere with the scoring of the segregation analysis and that mapping can be conducted relatively quickly. This procedure uses an *Arabidopsis* RFLP mapping set (ARMS) (3), which comprises of up to 13 markers which uniformly cover the five *Arabidopsis* chromosomes. These markers are able to differentiate between the ecotypes Landsberg *erecta* (Ler) and Columbia (Col), and Landsberg *erecta* and Enkheim.

To map a mutation that lies in the Ler, Enkheim, or Col ecotype, crosses need to be conducted between the appropriate lines that can be differentiated by the markers. The F1 must then be selfed to provide a segregating F2 population. A small set of F2 plants (~ 20) which are showing the recessive mutant traits are required. DNA is isolated from F3 pools from these lines and digested with EcoRI and transferred to nylon membranes. The filters can then be hybridized with the ARMS markers. The markers are grouped into two pools which allow the detection of every polymorphism in a heterozygous or homozygous state after only two hybridizations. The polymorphisms can then be scored and used for linkage analysis to provide map positions. Using 12–13 ARMS markers a mutation can be mapped to approximately 26 cM away from its closest marker.

Alternative strategies for fast molecular mapping including the use of RI lines (see Chapter 3) and various molecular markers including RAPDs (4, 5), CAPS (6), microsatellites (7), and AFLPs (8).

2.2 Influence of environment on phenotype

Environment can have a major influence on the expression of phenotype for particular mutant lines. One example of this is that of genes that influence pollen viability and cause male sterile phenotypes. The *eceriferum* (*cer*) mutants which exhibit bright green stems and siliques due to altered wax layers, also show degrees of sterility, both male and female depending on environmental conditions (9). Under relatively low humidity (60–80%) sterility is almost complete, whilst at high humidity ($> 90\%$) fertility is close to that of the wild-type. Depending on the nature of the mutation it may therefore be necessary to confirm that the observed phenotype is constant under a range of environmental conditions, namely temperature and humidity.

3. Microscopy

Microscopy can be used to help determine the nature of the phenotype at a tissue-specific level, this can include both fresh and fixed material analysis.

3.1 Fresh material characterization

Protocol 2 describes a number of approaches to analyse development by light microscopy using fresh *Arabidopsis* plant material.

Protocol 2. Light microscopy analysis of fresh *Arabidopsis* plant material

Equipment and reagents
- Whatman No. 1 filter paper
- Petri dishes, plastic rods
- Microscope slides and coverslips
- Glacial acetic acid

4: Arabidopsis *mutant characterization*

- 0.1% (w/v) aqueous neutral red solution in 20 mM phosphate buffer pH 7.8
- 0.001% (w/v) fluorol yellow 088 in 50% (w/v) polyethylene glycol, 40% (v/v) glycerol
- 2 μg/ml 4',6-diamidino-2-phenylindole (DAPI) in 7% (w/v) sucrose
- Pollen germination medium: 1% (w/v) agar, 20% (w/v) sucrose, 80 p.p.m. boric acid, 10 p.p.m. myo-inositiol, pH 6.0 with HCl
- Nail varnish
- Malachite green–acid fuschin stain: (added in order) 10 ml of 95% (v/v) ethanol, 1 ml of 1% (w/v) malachite green in 95% (v/v) ethanol, 50 ml distilled water, 25 ml glycerol, 5 g phenol, 5 g chloral hydrate, 5 ml of 1% (w/v) acid fuschin in distilled water, 0.5 ml of 1% (w/v) orange G in distilled water, and 4 ml glacial acetic acid
- 0.4 mg/ml fluorescein diacetate (FDA) solution in acetone, diluted with 7% (w/v) sucrose solution (aq) to 0.4 μg/ml

A. *Vacuole and vesicle staining* (10, 11)

1. Fresh plant material should be sectioned or squashed and incubated in 0.1% (w/v) aqueous neutral red solution for up to 5 min at room temperature.
2. View using bright-field illumination. The stain accumulates in the vacuoles and vesicles. Dead or damaged cells will contain stain throughout their cytoplasm.

B. *Lipid staining* (11, 12)

1. Stain sectioned plant material using 0.01% (w/v) fluorol yellow 088 in 50% PEG, 40% glycerol, for 1 h at room temperature.
2. Visualize using a blue filter.

C. *Nuclear DNA staining* (13)

1. Stain the sectioned tissue using a 2 μg/ml solution of DAPI solution, for 5 min at room temperature.
2. View using UV light and a conventional (UV) filter set.
3. DAPI specifically binds to double-stranded nucleic acids and fluoresces blue upon excitation with UV light.

D. *Pollen viability*

1. Cytoplasmic staining (14, 15).
   (a) Fix buds for 30 min in glacial acetic acid.
   (b) Rinse in distilled water, dissect out anthers, remove debris and excess water.
   (c) Add a drop of freshly prepared malachite green–acid fuschin stain and lightly squash using a coverslip.
   (d) Seal the edges of the coverslip with nail varnish. Leave anther squashes for 24 h at room temperature to take up the stain.
   (e) View under bright-field illumination. Viable pollen stains red/purple, whilst non-viable stains green/blue.

Protocol 2. *Continued*

2. Esterase activity staining (16).
   (a) Viability, as measured by esterase activity, can be determined by staining using FDA solution.
   (b) Tease apart the anthers in a drop of 0.4 µg/ml FDA solution and cover with a coverslip.
   (c) View after 10 min using UV light. Due to the presence of callose the stain is prevented from entering tetrads and therefore cannot be used to assess viability at this stage.

3. Germination studies (17).
   (a) Place two drops of molten pollen germination medium onto a microscope slide.
   (b) Allow the agar to solidify and place the slide (agar side up) on plastic rods above moist Whatman No. 1 filter paper in a Petri dish.
   (c) Dissect anthers from fresh, mature flowers and tease them apart on the surface of the agar medium to release the pollen.
   (d) Close the Petri dishes and incubate them under warm, white illumination (1800–2600 lux) at 25°C and 75% humidity for > 8 h.
   (e) Observe pollen under bright-field illumination. Viable pollen will have germinated to form pollen tubes.

3.2 Fixed material characterization

For light and electron microscopy, and also certain staining protocols, the fixation of fresh material is required. The aim of fixation is to stabilize tissue structure prior to subsequent treatment, while retaining the physical and biochemical characteristics of the living material. Hayat (18) gives a detailed review of approaches to fixation. An 'ideal' fixative penetrates rapidly and reacts with all cell components with minimal alteration, however it is necessary to choose fixatives based on their relative merits depending on the tissue to be fixed.

The bifunctional aldehydes are the most commonly used fixatives; they act by cross-linking between molecules to form a stable, three-dimensional network. Glutaraldehyde is the most effective fixative for stabilizing proteins, however it penetrates only slowly and for good preservation the maximum dimensions of tissue pieces should not exceed 0.5 mm. Formaldehyde on the other hand penetrates more rapidly, but its reactions are not so permanent; this means that it maintains immunoreactivity, but at the expense of ultra-structural preservation. For light microscopy, where tissue samples are relatively large, a combination of these two agents is used (see *Protocol 3*). It is essential that fixatives are made up freshly from fresh chemical stocks; self-

4: Arabidopsis *mutant characterization*

polymerization reduces the effectiveness of solutions, and degradation products from stored stocks cause tissue damage.

Acrolein is extremely effective as a fixative, it penetrates and reacts faster than most other fixatives; however it is a highly reactive and volatile, hazardous liquid, and an effective tear gas, and requires careful handling. It can be used in combination with glutaraldehyde and formaldehyde to fix difficult tissue such as seeds (19). Where membrane preservation is critical, potassium permanganate (2–3%, w/v) can be used as a fixative, however most other cell components are not stabilized, and are lost during processing.

For electron microscopy a primary aldehyde fixation is normally followed by osmium tetroxide treatment (Section 3.2.2, *Protocol 5*). This is necessary for good membrane preservation, as the aldehydes do not retain lipid well; the osmium also serves to add electron contrast.

To avoid tissue damage and distortion, the fixative should be used buffered, to a pH and a total osmolarity compatible with the tissue to be preserved. Fixation periods of several hours are normal, but microwave treatment can be used to achieve high quality fixation in a short time (20).

For most sectioning procedures acrylic or epoxy-based resins have replaced paraffin wax as embedding media. The use of semi-thin (0.5–1.0 μm) sections for light microscopy allows access to the full resolving power of the instrument. However, the presence of the embedding medium in the section (the plastic is not normally removed, as is the case when wax is used) limits their staining possibilities, and where full access to the tissue components is needed (as for *in situ* hybridization, see Section 4.3) a wax embedding procedure may be preferable.

Protocol 3. General fixation

Equipment and reagents

- 60°C oven
- Formaldehyde/glutaraldehyde fixative: 4% (w/v) formaldehyde, 5% (w/v) glutaraldehyde in 0.025 M phosphate buffer pH 7.2
- 0.025 M phosphate buffer pH 7.2
- Ethanol/water series: 10%, 20%, 40%, 60%, 80%, 90%, 95%, 100% (v/v)
- LR White resin (Agar Scientific)
- Gelatine capsules

Method

1. Fix material for 2–5 h at room temperature in freshly prepared formaldehyde/glutaraldehyde fixative. Large tissue pieces (leaf, stem) should be cut up with a very sharp blade in a drop of fixative.

2. Rinse material in cold 0.025 M phosphate buffer at 4°C for 15 min.

3. Dehydrate material in an ethanol/water series at 4°C in the following steps: 10%, 20%, 40%, 60%, 80%, 90%, 95% (v/v) for 10 min each, then 100% for 15 min, then twice in 100% for 30 min each (or once for 30 min and then in fresh 100% overnight).

Protocol 3. *Continued*

4. If material has been fixed in groups of tissue, e.g. bud clusters, the material can then be dissected into the individual units to be embedded.
5. Material should be infiltrated with two changes (1 h; room temperature) of LR White resin, and overnight at 4°C in fresh resin.
6. Place tissues into gelatine capsules with fresh resin, replace the lids, and polymerize at 60°C for 24 h.

3.2.1 Light microscopy

For light microscopy sections of 0.5–1.0 μm thickness should be cut using an ultramicrotome. For general visualization of tissue, sections can be stained using toluidine blue and sodium carbonate (both 0.1%, w/v) for a few minutes, followed by rinsing in distilled water. Specific staining (see *Protocol 4*) can be used to differentiate between the tissue types.

Protocol 4. Specific staining for fixed tissues

Equipment and reagents

- Copper rod, spirit burner
- Microscope slides and coverslips
- 0.05% (w/v) aniline blue in 0.067 M phosphate buffer pH 8.5
- Ethanol:acetic acid (3:1)
- 70% (v/v) ethanol
- 45% (v/v) acetic acid
- 1% propionic–carmine: 1% (w/v) carmine in propionic acid
- Nail vanish
- 0.25 mg/ml azure blue in 0.1 M sodium citrate pH 4.0

A. *Callose staining* (11, 21)

1. Stain 0.5–1.0 μm sections with 0.05% (w/v) aniline blue for 5 min at room temperature.
2. View under UV illumination.

B. *Meiosis analysis in buds* (22)

1. For Landsberg *erecta*, harvest the first buds which appear at the rosette stage 18–20 days after sowing (1500–2200 lux; 21°C ± 2°C).
2. Fix buds in ethanol:acetic acid (3:1) for 24 h. Store in 70% (v/v) ethanol at 4°C until required.
3. Dissect buds (1–2 mm in length) in a drop of 45% (v/v) acetic acid. Remove the floral debris and excess acetic acid to leave the white anthers (≤ 400 μm).
4. Stain with 1% (w/v) propionic–carmine solution.
5. Release dividing cells by gently tapping with a copper rod.

4: Arabidopsis *mutant characterization*

6. Place a coverslip onto the sample and pass the slide briefly through the flame of a spirit burner. Apply gentle pressure to the coverslip and seal with nail varnish.
7. View samples using a green filter.

C. *RNA staining (11)*
1. Embed tissues in paraffin (see Section 4.3.1).
2. Stain tissue using 0.25 mg/ml azure blue in 0.1 M sodium citrate pH 4.0, for 2 h at 50°C.
3. Visualize using bright-field.

3.2.2 Transmission electron microscopy

Tissue samples fixed for electron microscopy should be no larger than 0.5 mm in any dimension; this ensures that fixative penetration occurs as fast as possible through all surfaces. Material can be cut up in a drop of fixative on a wax surface using a new degreased scalpel or razor blade. Avoid handling the tissues; the various solution changes involved in the proceeding protocols should be achieved by changing the solutions in the tissue containers rather than by moving the tissue from one container to another. Full procedures are given in, e.g. Hunter (23). Modifications to the standard fixation protocol for TEM fixation (Section 3.2, *Protocol 3*) are described in *Protocol 5*.

Protocol 5. Modifications to general fixation (*Protocol 3*) for electron microscopy

Equipment and reagents
- 60°C oven
- Electron microscopy specimen grids
- Dental wax
- 2.5% (w/v) glutaraldehyde in 0.025 M phosphate buffer pH 6.8
- 0.025 M phosphate buffer pH 6.8
- Buffered 2% (w/v) osmium tetroxide
- Ethanol/LR White resin: (3:1, 1:1, 1:3)
- Ethanol/water series: 10%, 20%, 40%, 60%, 80%, 90%, 95%, 100% (v/v)
- 100% LR White resin
- Gelatine capsules
- Sodium hydroxide pellets
- 2% (w/v) uranyl acetate
- 1% (w/v) Reynolds lead citrate (Agar Scientific)

Method
1. Fix material in 2.5% (w/v) glutaraldehyde in 0.025 M phosphate buffer pH 6.8, overnight at 4°C.
2. Rinse in buffer, then carry out a post-fixation treatment for 3 h at 4°C in buffered 2% (w/v) osmium tetroxide.
3. Rinse in cold 0.025 M phosphate buffer pH 6.8 for 5 min at 4°C.
4. Dehydrate according to *Protocol 3*, step 3.
5. Infiltrate with 3:1 ethanol/LR White resin dropwise over 30 min at room temperature, then 1:1 and 1:3 ethanol/LR White resin for > 1 h

Protocol 5. *Continued*

   each (room temperature). Incubate twice in fresh 100% resin overnight at 4°C.
6. Place tissue into fresh resin in gelatine capsules, cover, and polymerize at 60°C for 24 h.
7. For electron microscopy, cut sections between 50–90 nm (silver grey-pale gold) in thickness. Mount on grids.
8. Place drops of 2% (w/v) uranyl acetate on dental wax in a Petri dish and place each grid on one drop. Stain for 5 min at room temperature.
9. Rinse in distilled water.
10. Transfer grids to drops of 1% (w/v) Reynold's lead citrate (23) on dental wax in Petri dishes containing sodium hydroxide pellets to reduce CO_2 levels.
11. Wash grids in distilled water and observe.

3.2.3 Scanning electron microscopy

The scanning electron microscope (SEM) can reveal morphological detail at high magnification, but is particularly useful at low magnification where its great depth of field reveals otherwise unobtainable images. To withstand the low pressures inside the microscope, samples must be dehydrated (though cryo-techniques allow examination of fresh, frozen samples), and this must be carried out with minimum shrinkage or distortion of the material. Freeze-drying can be used, but critical point drying offers the best way of avoiding the surface tension stresses which occur on air drying (see *Protocol 6*). The procedure involves the replacement of water in the sample with another miscible liquid (e.g. acetone), and then the replacement of this with a transitional fluid (e.g. carbon dioxide, Freon-13). The sample is then subjected to a temperature and pressure increase so that the fluid passes its critical point, at which the densities of liquid and gas are the same and the surface tension is zero, then with the temperature above the critical temperature the pressure is lowered so that the gas phase is slowly released. Once dried, samples are mounted on specimen stubs, given a conductive coat of gold or some other metal, and can then be examined directly in the SEM.

Protocol 6. Sample preparation for scanning electron microscopy (SEM)

Equipment and reagents

- Critical point drying apparatus
- Gold coating (c. 250 Å, 25 nm) or by a sputter coater
- 2.5% (w/v) glutaraldehyde in 0.025 M phosphate buffer pH 6.8
- SEM stubs, double-sided adhesive tape
- Buffered 2% (w/v) osmium tetroxide
- Ethanol/water series: 10%, 20%, 40%, 60%, 80%, 90%, 95%, 100% (v/v)

Method

1. Fix the tissues in glutaraldehyde followed by osmium tetroxide as for transmission electron microscopy (see *Protocol 5*).
2. Dehydrate in an ethanol or acetone series (see *Protocol 3*), depending on the transition fluid to be used (24).
3. Process with a critical point drying apparatus.
4. Attach samples to SEM stubs using double-sided adhesive tape.
5. Coat with a thin film of gold (c. 250 Å, 25 nm) either by evaporation, or by using a sputter coater.
6. Examine in the SEM.

4. Analysis of plant gene expression
4.1 RNA isolation

The general rules for RNA work apply, gloves must be worn at all times, glassware must be baked for at least 4 h at 180°C. Plasticware must be DEPC treated (0.2% (w/v) DEPC in water for 4 h, 37°C). All solutions must also be treated with DEPC (0.4% (w/v) DEPC in water for 4 h, 37°C) before being autoclaved. Some chemicals (e.g. Tris) are not compatible with DEPC, these must be prepared in DEPC-treated water.

Protocol 7. *Arabidopsis* total RNA isolation (26)

Equipment and reagents

- Mortar and pestle, liquid nitrogen, centrifuge tubes, microcentrifuge tubes
- RNA extraction buffer: 4 M guanidinium thiocyanate, 20 mM sodium citrate pH 7.0, 0.5% (w/v) Sarcosyl, 0.1 M 2-mercaptoethanol[a]
- 2 M sodium acetate pH 4.0
- Equilibrated phenol pH 8.0
- Chloroform
- Equilibrated phenol (pH 8.0):chloroform: isoamyl alcohol (25:24:1)
- Isopropanol
- 75% (v/v) ethanol
- DEPC-treated double distilled water
- 8 M lithium chloride

Method

1. Harvest and freeze plant material in liquid nitrogen. Use directly or store at –80°C until required.
2. Grind frozen plant material in liquid nitrogen in a pestle and mortar.
3. Transfer powder to a centrifuge tube and add 3 ml of RNA extraction buffer.
4. Homogenize for 1 min.

Protocol 7. *Continued*

5. Add the following to the homogenate, vortexing briefly after each addition:
   (a) 0.1 vol. of 2 M sodium acetate pH 4.0.
   (b) 1 vol. of equilibrated phenol pH 8.0.
   (c) 0.2 vol. of chloroform.
6. Shake vigorously for 1 min and place on ice for 10 min.
7. Centrifuge for 15 min at 10 000 g.
8. Transfer the aqueous phase into a fresh tube and add an equal volume of equilibrated phenol (pH 8.0):chloroform:isoamyl alcohol (25:24:1).
9. Repeat steps 6 and 7.
10. Transfer the aqueous phase into a fresh tube that contains 1 vol. of isopropanol. Precipitate at −20 °C for at least 1 h.
11. Recover the pellet by centrifugation (10 000 g, 15 min).
12. Wash the pellet with 75% (v/v) ethanol, air dry, and resuspend in DEPC-treated water (30 μl water/g plant tissue used).[b] Transfer samples to microcentrifuge tubes.
13. Add 1 vol. of 8 M lithium chloride to the sample and leave to precipitate at −20 °C overnight.
14. Recover pellets by centrifugation (13 000 g, 1 h, 4 °C). Wash pellets twice with 75% (v/v) ethanol.
15. Air dry pellets and resuspend in 100–200 μl of DEPC-treated sterile water.

[a] 2-mercaptoethanol must be added immediately before use.
[b] RNA must be in a relatively high concentration for precipitation with lithium chloride.

RNA purification is usually complicated by the presence of endogenous RNases in plant cells. To avoid RNA degradation during isolation a variety of RNase inhibitors can be used (25). In the total RNA isolation protocol presented here (see *Protocol 7*), RNase action is inhibited by guanidinium isothiocyanate which is a strong denaturant. This method gives very good quality total RNA, but is relatively time-consuming. Alternatively, commercially available isolation kits, e.g. RNeasy (Qiagen), Trizol (Gibco BRL), can be used although problems can be encountered with isolations from some plant tissues. In our laboratory we have used both the RNeasy (Qiagen) and Trizol (Gibco BRL) kits and found them very simple and fast to use. In our hands they tended to give higher RNA yields, but generally of lower quality than the isolation procedure in *Protocol 7*. However, an additional purification

4: Arabidopsis mutant characterization

step, i.e. extra phenol:chloroform purification improved the quality of the resulting RNA.

Nuclear mRNA can be separated from total RNA using oligo-T traps. There are also various mRNA isolation kits available commercially. We have successfully used the Qiagen mRNA isolation kit which requires total RNA to be isolated and then the poly(A^+) separated. An alternative to the commercially available kits is the direct isolation of poly(A^+) RNA (27).

4.2 Northern analysis

It is essential that during electrophoresis of RNA, secondary structures or hybridization to other RNA molecules does not occur. Thus the RNA samples are denatured prior to loading and run under conditions that prevent reassociation. This can be achieved by reacting the free acids of the bases with formaldehyde. The Schiff bases formed are not capable of hydrogen bonding which reduces the likelihood of secondary structures. It is possible to use other denaturants such as glyoxal or methyl mercuric hydroxide but they are expensive and more toxic than formaldehyde. A method for RNA gel electrophoresis and Northern blotting is presented in *Protocol 8*.

Protocol 8. RNA electrophoresis and blotting (28)

Equipment and reagents

- 75°C incubator
- Agarose
- Nylon membrane (Hybond-N)
- Double distilled DEPC-treated water
- 10 × MOPS buffer:[a] 0.4 M morpholino-propanesulfonic acid, 0.1 M sodium acetate, 10 mM EDTA, adjust to pH 7.2 with NaOH
- 37% (w/v) formaldehyde
- 2 × loading buffer:[b] 2 × MOPS, 8% (w/v) formaldehyde, 5% (v/v) glycerol, bromophenol blue
- 0.5 mg/ml ethidium bromide stock[c]
- 20 × SSC: 3 M NaCl, 0.3 M Na_3 citrate
- 10 × SSC
- 2 × SSC, 0.1% (w/v) SDS
- 0.1 × SSC, 0.1% (w/v) SDS

Method

1. Dissolve 1.1 g of agarose in 85 ml of distilled, DEPC-treated water.
2. Cool to approx. 70°C and add 10 ml of 10 × MOPS buffer.
3. Add 16 ml of 37% (w/v) formaldehyde to the gel solution. Stir thoroughly and pour gel.
4. Let it solidify for at least 30 min at room temperature.
5. Prepare samples for loading; use 6–8 μg of total *Arabidopsis* RNA for electrophoresis. Add 1 μl of loading buffer per 1 μl of sample.
6. Denature samples by placing them at 75°C for 10 min. Then cool on ice for 30 sec.
7. Using 1 × MOPS, with the addition of 160 ml of 37% (w/v) formaldehyde per litre of running buffer, as running buffer, load samples

Protocol 8. *Continued*

and run gel until the bromophenol blue migrates to three-quarters of the gel. Electrophoresis should be conducted with buffer circulation and cooling. If that is not possible the buffer should be changed at least once during the run.

8. Stain the gel with ethidium bromide to visualize the RNA.[d] Rather than staining the entire gel post-electrophoresis, ethidium bromide can also be added directly to the samples prior to loading (0.5 μl of 0.5 mg/ml per sample), although this may reduce the efficiency of the RNA transfer to the membrane.

9. Transfer the gel to a box containing at least five times the volume of the gel of sterile distilled water. Wash for 30 min to remove the formaldehyde.

10. Transfer the RNA from the gel onto a nylon membrane by capillary transfer (25) using 10 × SSC.

11. Hybridization and post-hybridization washes are essentially the same with that of DNA. Wash at 65°C with 2 × SSC, 0.1% (w/v) SDS, followed by 0.1 × SSC, 0.1% (w/v) SDS for 20 min each. However, since DNA–RNA and RNA–RNA binding is stronger than DNA–DNA binding it is possible to use higher hybridization and washing temperatures to decrease background without loss of signal, although we have not found higher temperatures necessary.

[a] MOPS buffer cannot be autoclaved. It has to be filter sterilized. It should be colourless, if yellow colour appears discard. Store at 4°C.
[b] Keep at –20°C and discard if colour changes.
[c] Care must be taken when handling ethidium bromide since it is a carcinogen and mutagen.
[d] Ethidium bromide predominantly stains double-stranded molecules. Thus ssRNA stains very weakly in comparison with dsDNA.

4.3 *In situ* hybridizations

In situ hybridizations are used for the analysis of temporal and spatial RNA accumulation and have been extremely useful in developmental studies. This is particularly true for animal tissues, although since the late 1980s there has been an increasing number of *in situ* expression analysis in plants. *Protocol 9* describes a procedure for the fixation and embedding of plant material for the subsequent identification of RNA localization patterns by *in situ* hybridization in various tissues of *Arabidopsis* and *Antirrhinum*; they should, however be applicable to most plant species.

Until recently *in situ* hybridization detection was done by autoradiography usually using photographic emulsion. However with the development of non-radioactive detection methods there are a growing number of laboratories finding this method of detection preferable. We found non-radioactive *in situ*

4: Arabidopsis mutant characterization

procedures much faster, cheaper, safer, and sensitive enough for our work. Thus the focus of this section will be on non-radioactive methods, although with the exception of the probe labelling and detection, the procedures described here can be used for both radioactive and non-radioactive probes. For radioactive labelling and detection see refs 27 and 29.

4.3.1 Fixation

Successful fixation of plant material for *in situ* hybridization has to achieve preservation of tissue morphology and *in vivo* fixation of RNA molecules. Generally fixation by cross-linking, e.g. using aldehydes, have been shown to be better for *in situ* hybridizations than fixation by precipitation (27).

In *Protocol 9* we present two procedures, one using formaldehyde the other using glutaraldehyde as fixatives. It has been suggested (27) that formaldehyde might be a better fixative for tissues where the penetration is more difficult (e.g. embryo tissues inside the seed). In our laboratory we routinely use formaldehyde to fix *Arabidopsis* material, although glutaraldehyde has been successfully used. Buffered formalin has been recently proposed as a successful alternative to formaldehyde and glutaraldehyde. Under-fixation is usually obvious since it results in loss of tissue structure. It has been claimed that over-fixation results in signal decrease (27). Ethanol and Histoclear (Cell-Path) are used for the dehydration and clearing steps, although if needed Histoclear could be substituted by xylene or inhibisol. Changes in ethanol and Histoclear steps should be done gradually to avoid cytoplasm shrinking and cellular distortion. When vacuum is used to facilitate the penetration of solutions, changes in pressure should be gradual to avoid tearing of the tissue.

After fixation, clearing, and embedding, the tissue can be stored in a dry box at 4°C for at least nine months, and probably for even longer.

Protocol 9. Fixation, clearing, and embedding of plant tissue for *in situ* hybridizations

Equipment and reagents

- Vacuum desiccator
- Sample embedding moulds
- 4% (w/v) formaldehyde[a]
- Ethanol: 20%, 30%, 50%, 60%, 70%, 80%, 85%, 95%, 100% (v/v)
- 50%, 75%, 100% (v/v) Histoclear in ethanol (Cell-Path)
- 0.85% (w/v) saline solution (NaCl)
- Paraffin wax chippings (pastillated paramat, BDH)
- 1% (w/v) glutaraldehyde in 0.02 M Hepes buffer pH 7.2
- 0.02 M Hepes buffer

A. *Formaldehyde fixation (30)*

1. Fix tissue in fresh 4% (w/v) formaldehyde[b] for 30 min under vacuum.[c] Discard and replace the formaldehyde solution and leave overnight at 4°C.

2. Wash the tissue for 30 min in 0.85% (w/v) saline solution on ice.

Protocol 9. *Continued*

3. Remove the saline and add 50% (v/v) ethanol. Incubate for 45 min under constant vacuum at room temperature. Repeat this for the following ethanol solutions: 60%, 70%, 85%, 95%, 100% (v/v). Leave overnight (without vacuum, 4°C) in fresh 100% ethanol.
4. Incubate in fresh 100% ethanol (room temperature, 2 h).
5. Remove the ethanol and add 5 ml of 50% (v/v) Histoclear in ethanol (room temperature, 1 h). Repeat using 75% (v/v) Histoclear solution (in ethanol). Replace with 100% Histoclear (room temperature, 1 h). Repeat the 100% infiltration step.
6. Without removing the Histoclear add an equal volume of paraffin wax chippings. Leave overnight at room temperature.
7. Incubate the tissue at 42°C to solubilize any remaining wax. Add approx. half the amount of wax used in step 6 and incubate for another 3–4 h (42°C).
8. Pour off the wax/Histoclear solution. Add 10 ml of melted[d] wax. Incubate at 60°C for at least 8 h. Repeat at least six times.
9. Pour the molten wax and tissue sample into a mould. Allow to cool overnight. Store at 4°C.

B. *Glutaraldehyde fixation (27)*

1. Cut the tissue and immediately place it in a vial containing 5 ml of 1% (w/v) glutaraldehyde in 0.02 M Hepes buffer pH 7.2.
2. Fix for 3 h at room temperature. Degas (in a desiccator) for 10 min after 1 h and then 2 h. When degassing, gradually increase and decrease the vacuum.
3. Wash the tissue with 5 ml of 0.02 M Hepes buffer (30 min, room temperature). Repeat.
4. Remove the buffer and add 5 ml of 20% (v/v) ethanol. Incubate at room temperature at constant vacuum for 30 min. Repeat this for the following ethanol solutions: 20%, 30%, 50%, 60%, 70%, 80%, 95%, 100% (v/v). Leave the tissue overnight in 100% ethanol, without vacuum, 4°C.
5. Continue from part A, step 4.

[a] Formaldehyde can be freshly made from paraformaldehyde by dissolving 4 g of paraformaldehyde in 100 ml of PBS pH 7.2. Paraformaldehyde does not dissolve well at neutral pH; it is therefore advisable to increase the pH to 11 (using NaOH), dissolve the paraformaldehyde in warm PBS, and then adjust to pH 7.2 (using H_2SO_4).
[b] The tissue to fluid ratio should never be greater than 1:10. The tissue pieces should be small to facilitate fixation.
[c] Most plant tissues have a cuticle and will thus float on the surface of the fixative. To allow penetration of the fixative the tissue must be low pressure vacuum infiltrated.
[d] Do not allow temperatures to exceed 62°C at any point as this will damage the polymer structure of the paraffin.

4.3.2 Pre-hybridization treatments

Slides and coverslips have to be cleaned and coated to prevent loss of sections. Slide preparation is given in *Protocol 10*. Depending on the size of the plant material we use 6–8 μm serial sections. Sections are first spread on a 45°C water-bath, mounted on slides, and left at 42°C overnight.

Protocol 10. Slide preparation

Equipment and reagents
- Slides and coverslips
- 180°C oven
- 40% (w/v) chromium trioxide
- 1 mg/ml poly-L-lysine (Sigma)

Method
1. Clean slides and coverslips by incubating in 40% (w/v) chromium trioxide overnight or longer.
2. Wash with distilled water for 10 min. Repeat at least ten times.
3. Bake at 180°C for at least 4 h.
4. Coat the slides with 1 mg/ml poly-L-lysine.[a]
5. Leave slides to air dry in a dust-free place.

[a] This can be done by spreading a drop of poly-L-lysine on the slide using a coverslip. If drops of liquid instead of a thin film layer are forming the slides are not clean enough.

The pre-hybridization treatments (*Protocol 11*) together with the quality of the probe are critical for the success of the whole procedure. The pre-hybridization steps have two main goals; first to make the target DNA more accessible, and secondly to decrease background signal. To make the target DNA accessible after cross-linking fixation a partial protease digestion is conducted. Proteinase K and pronase E are most commonly used. We found pronase E gives more reproducible results, yet regardless of the protease used, it is recommended to calibrate the optimum deproteinization time for your own tissue. In addition proteases from different batches may vary significantly. Thus the optimal protease incubation time may also vary depending on the supplier. It has been suggested that tissue differences are less important than fixation differences in determining the optimal protease treatment (27). Over-digestion with proteases will cause loss of signal and possibly tissue damage. Acetic anhydride is used to reduce background in the sections and slides by acetylating positive charges. Positive charges on the slides may otherwise result in indiscriminate binding of the probes which are acid.

Protocol 11. Tissue pre-hybridization treatments (30)

Equipment and reagents

- Desiccator, staining dish
- Copling jar
- Inhibisol
- Ethanol 100%, 95%, 85%, 70%, 60%, 50%, 30% (v/v)
- DEPC-treated distilled water
- Phosphate-buffered saline (PBS) pH 7.2 (Sigma): 1.3 M NaCl, 0.03 M Na$_2$HPO$_4$, 0.03 M NaH$_2$PO$_4$
- Pronase E (Sigma): prepare a 40 mg/ml aqueous stock, pre-digest by incubating for 4 h at 37°C. Store at –20°C in 1 ml aliquots. Working stock: 0.125 mg/ml in 50 mM Tris–HCl pH 7.5, 5 mM EDTA.
- 0.2% (v/v) glycine in PBS
- 4% (v/v) fresh formaldehyde in PBS
- Acetic anhydride
- 0.1 M triethanolamine[a] pH 8.0

Method

1. Incubate the slides in inhibisol for 10 min to remove the paraffin. Repeat.
2. Hydrate the sections by passing them through the following solutions: 100%, 100%, 95%, 85%, 70%, 60%, 50%, 30% (v/v) ethanol leaving them for 1 min in each solution at room temperature.
3. Incubate for 2 min at room temperature in DEPC-treated distilled water. Repeat, replacing the water with PBS pH 7.2.
4. Incubate the sections in 0.125 mg/ml pronase E in a Copling jar for 10 min at room temperature to remove proteins.[b]
5. Remove the pronase E and transfer the slides to a staining dish containing 0.2% (v/v) glycine in PBS (2 min) to stop the deproteinization.
6. Wash in PBS pH 7.2 for 2 min at room temperature.
7. Incubate in 4% (v/v) fresh formaldehyde in PBS for 10 min at room temperature.
8. Wash in PBS pH 7.2 for 2 min. Repeat.
9. Add acetic anhydride to a fresh staining dish, transfer the slides, and add 0.1 M triethanolamine pH 8.0. The final acetic anhydride concentration should be 0.25% (v/v). Incubate at room temperature for 10 min.
10. Wash the slides with PBS in a staining dish for 2 min.
11. Dehydrate the sections by passing them through the following solutions: 30%, 50%, 60%, 70%, 85%, 95%, 100% (v/v) ethanol as in step 2. Wash with fresh 100% ethanol. Dry in a desiccator at room temperature.[c]

[a] Triethanolamine cannot be autoclaved, prepare using sterile water in a sterile bottle.
[b] The optimum deproteinization times should be determined for each tissue and batch of pronase E.
[c] If needed at this stage the sections can be stored in a dry box for a few hours at room temperature.

4.3.3 Hybridization

Single-stranded RNA probes are usually used for *in situ* hybridizations since RNA probes allow the removal of unbound probe after hybridization without signal loss. In addition RNA–RNA duplexes show higher stability than DNA–RNA duplexes allowing higher stringency washes to be used. Single-stranded RNA probes can be labelled both radioactively and non-radioactively using an *in vitro* transcription system. Radioactive labelling will not be discussed here (for protocols and discussion on radioactive probes see refs 27 and 29). The most commonly used label for non-radioactive ssRNA probes for *in situ* hybridizations is DIG-11-UTP (Boehringer Mannheim). The manufacturer suggest using a ratio of 3.5 labelled to 6.5 non-labelled UTP during *in vitro* transcription, since the size of the digoxigenin molecule means it can inhibit the activity of the transcription enzymes, however we have found that using non-labelled UTP results in lower signals. *Protocol 12* gives a DIG-11-UTP labelling procedure.

Protocol 12. DIG-11-UTP RNA probe labelling by *in vitro* transcription

Reagents

- Transcription buffer (Pharmacia)
- RNase inhibitor (Pharmacia)
- Nucleotide mix (containing 5 mM ATP, GTP, CTP)
- 1 mM Dig-11-UTP
- 12.5 µl of linearized DNA template in water (0.5–1.0 µg DNA)
- DEPC-treated distilled water
- DNA-dependent RNA polymerase (10 U/µl) (Pharmacia)
- 10 µl of tRNA (20 mg/ml)
- RNase-free DNase (10 U/µl) (Pharmacia, FPLC pure)
- 5 M NH$_4$Ac
- Ethanol 100%, 70% (v/v)

Method

1. Add the following to an 1.5 ml microcentrifuge tube in this order:
   - 5 µl of transcription buffer
   - 1 µl of RNase inhibitor
   - 2.5 µl of nucleotide mix (containing 5 mM ATP, GTP, CTP)
   - 2.5 µl of 1 mM Dig-11-UTP
   - 12.5 µl of linearized DNA template in water (0.5–1.0 µg DNA)
   - 1 µl (10 U) DNA-dependent RNA polymerase
2. Mix and incubate at 37°C for 2 h.
3. Stop the reaction by adding 75 µl of DEPC-treated water, 10 µl of tRNA (20 mg/ml), and 1 µl (10 U) of RNase-free DNase. Leave at 37°C for 15 min.
4. Add 75 µl of 5 M NH$_4$Ac and 600 µl of ice-cold ethanol to precipitate

Protocol 12. *Continued*

  the RNA transcripts. Incubate at −80 °C for 30 min or at −20 °C for at least 2 h.

5. Collect the RNA pellet (13 000 g, 10 min).
6. Wash the pellet with ice-cold 70% (v/v) ethanol and recentrifuge (13 000 g, 10 min).
7. Remove the supernatant and air dry.
8. Resuspend in 50 µl DEPC-treated water.
9. Remove 0.2 µl to check the probe by a dot blot (see *Protocol 15A*). Store the rest at −20 °C.

4.3.4 Probe size and concentration

The optimal concentration of probe depends on the probe complexity. About 0.3 µg/ml of probe per kb of probe complexity is required (27). The optimal probe size for *in situ* hybridization has been estimated at 100–200 bp (27) or 250 bp however, we have successfully used probes as large as 600 bp. If very large probes are to be used they can be partially hydrolysed after synthesis by alkaline hydrolysis (see *Protocol 13*).

Protocol 13. Regulation of RNA probe length by alkaline hydrolysis

Equipment and reagents

- 60 °C incubator
- 2 × carbonate buffer: 80 mM NaHCO$_3$ and 120 mM NaCO$_3$, pH 10.2
- 10% (v/v) acetic acid
- 3 M sodium acetate pH 6.0
- Ethanol 100%, 70% (v/v)
- TE: 10 mM Tris–HCl pH 8.0, 1 mM EDTA

Method

1. Add 50 µl of 2 × carbonate buffer.
2. Incubate at 60 °C for the appropriate time.[a]
3. Stop the reaction and precipitate by adding 10 µl of 10% (v/v) acetic acid, 12 µl of 3 M sodium acetate pH 6.0, and 300 µl of ice-cold 100% ethanol. Leave at −80 °C for 30 min or at least 2 h at −20 °C.
4. Precipitate (13 000 g, 10 min). Wash with 70% (v/v) ethanol.
5. Resuspend in 50 µl of TE. Store at −20 °C.

[a] The incubation time (t) depends on the original transcript length (L$_0$) and the required final length (L). The incubation time can then be calculated from the formula:
$$t = (L_0 - L)/(K\, L_0\, L)$$
where K = 0.11. Calculated time is in minutes (according to ref. 27).

4: Arabidopsis *mutant characterization*

4.3.5 Hybridization

Protocol 14 describes hybridization conditions and post-hybridization washes for homologous probes. Hybridization is conducted at 50°C in a buffer containing 50% (v/v) formamide. When using non-homologous probes the hybridization temperature must be reduced (1% sequence divergent lowers the T_m by approximately 1°C). However, when the homology is lower than 80% it may be very difficult to get a low enough noise-to-signal ratio for the signal to be practically detectable above background.

To decrease the background a combination of RNase A digestion and low salt, high temperature, formamide washes are used. RNase A digests only ssRNA therefore removing the unbound probe without decreasing the signal. Formamide permits the washes to be stringent enough without having to increase the washing temperature above 50°C which could damage the tissue.

Protocol 14. Hybridization and post-hybridization washes (27)

Equipment and reagents

- Staining dishes, humidified box for hybridization, coverslips
- 20 × SSPE: 3.6 M NaCl, 0.2 M NaH$_2$PO$_4$, 0.02 M EDTA pH 7.7
- Hybridization buffer: 5 ml formamide, 1 ml of 20 × SSPE, 500 µl of 100 × Denhardt's solution, 2 ml of 50% (w/v) dextran sulfate, 100 µl of 10 mg/ml tRNA,[a] 500 µl of 10% (w/v) SDS, 840 µl DEPC water
- Wash buffer: 2 × SSC, 50% (v/v) formamide
- 100 × Denhardt's solution: 2% (w/v) BSA, 2% (w/v) Ficoll 400, 2% (w/v) polyvinyl pyrrolidone
- 20 × SSC: 3 M NaCl, 0.3 M Na$_3$ citrate (dilute this stock solution to 2 × SSC)
- NTE: 0.5 M NaCl, 10 mM Tris–HCl pH 7.5, 1 mM EDTA
- NTE containing 20 µg/ml RNase A
- PBS pH 7.2 (see *Protocol 11*)

A. Hybridization

1. Take 4 µl of the probe[b] per slide and add 16 µl of hybridization solution. Denature at 80°C for 5 min.
2. Cool on ice for 2 min.
3. Apply a total volume of 20 µl per slide and add a coverslip.
4. Incubate the slides in a humidified box at 50°C for at least 17 h.

B. Post-hybridization washes

1. Place the slides in a staining dish containing 2 × SSC to remove the coverslips.
2. Transfer the slides to a fresh staining dish containing wash buffer. Incubate at 50°C for 1 h. Repeat.
3. Remove the wash solution and add NTE. Incubate at 37°C for 10 min.
4. Replace the NTE with fresh NTE containing 20 µg/ml RNase A at 37°C for 30 min.

Protocol 14. *Continued*

5. Wash briefly with NTE.
6. Replace NTE with wash buffer. Incubate at 50°C for 1 h.
7. Wash briefly with 2 × SSC at room temperature.
8. Remove the SSC and wash in PBS buffer for 5 min. Replace with fresh PBS buffer and store at 4°C.

[a] Added immediately before hybridization.
[b] Probes should be checked for labelling efficiency, prior to use for *in situ* hybridizations, by a dot blot and detection (see *Protocol 15A*).

4.3.6 Detection and staining

Here we will only consider the detection of digoxigenin labelled probes, for radioactive detection and staining see Cox and Goldberg (27). Unlike autoradiography where the signal detected comes directly from the probe, in the DIG system the hybridized probes are immunodetected with an alkaline phosphatase-conjugated anti-digoxigenin antibody, and then visualized with the colorimetric substrates NBT and X-phosphate (Boehringer Mannheim). *Protocol 15* describes a method for detection and staining of DIG labelled probes in *in situ* hybridization experiments. The aim of this procedure is to keep non-specific binding of the antibody conjugate as low as possible, whilst maximizing signal from the probe. Background antibody conjugate binding is minimized by pre-incubation of the slides with BSA and 'blocking reagent' (proteolytic caesin in powder form) (Boehringer Mannheim). Further decreases in the background signal are achieved by rigorous washes with buffer 3 (see *Protocol 15*). Buffer 3 contains detergent (Triton X-100 or Tween 20) which helps in the removal of non-specific bound antibody. We have never had problems with high antibody-produced background using this procedure. However, should such problems occur it is possible to add sheep serum in buffer 3 to a final concentration of 3% (v/v) to further decrease background. A 1:3000 dilution of the antibody conjugate is suggested in this protocol. However, we have found that a 1:4000 dilution can also be used provided that the incubation time is doubled.

Protocol 15. Detection of DIG labelled probes for *in situ* hybridization (30)

Equipment and reagents

- Staining dish
- Buffer 1: 100 mM Tris–HCl, 150 mM NaCl pH 7.5
- Buffer 2: buffer 1 containing 0.5% (w/v) blocking reagent (Boehringer Mannheim)[a]
- Buffer 3: buffer 1 containing 1% (w/v) BSA, 0.3% (v/v) Triton X-100 (or 1% (v/v) Tween 20)[b]
- Anti-digoxigenin–alkaline phosphatase conjugate (Boehringer Mannheim)

4: Arabidopsis *mutant characterization*

- Buffer 4: 100 mM Tris–HCl, 100 mM NaCl pH 9.5
- Nylon membrane (for dot blot)
- Nitro blue tetrazolium salt (NBT): 75 mg/ml in 75% (v/v) dimethyl formamide
- 5-bromo-4-chloro-3-indoyl phosphate (X-phosphate): 50 mg/ml in dimethyl formamide
- TE: 10 mM Tris–HCl pH 8.0, 1 mM EDTA

A. Dot blot analysis of probe prior to hybridization

1. Spot 1 µl of a 1:5 dilution of the probe onto nylon membrane and UV cross-link (30 mJ).
2. Wet in buffer 1.
3. Incubate in buffer 2 for 30 min, agitating gently.
4. Wash in buffer 1.
5. Incubate in 10 ml of buffer 1 containing 1 µl of anti-digoxigenin–alkaline phosphatase conjugate, agitating gently.
6. Rinse in buffer 1 with agitation, twice for 15 min.
7. Wash briefly in buffer 4.
8. Incubate for 10 min in 10 ml of buffer 4 containing 14 µl of NBT and 10 µl of X-phosphate.
9. Signal should be detected within a few minutes, reaching a maximum after 12 h.
10. Wash membrane in buffer 1 for 5 min to stop the reaction.
11. Probes should only be used for *in situ* hybridizations if they give strong DIG staining.

B. DIG detection for in situ hybridizations

1. Wash the slides in buffer 1 in a staining dish for 5 min.
2. Block background antibody binding, by incubating slides in a staining dish on a shaker with: buffer 2 for 1 h, and then in buffer 3 for 1 h at room temperature.
3. Transfer slides to a tray.[c] Add buffer 3 containing anti-digoxigenin–alkaline phosphatase conjugate in a 1:3000 dilution. Add the antibody conjugate just before use. Incubate at room temperature for 2 h with agitation.
4. Transfer slides to a staining dish and wash with buffer 3 for 20 min. Repeat three times.
5. Rinse with buffer 1.
6. Equilibrate in buffer 4 for 5 min.
7. Detect the hybridization signal by incubating in buffer 4 containing 2 µl of NBT and 1.5 µl of X-phosphate per ml of buffer 4. Do not shake. Signal should be detectable in 6–72 h.

Protocol 15. *Continued*

8. Stop the reaction by transferring the slides in a staining dish containing TE. At this stage sections can be stored for a few hours before tissue staining and mounting.

[a] Make fresh just before use, dissolve at 60°C for 30 min. The solution remains turbid.
[b] Make fresh just before use.
[c] The sections are transferred to a tray to minimize the volume of antibody-containing solution used.

4.3.7 Staining and photography

Background tissue staining with Calcofluor (Sigma 'Fluorescent Brightner 28') (*Protocol 16*) is convenient since it does not interfere with the signal as it is only visible only under UV light. However, it is also possible to use other staining procedures for plant tissue (e.g. nuclear fast red, acridine orange) provided they do not interfere with the dark-purple/dark-blue colour of the signal. When Calcofluor is used it is essential that the sections are viewed under UV light for a minimal time, since prolonged UV exposure causes fading of the fluorescence. Photographs can be taken using bright or dark-field simultaneously with UV light. High quality photographs can be taken using double exposure: first bright/dark-field then UV.

Protocol 16. Staining and detection of sections

Reagents

- Ethanol: 50%, 70%, 95%, 100% (v/v)
- Euparal (Merck)
- 0.1% (w/v) Fluorescent Brightner 28 (Calcofluor) (Sigma)

Method

1. Rinse the sections with distilled water.
2. Take the sections through the following solutions: 50%, 70%, 95%, 100% (v/v) ethanol for 2 min each. Then back through to water. This removes background colour staining.[a]
3. Stain by dipping the sections in 0.1% (w/v) Calcofluor for 5 min.
4. Rinse with distilled water. Repeat twice.
5. Air dry slides in a dust-free place.
6. Mount slides with Euparal. Leave overnight and store in a dry box. Avoid prolonged exposure to light.
7. Photograph the sections using bright-field and UV light.

[a] This may also reduce the signal. To avoid this ensure that while rehydrating in step 2 the sections stay in ethanol for a minimum time.

References

1. Kosambi, D. D. (1944). *Ann. Eugen.*, **12**, 172.
2. Koornneef, M. and Stam, P. (1992). In *Methods in Arabidopsis research* (ed. C. Koncz, N.-H. Chua, and J. Schell), p. 83. World Scientific Pub. Co.
3. Fabri, C. O. and Schäffner, A. R. (1994). *Plant J.*, **5**, 149.
4. Reiter, R. S., Williams, J. G. K., Feldmann, K. A., Rafalski, J. A., Tingley, S. V., and Scholnik, P. A. (1992). *Proc. Natl. Acad. Sci. USA*, **89**, 1477.
5. Williams, J. G. K., Kubelik, A. R., Livak, K. J., Rafalski, J. A., and Tingley, S. V. (1990). *Nucleic Acids Res.*, **18**, 6531.
6. Konieczny, A. and Ausubel, F. M. (1993). *Plant J.*, **4**, 403.
7. Bell, C. J. and Ecker, J. R. (1994). *Genomics*, **19**, 137.
8. Vos, P., Hogers, R., Bleeker, M., Rijans, M., Van der Lee, T., Hornes, M., *et al.* (1995). *Nucleic Acids Res.*, **23**, 4407.
9. Koornneef, M., Hanhart, C. J., and Thiel, F. (1989). *J. Hered.*, **80**, 118.
10. Mahlberg, P. (1972). *Can. J. Bot.*, **50**, 857.
11. Regan, S. M. and Moffatt, B. A. (1990). *Plant Cell*, **2**, 877.
12. Brundrett, M. C., Kendrick, B., and Peterson, C. A. (1991). *Biotechnic. Histochem.*, **66**, 111.
13. Coleman, A. W. and Goff, L. J. (1985). *Stain Technol.*, **60**, 145.
14. Alexander, M. P. (1969). *Stain Technol.*, **44**, 117.
15. Dawson, J., Wilson, Z. A., Aarts, M. G. M., Braithwaite, A. F., Briarty, L. G., and Mulligan, B. J. (1993). *Can. J. Bot.*, **71**, 629.
16. Heslop-Harrison, J. and Heslop-Harrison, Y. (1970). *Stain Technol.*, **45**, 115.
17. Pickert, M. (1988). *Arabidopsis Inf. Serv.*, **26**, 39.
18. Hayat, M. (ed.) (1981). *Fixation in electron microscopy.* p. 501. New York, London, Academic Press.
19. Mollenhauer, H. and Totten, C. (1971). *J. Cell Biol.*, **48**, 387.
20. Henmann, H. (1992). *Histochemistry*, **98**, 341.
21. Smith, M. M. and McCully, M. E. (1978). *Protoplasma*, **95**, 229.
22. Vieira, M., Briarty, L. G., and Mulligan, B. J. (1990). *Ann. Bot.*, **66**, 717.
23. Hunter, E. (ed.) (1993). *Practical electron microscopy.* p. 173. Cambridge, Cambridge University Press.
24. Gabriel, B. (ed.) (1982). *Biological scanning electron microscopy.* New York, Van Nostrand Reinhold Company.
25. Sambrook, J., Fritsch, E. F., and Maniatis, T. (ed.) (1989). *Molecular cloning: a laboratory manual* (2nd edn). Cold Spring Harbour.
26. Chomczynski, P. and Sacchi, N. (1987). *Anal. Biochem.*, **162**, 156.
27. Cox, K. and Goldberg, R. B. (1988). In *Plant molecular biology: a practical approach* (ed. C. Shaw), p. 1. IRL press, Oxford.
28. DuPont. (1992). *Hybridization and detection protocols.*
29. Cox, K. H., DeLeon, D., Angerer, J., and Angerer, R. C. (1984). *Dev. Biol.*, **101**, 485.
30. Bradley, D., Carpenter, R., Sommer, H., Hartley, N., and Coen, E. (1993). *Cell*, **72**, 85.

5

Classical and molecular cytogenetics of *Arabidopsis*

G. H. JONES and J. S. HESLOP-HARRISON

1. Introduction

In many species of importance, either as genetical research tools or because of their economic value, cytogenetics has played a prominent role in providing a framework for genetical and molecular analyses of their genomes. The small genome size of *Arabidopsis thaliana* L. which confers particular advantages for molecular analysis has been a distinct hindrance to cytogenetical analysis. It has been unkindly suggested that *Arabidopsis* chromosomes resemble 'mouse droppings seen through the wrong end of a telescope'! While this is perhaps too harsh a judgement, the chromosomes of *Arabidopsis* are certainly very small by comparison with most eukaryotes, and this limitation has without doubt restricted the examination of chromosome organization and behaviour during mitosis and meiosis, and their numerical and structural variants, in short—cytogenetics. Despite these difficulties, *Arabidopsis* has attracted the attention of cytogeneticists over a long period of time and slow but erratic progress has been made in characterizing its chromosome organization and behaviour both in normal and abnormal situations.

Steinitz-Sears (1) cited a number of early studies (2–4) as having established the chromosome number of *Arabidopsis* as 2n = 10, but hers was the first study which attempted to characterize the *Arabidopsis* karyotype in terms of chromosome lengths, arm ratios, and nucleolus organizing regions (NORs), albeit incompletely and, in part, incorrectly. In addition she described the five primary trisomics of *Arabidopsis* and presented a preliminary chromosome assignment of three of them. All five chromosomes were described as bi-armed, ranging in length from 0.8–2.1 μm at mitotic metaphase, giving a range of 2.6 × in chromosome lengths which is rather larger than the ranges determined in later studies of mitotic chromosomes (1.9 ×) (5) and meiotic pachytene bivalents (1.6 ×) (6) (1.5 ×) (7). All but two of the chromosomes were said by Steinitz-Sears (1) to be distinguishable but curiously, only one nucleolus organizing chromosome was detected and this was, incorrectly, identified as the largest chromosome of the complement with the NOR located in its long arm.

A fuller karyotype description and trisomic analysis followed a few years later (8). This study utilized meiotic diplotene pollen mother cells, with the advantages that they revealed the positions of densely stained pericentromeric heterochromatic blocks, identified NORs by their association with nucleoli, and conclusively identified the three homologous chromosomes in trisomics by virtue of their association in trivalents. This study correctly deduced, for the first time, the presence of two nucleolus organizing chromosome pairs in *Arabidopsis*, but persisted in regarding one of these (linkage group 4) as the longest chromosome of the genome possessing a median centromere, which we now know to be erroneous.

The next major advance was the application of Giemsa C-banding to mitotic metaphase chromosomes (6) combined with detailed observations and measurements of chromosome dimensions and centromere and NOR locations. This study again detected only one NOR (chromosome 5 = linkage group 4) which was described as having a median centromere, although they recognized that its length was extremely variable due to stretching of the NOR-bearing short arm. Some of this variability could also have been due to the destructive action of Giemsa C-banding which can disrupt chromosome morphology. This C-banded karyotype was subsequently aligned to the genetic linkage map of Koornneef *et al.* (9) by Schweizer *et al.* (10) also incorporating Steinitz-Sears and Lee Chen's (8) observation of two NORs, assigned to linkage groups 2 and 4.

In recent years much clearer images of *Arabidopsis* mitotic chromosomes have been obtained by staining with the fluorochrome DAPI and the locations of NORs (45S rDNA) and paracentromeric heterochromatin blocks have been confirmed by fluorescent DNA–DNA *in situ* hybridization (FISH) (11–13). Fransz *et al.* (7) developed a nuclear spreading and DAPI staining technique for light microscopic analysis of *Arabidopsis* meiotic pachytene bivalents, which combined with FISH has proved useful for karyotyping and for high resolution physical mapping of repetitive and low copy-number probes. The three longest chromosomes of *Arabidopsis thaliana* are metacentric or nearly so and show a gradation of overall lengths. It seems to be generally agreed that these are designated numbers 1, 3, and 5 in accordance with their corresponding linkage groups, despite the non-concurrence with their physical length rankings. Chromosomes 1 and 5, the two longest ranking chromosomes, are distinguished by the presence of one or two pericentromeric 5S rDNA loci on chromosome 5. The somewhat shorter and sub-metacentric chromosome 3 may also carry a 5S rDNA locus, but this is variable in occurrence and position among different ecotypes. The remaining shorter chromosomes, designated 2 and 4 (again to concur with linkage groups) are both acrocentric, and of similar overall lengths, with distinctly unequal arms; NORs have been located to the short arms of both these chromosomes by

5: Classical and molecular cytogenetics of Arabidopsis

FISH using a 45S rDNA probe (7, 13). A high resolution karyotype of *Arabidopsis* produced by synaptonemal complex (SC) surface spreading has confirmed that two NORs are located subterminally on the two shortest SCs (6). Unfortunately the published SC karyotype does not locate centromeres as these were not preserved in the SC preparations used to prepare the karyotype. Centromeres show variable preservation in SC spreads and so it is possible that an SC karyotype including centromeres could be produced in the future.

This brief historical survey reveals that the small sizes of *Arabidopsis* chromosomes resulted in considerable difficulties in producing a consistent and reliable karyotype and there was some initial confusion especially concerning the number and the identities of NOR-bearing chromosomes which extended to the relative sizes and arm ratios of these chromosomes and the chromosomal locations of the NORs. These difficulties are now largely resolved and a relatively clear karyotype has emerged, which is consistent for both mitotic and meiotic chromosomes.

The genus *Arabidopsis* comprises about 20 species, mostly of Asian origin, but the species *A. thaliana* extends into Europe and North America. Cytogenetic and molecular comparisons of chromosome and DNA organization in these species are potentially of great interest and value, both to geneticists and taxonomists. Early indications confirm this view (14, 15).

One of the foremost objectives of cytogenetics applied to *Arabidopsis* is to construct an accurate, reliable, and consistent karyotype which can form the basis for a physical map of the genome. *In situ* hybridization of defined DNA sequences to mitotic metaphase chromosomes may contribute to this, but the small sizes of *Arabidopsis* chromosomes limit the resolution of a physical map produced in this way. The application of *in situ* hybridization to interphase nuclei (interphase cytogenetics) will almost certainly prove valuable in this respect since in many species the extended decondensed chromatin of individual chromosomes is known to retain its linear organization within spatially limited chromosome domains. Thus, DNA sequences which are inseparable by FISH on condensed mitotic metaphase chromosomes can be separated and ordered in interphase nuclei (16). Another potentially valuable approach to high resolution physical mapping is *in situ* hybridization to pachytene bivalents or to SCs (7, 17). Bivalents and SCs at the pachytene stage of meiosis are about ten times longer than corresponding mitotic metaphase chromosomes, offering a correspondingly higher resolution for the separation and ordering of closely adjacent genes and DNA sequences on the physical map. Even higher resolution separation and ordering of DNA sequences can be achieved by FISH to spread DNA/chromatin fibres (18), but this approach is especially appropriate to very closely adjacent sequences.

Another important role for cytogenetics lies in identifying and characterizing variants of chromosome number, structure, and behaviour. Trisomics and telotrisomics have been extensively employed in *Arabidopsis* for associating chromosomes with their linkage groups and for locating centromeres on genetical linkage maps (9). Autotetraploid lines of *Arabidopsis* exist which have been confirmed by mitotic cell chromosome counts or by counting heterochromatic chromocentres in interphase nuclei (19, 20). However, there have been few studies of meiotic chromosome behaviour in aneuploid or polyploid *Arabidopsis* although it is evidently feasible to analyse chromosome association at diplotene, diakinesis, or metaphase I (8, 21) and it should also be possible to analyse their synaptic behaviour in SC spread preparations, and their subsequent disjunctional properties. Similarly there have been few cytogenetic studies of structurally altered chromosomes (translocations, inversions, deletions) in *Arabidopsis* although here again there is considerable scope for defining the breakpoints of rearrangements and investigating the synapsis and subsequent disjunction of rearranged chromosomes during meiosis. Cytogenetics also has an important role in defining the phenotypes of meiotic mutants originating from programmes to isolate and characterize genes involved in meiosis in *Arabidopsis* (22, 23).

2. Mitotic chromosome analysis by light microscopy

The most convenient material for chromosome preparations is young flower buds. They have a high metaphase index, and large numbers can be obtained from one plant. Root tips from seedlings, and presumably adult plants, can also be used, but their small size makes them difficult to handle, and each root often contains less than ten metaphases. *Protocol 1* describes a method that gives high quality chromosome preparations suitable for morphological studies or *in situ* hybridization. Three stages are involved. First the plant material is pre-treated to accumulate more metaphases and prometaphases than would be present in untreated tissues. The metaphase arresting agent 8-hydroxyquinoline is recommended because it preserves good chromosome morphology while other inhibitors such as colchicine shorten the small chromosomes too much. Then the plant material is treated by enzyme digestion to dissolve much of the cell wall, freeing chromosomes and interphase nuclei from their cells. The preparations are then squashed to give well spread chromosomes which are free of cytoplasm, but still with complete metaphases. Other methods may also be used (13).

Protocol 1. Mitotic metaphase chromosome preparation

Equipment and reagents

- Plasticware for fixation[a]
- Microcentrifuge tubes, e.g. Eppendorf
- Microscope slides[b] and 18 × 18 mm coverslips dipped in 100% ethanol and polished with a clean tissue before use
- Polystyrene 1–5 cm^3 tubes, e.g. Sterilin
- Dissecting instruments
- Stereo dissecting microscope
- Diamond or carbide-tipped pen
- Drying rack for slides

5: Classical and molecular cytogenetics of Arabidopsis

- Dry ice, liquid nitrogen, or metal block in −80°C freezer
- 100 mm filter paper (Whatman No. 1)
- Murashige and Skoog agar medium[c] without sucrose (Sigma)
- Plant material
- Solution of 0.2 mM 8-hydroxyquinoline in distilled water[d]
- Tissue
- 45% (v/v) acetic acid
- 60% (v/v) acetic acid
- 100% ethanol
- Glacial acetic acid
- Enzyme buffer: 0.1 M citric acid (solution A), 0.1 M tri-sodium citrate (solution B)
- Cellulase (from *Aspergillus niger*; Calbiochem, 21947)
- Pectinase (from *Aspergillus niger*, solution in KCl and sorbitol; Sigma, P-9179)

A. Accumulation of metaphases and fixation

1. Use either root tips from seeds germinated on Murashige and Skoog agar medium without sucrose, or buds from plants grown in soil. If using root tips, use the entire seedling. If ample flower material is available, it is convenient to use the tip of the floral stem from open flowers upwards. If more flowers are required from a plant, particularly if the plant for analysis is special, remove individual flowers and buds at stages between anthesis and a bud length of 1.5 mm. Fill the tubes about half-full with 8-hydroxyquinoline solution, warm to the temperature of plant growth, and shake briefly to aerate. Place the plant material into the tubes, and shake again, ensuring material is submerged.

2. Leave in a chemical-free environment at the same temperature for 1–2 h. Transfer to 4°C, still chemical-free, for 1–2 h.

3. Remove the plant material, touch it against dry tissue to remove excess 8-hydroxyquinoline, and transfer the material to freshly made 3:1 ethanol:glacial acetic acid fixative at 4°C for a minimum of 4 h fixation. Use for immediate preparation or transfer to fresh fixative and maintain at −20°C for long-term storage of up to three months.

B. Enzyme digestion

1. Make an enzyme stock buffer by mixing 4 ml of solution A and 6 ml of solution B. Dilute 1:10 with distilled water for use.

2. Wash fixations twice in diluted enzyme buffer for 10 min each.

3. Make an enzyme solution of 2% (w/v) cellulase and 20% (w/v) pectinase in enzyme buffer. Aliquot into 1.5 ml microcentrifuge tubes and store for up to one year at −20°C.

4. Transfer buds or root tips to the enzyme buffer and incubate at 37°C for 90–120 min (buds) or 30–60 min (root tips). The optimum time must be found by experiment and ± 15 min will make a substantial difference to the preparation quality.

5. Transfer to the diluted enzyme buffer. Material can be left for 1–2 h in the buffer at room temperature, or overnight at 4°C.

Protocol 1. *Continued*

C. *Squash preparations*

1. Transfer a few root tips or buds from the enzyme buffer to 60% (w/v) acetic acid.

2. Place a 15 µl drop of 60% (w/v) acetic acid onto a cleaned microscope slide. The size of drop may need adjustment, but is critical to good preparations.

3. Under a dissecting microscope, either dissect out one pistil from a bud, or the terminal 1 mm of a root tip, and place in the drop on the slide. The material will be soggy following enzyme treatment, so two pairs of forceps should be used for transfer.

4. Put a 18 × 18 mm glass coverslip over the centred material and allow it to settle for a few seconds. Tap, quite hard, many times over the material with a dissecting needle. Fold two 100 mm circles of filter paper in half, put the edge of the slide in the fold, on a bench, and press your thumb onto the filter paper over the coverslip with the maximum force possible.

5. Place the slide on a block of dry ice (preferable), or on a piece of metal in a −80 °C freezer, for 5–10 min (not more), or immerse in liquid nitrogen until frozen. Flick off the coverslip with a razor blade.

6. Air dry the slide for 2–8 h, or for 10 min at 37 °C.

7. Scratch two lines on the bottom of the slide to show the edges of the coverslip[e] for reference when adding solutions and coverslips for chromosome examination (*Protocols 2–4*).

[a] In general, it is best to use new plasticware for pre-treatments. Any contamination with chemicals (particularly acids and alcohol) will reduce the number of metaphases in the preparation. However, some tissue culture plasticware is also unsuitable because of residues of ethylene oxide used for sterilization, or from mould release agents.

[b] Microscope slides with frosted ends can be used but extra care should be taken because glass dust left from sand-blasting will prevent the coverslip pressing down flat onto the chromosomes.

[c] Murashige, T. and Skoog, F. (1962). A revised medium for rapid growth and bioassays with tobacco tissue cultures. *Physiol. Plant.*, **15**, 473.

[d] The crystals take about 8 h to dissolve with stirring, and the solution is pale yellow. It can be stored at 4 °C for several months.

[e] If the original edges of the coverslip cannot be seen easily, breathe gently on the top of the slide. Scratch reference numbers or codes into the slide ends using a diamond or carbide-tipped pen. With care, pencil writing on frosted-end slides will survive the *in situ* hybridization procedure, but few other methods will.

Protocol 2 describes staining preparations for chromosome examination. Preparations can be examined to check that chromosomes are present, count their number, and examine their morphology. If *in situ* hybridization is to follow, the preparation should be examined to check the quality. *Arabidopsis*

5: Classical and molecular cytogenetics of Arabidopsis

chromosomes are too small to see easily without staining, so are stained with a fluorochrome (see *Protocol 2*), whilst chromosomes of species with larger chromosomes can be examined dry by phase-contrast microscopy before *in situ* hybridization. The fluorochromes propidium iodide or DAPI (4′,6-diamidino-2-phenylindole) are recommended. However many microscope lenses and some immersion oils are not UV transparent and hence unsuitable for DAPI even with the correct filter sets. If no fluorescent microscope is available, Giemsa can be used to stain the chromosomes.

Protocol 2. Staining preparation for examination of chromosomes

Equipment and reagents

- Compound microscope with epifluorescence if fluorescent stains are used (strongly recommended)
- 22 × 22 mm glass coverslips
- Large thin coverslips, No. 0, 20 × 40 mm
- Filter paper
- 100%, 90%, 70% (v/v) ethanol
- Nail varnish

For semi-permanent mounts
- McIlvaine's buffer pH 7.0: 18 ml of 0.1 M citric acid plus 82 ml of 0.2 M Na_2HPO_4
- DAPI (4′,6-diamidino-2-phenylindole) (Sigma) staining solution: stock 0.1 mg/ml in water, stored at −20°C, diluted to 2 μg/ml in McIlvaine's buffer for use

or
- Propidium iodide staining solution: stock 100 μg/ml in water, stored at −20°C, dilute to 2.5 μg/ml in 4 × SSC–Tween made from

20 × SSC (3 M sodium chloride, 0.3 M sodium citrate) with 0.2% (v/v) Tween 20 (polyoxysorbitan monolaurate) (Sigma) for use

or
- Giemsa staining solution: make fresh solution of 4% (v/v) Giemsa (Sigma) in Sorensen's buffer (0.03 M KH_2PO_4 plus 0.03 M Na_2HPO_4 pH 6.8)

To make preparations semi-permanent
- Antifade mounting solution, e.g. Citifluor AF1 (Agar), Vectashield (Vector Laboratories)
- Coverslips of 20 × 40 mm rectangles cut from autoclavable plastic bags used for disposal of biohazardous waste, e.g. Sterilin
- 4 × SSC–Tween (see propidium iodide staining)

A. *Fluorochrome staining*

1. Place a drop (50–100 μl) of staining solution onto the chromosome preparation.[a]

2. For temporary preparations for chromosome counts, or before *in situ* hybridization place a 22 × 22 mm coverslip over the preparation, wait 1–10 min for staining, and observe. The edges of the coverslip can be sealed with nail varnish, and the slide kept for one to two weeks at 4°C.

3. For *in situ* hybridization or to store the slide indefinitely in a dry state, remove the coverslip immediately by dipping into 70% (v/v) ethanol and allowing it to fall off. Transfer slide through 90% and 100% ethanol for 5 min each and allow to air dry.

4. For semi-permanent preparation to examine chromosome morphology or following *in situ* hybridization place a plastic coverslip over

Protocol 2. *Continued*

　the preparation and wait 1–10 min for staining. Wash the slide briefly in 4 × SSC–Tween until the coverslip falls off, and shake to remove drops. If both propidium iodide and DAPI staining are required, repeat from step 1 with the other stain: DAPI is best used first as it is less soluble than propidium iodide.

5. Put a generous drop (50–100 µl) of antifade solution onto the preparation, and cover it with a large, thin coverslip (No. 0, 20 × 40 mm). Press the slide very hard between four sheets of filter paper. A minimum thickness of antifade solution is necessary to produce the best images. Observe the slide. Normally, the pressing should be repeated before each observation. Store the slide horizontally at 4°C in the dark; it will normally remain useful for one to two years without sealant.

B. *Giemsa staining*

1. Immerse the slides in 50 ml staining solution for 7–12 min at room temperature (depending on staining intensity required).
2. Wash in distilled water by flooding away the stain solution; removing slides through the surface film often results in stain precipitate from the surface adhering to the preparations. Mount in microscopy mountant such as euparal or DPX before observation by transmitted light microscopy.

[a] Use a complementary colour of fluorochrome to probe detection systems after *in situ* hybridization: for example, do not use propidium iodide with Cy3, or rhodamine labelled probes nor DAPI with AMCA labels.

3. *In situ* hybridization to mitotic chromosome preparations

Chromosome *in situ* hybridization is a powerful technique that allows localization of DNA sequences along metaphase chromosomes and within interphase nuclei, hence combining cytological information with molecular data about DNA sequences. *In situ* hybridization is valuable for the physical mapping of sequences and investigation of their long-range organization in the genome. *In situ* hybridization is the method of choice to map repetitive DNA sequences to a linkage group; 18S–5.8S–25S rDNA sequences, repeated some 570 times in the *A. thaliana* genome, were the first to be mapped by *in situ* hybridization (11), and the 5S rDNA sites were mapped more recently (7, 13). Similarly, a repetitive sequence in *A. thaliana*, pAL1, isolated and characterized by Southern hybridization (24) was shown to locate at all ten paracentromeric regions by *in situ* hybridization. Sequences at one site where the target is above 10 kb in size can be routinely localized, while methods are

5: Classical and molecular cytogenetics of Arabidopsis

improving to enable smaller cosmid or plasmid clones to be localized (7, 25). *Protocols 3* and *4* are methods for pre-treatment and *in situ* hybridization based on those given by Schwarzacher *et al.* (26) that have been modified and condensed specifically for application to chromosome spreads of *A. thaliana*. Chromosome preparations must be free of RNA and well fixed so material does not come off the slide during the hybridization and the stringent washes.

The quality of the chromosome preparation is very important for *in situ* hybridization. There should be many metaphases on the slide with clearly separated chromosomes and little cytoplasm. These factors can be altered by fixation (times of 8-hydroxyquinoline treatment), enzyme mixture digestion time, and squashing force or tapping technique. Slide selection should be very rigorous; in our laboratory, we routinely discard half of all fixation batches as being unsuitable and 70% of slide preparations from good fixations.

Protocol 3. Pretreatment of slides for *in situ* hybridization to somatic metaphase preparations

Equipment and reagents

- Humid chamber
- Oven 37°C
- Coplin jars
- Plastic coverslips cut from autoclavable waste disposal bags
- Dehydrated chromosome preparations (see *Protocols 1* and *2*)
- Heated stirrer
- DNase-free RNase A (Sigma): make a stock by dissolving 10 mg/ml RNase A in 10 mM Tris–HCl pH 7.5 and 15 mM NaCl; place in a closed tube in boiling water for 15 min to destroy any DNase activity, aliquot, and store at –20°C

- 20 × SSC stock: 3 M NaCl, 0.3 M Na citrate, adjusted to pH 7.0 (working solution 2 × SSC)
- Pepsin (porcine stomach mucosa, activity: 3200–4500 U/mg, Sigma): make a stock solution of 500 mg/ml in 0.1 M HCl and store at –20°C
- 0.01 M HCl
- Paraformaldehyde (microscopy grade, e.g. Sigma)
- 70%, 90%, and 100% (v/v) ethanol
- Distilled water
- 0.1 M NaOH

Method

1. Place the dehydrated chromosome preparation into an oven at 37°C overnight.
2. Dilute the stock RNase A 1:100 in 2 × SSC, add 200 µl to each slide preparation, cover with a plastic coverslip, and incubate for 1 h in a humid chamber at 37°C.
3. Prepare fresh paraformaldehyde solution by adding 4 g paraformaldehyde to 80 ml water and heating to 60°C while stirring for 10–15 min. Add 20 ml 0.1 M NaOH to clear the solution.
4. Wash the slides three times in 2 × SSC (5 min each) in Coplin jars using enough solution to fully immerse the slides.
5. Place the slides in 0.01 M HCl in Coplin jars.
6. Add 200 µl of 40 U/ml pepsin in 0.01 M HCl to each slide and cover

Protocol 3. Continued

with a plastic coverslip. Typically incubate for 10 min at 37°C in a humid chamber. The time (up to 1 h) and pepsin concentration (up to 400 U/ml) can be altered (or omitted) depending on the amount of cytoplasm on the slides.[a]

7. Wash the slides three times in 2 × SSC for 5 min each. The coverslip usually falls off in the first wash.
8. Place in paraformaldehyde solution (see step 3) for 10 min at room temperature.
9. Wash the slides three times in 2 × SSC for 5 min each.
10. Dehydrate through an ethanol series (70%, 90%, 100%, v/v) for 5–10 min at each stage and air dry.

[a] Cytoplasm varies depending on factors such as plant genotype, growth conditions, fixation, and spreading techniques.

Protocol 4. *In situ* hybridization

Equipment and reagents

- Programmable temperature cycler (e.g. Hybaid)
- Epifluorescence microscope
- Microcentrifuge
- Humid chamber
- Plastic coverslips (see *Protocol 3*)
- DNA probe, typically a cloned DNA fragment
- Formamide[a] (Sigma, Roche)
- Labelled nucleotide (one or more of the following):
  (a) Digoxigenin-11-dUTP (1 mM solution; Roche)
  (b) Biotin-11-dUTP (powder; Sigma) or biotin-14-dATP (0.4 mM solution; Sigma)
  (c) Fluorescein-12-dUTP (1 mM solution; Roche) or fluorescein-11-dUTP (Fluoro-Green, 1 mM solution; Amersham)
  (d) Rhodamine-4-dUTP (FluoroRed, 1 mM solution; Amersham)
  (e) Coumarin-4-dUTP (FluoroBlue, 1 mM solution; Amersham)
  (f) Cy5-dCTP (Fluorolink, 25 nM solution; Amersham)

  Prepare stock solutions and store at –20 °C. Digoxigenin: take 18.5 ml of 1 mM solution of TTP in Tris–HCl pH 7.5 and mix with 11.5 ml digoxigenin-11-dUTP. Biotin: make a 0.4 mM solution in Tris–HCl pH 7.5. Direct labelled nucleotides: make 10 ml of 0.5 mM solution of TTP and mix with 25 ml of 1 mM solution of fluorochrome-conjugated dUTP.

- Unlabelled nucleotides: dATP, dCTP, and dGTP (lithium salt, 100 mM solution in Tris–HCl pH 7.5; Roche). Make a 0.5 mM stock solution of each nucleotide in 100 mM Tris–HCl pH 7.5 and mix 1:1:1. Store at –20 °C for six months maximum.
- Klenow enzyme (DNA polymerase I, large fragment, labelling grade), 6 U/μl (Roche)
- Hexanucleotide reaction mixture (in 10 × buffer; Roche) or 'random' hexanucleotides or primers[b]
- 3 M sodium acetate or 4 M lithium chloride
- 0.3 M EDTA pH 8.0
- Ice-cold 100% ethanol
- Ice-cold 70% (v/v) ethanol
- TE buffer: 0.01 M Tris–HCl pH 8.0 and 0.001 M EDTA
- 50% (w/v) dextran sulfate in filter (0.22 μm) sterilized water; store as aliquots at –20 °C
- Salmon sperm DNA: autoclaved to fragments of 100–300 bp (typically 5 min at 100 kgm^{-2} in a household pressure cooker); store at –20 °C
- 5% (w/v) BSA block in 4 × SSC–Tween
- Low stringency washes: 500 ml of 2 × SSC, 200 ml of 0.1 × SSC at 42 °C
- 200 ml stringent formamide wash: 20% (v/v) in 0.1% SSC at 42 °C
- 1 litre of 4 × SSC–Tween
- Freshly made FITC–anti-digoxigenin and Cy3–streptavidin, or appropriate detection colours and molecules, in BSA block

5: Classical and molecular cytogenetics of Arabidopsis

A. *Method for random primer labelling*[c]
1. Prepare DNA for labelling. Cloned DNA should be either linearized whole plasmid or cleaned insert.
2. Take 50–200 ng of DNA for labelling and mix with sterile distilled water to make a volume of 12.5 µl in an 1.5 ml microcentrifuge tube. (Firmly tape or clip the lid to prevent opening during step 3.)
3. Denature the DNA by placing the tube in boiling water for 10 min, put on ice for 5 min, and microcentrifuge briefly.
4. Add in the following order:-
   - 3 µl of unlabelled nucleotide mixture
   - 1.5 µl of labelled nucleotide mixture
   - 2 µl of hexanucleotide reaction buffer

   Briefly vortex the mixture.
5. Add 1 µl of Klenow enzyme and mix gently.
6. Incubate at 37°C for 6–8 h or overnight.
7. Add 2 µl of 0.3 M EDTA to stop the reaction, vortex, microcentrifuge, and add 0.1 vol. of sodium acetate or lithium chloride.
8. Add 2/3–3/4 vol. of cold 100% ethanol and mix gently.
9. Incubate at –20°C overnight or at –80°C for 1–2 h.
10. Microcentrifuge the tubes containing precipitated DNA at –10°C for 30 min at 12 000 g.
11. Discard the supernatant carefully. Add 0.5 ml of cold 70% (v/v) ethanol and spin at –10°C for 5 min at 12 000 g.
12. Discard the supernatant and leave the pellet to air dry.
13. Resuspend the DNA in 10–30 µl of TE. The probe can be stored indefinitely at –20°C.

B. *Preparation of hybridization mixture*
1. Add the following to a 1.5 ml microcentrifuge tube: 20 µl of formamide, 8 µl of 50% (w/v) dextran sulfate, 4 µl of 20 × SSC, 0.5–1 µl of 10% (w/v) SDS, 1–5 µl of probe DNA (final concentration 0.5–2 µg/µl), 1–5 µl of salmon sperm DNA (final concentration 5–50 µg/µl, 10–200 × of probe concentration). Adjust to a final volume of 40 µl with distilled water, briefly vortex, then centrifuge the mixture.
2. Denature the hybridization mixture at 70°C for 10 min and then transfer to ice for 5 min.

C. *Denaturation of preparations and hybridization*
1. Place 40 µl of the hybridization mixture from part B onto each chromosome preparation.

Protocol 4. *Continued*

2. Cover with a plastic coverslip and place into the humid chamber of a programmable temperature cycler.
3. Programme to denature for 5 min at 85°C, then cool slowly, stopping for 1 min at 75, 65, 55, and 45°C. Leave overnight at 37°C for hybridization.[d]

D. *Washes and detection of probe hybridization sites*

1. Remove the slides from the humid chamber, carefully remove the coverslips, and place slides in a Coplin jar containing 2 × SSC at 42°C.
2. Pour off the 2 × SSC and replace with stringent formamide wash. Incubate for 5 min at 42°C, shaking gently. Repeat this wash.[e]
3. Wash the slides in 0.1 × SSC for 2 × 5 min at 42°C, and then in 2 × SSC for 2 × 5 min at 42°C.
4. Transfer to room temperature and incubate the slides in 4 × SSC–Tween for 5 min. If the hybridization used probes directly labelled with fluorochromes, no detection is required, so proceed to part E.
5. Add 200 μl of BSA block to each slide, cover with a plastic coverslip, and incubate for 5 min at room temperature.
6. Remove the coverslip, drain the slide, and add 30 μl of the anti-digoxigenin and/or avidin solution. Cover and incubate the slides in a humid chamber for 1 h at 37°C.
7. Wash the slides in 4 × SSC–Tween for 3 × 8 min at 37°C.

E. *Counterstaining and mounting of slides and examination of preparations*

1. Repeat *Protocol 2*A.
2. Observe the slides under an epifluorescent microscope using suitable filter sets.[f]
3. Photograph suitable preparations (see Section 3.1). Although mounted in antifade solution, the solution is not completely effective, and, in particular, observation of DAPI fluorescence can fade *in situ* hybridization signal in 10–20 sec. Hence photograph as soon as suitable preparations are found, without extensive observation, and photograph the longest wavelengths first (e.g. red, then green, and finally blue).

[a] A good 'molecular biology' grade (but not the highest grade, which is very expensive). Immediately after purchase divide into 40 ml aliquots and store at −20°C.
[b] To prepare 50 ml of hexanucleotide reaction buffer, if ready made buffer is not available from manufacturer, mix 0.5 M Tris–HCl pH 7.2, 0.1 M $MgCl_2$, 1 mM dithierythritol, 2 mg/ml BSA (bovine serum albumin), 62.5 A_{260} units/ml 'random' hexanucleotide. Store at −20°C.

5: *Classical and molecular cytogenetics of* Arabidopsis

> ^c Any suitable alternative labelling protocol can be used, including nick translation and PCR incorporation of labelled nucleotides. Kits are often convenient and economical for non-molecular laboratories.
> ^d For single copy probes, higher levels of signal may be seen by hybridization for up to 72 h.
> ^e If probes of low homology are used, decrease the stringency by increasing to 1 × SSC or 2 × SSC.
> ^f Preparations stabilize for the first 12 h after mounting, so avoid extensive observation during this time. Single wavelength filter sets are most valuable, although multiple band pass filters are useful for some applications, enabling simultaneous visualization of red, green, and blue fluorescence.

3.1 Photography of *in situ* hybridizations

We have a considerable preference for colour print film, and use Fujicolor 400. These films have much wider exposure latitude, are very high contrast (called 'amateur' in photographic terminology), and are cheaper to print than slide films. Exposure times vary from 2 sec (some DAPI exposures) to 4 min (single copy probes). CCD or other video cameras or confocal microscopes can also be used for imaging, but rarely are as good in resolution or convenience compared to 35 mm film, and are often extremely expensive.

Image processing is often useful to enhance images. Scanning of negatives using the Kodak PhotoCD system is very effective and gives high resolution scans for $1–2 per image without capital investment. The resulting scans are much higher quality than those from scanning prints (which are already processed), using cheap negative scanners or from cheap video cameras and capture boards.

Adobe Photoshop (for PC and Apple computers) is a standard package (with high academic discounts) for image processing, enabling contrast adjustment, overlaying of different images, and printing. Several free-ware, shareware, or lower cost programs with similar capabilities are available, and often distributed free with computer magazines. NIH image is a major research package, available free via the Internet (Home page: URL::http://rs.binfo.nih.gov/nih-image/). Extensive user-written add-ins are freely available, and a useful newsgroup gives continuous information about the program, additions, and uses. A single full frame of 400ASA film exposed with a × 100 oil immersion lens contains about 4.5 MB of colour information (1000*1500 pixels). An *Arabidopsis* metaphase will fill about 1/8th of the frame representing about 600 KB. No information is gained by processing at higher resolutions but information is lost at lower resolutions. A full page A4 colour plate, printed to a standard used by many scientific journals, requires about 4.5 MB.

4. Meiotic chromosome analysis by light microscopy

Traditionally, at least over the last sixty years, meiotic analysis has relied on light microscopical (LM) studies of fixed, squashed, and stained preparations of pollen mother cells, and this is as true of *Arabidopsis* as of other higher eukaryotes. The main technical difficulty in *Arabidopsis* LM meiotic studies (as in mitotic studies) derives from the extremely small size of the chromosomes, their low DNA content, and the consequent difficulty of obtaining a sufficiently intense staining reaction for analysis by LM. A lesser problem is the rather low number of pollen mother cells in each anther. One bonus of the reproductive system shared by *Arabidopsis* and other Angiosperms is the high degree of synchronization of meiosis within anthers and to a lesser degree between anthers within the same bud.

Protocols 5 and *6* present two approaches for LM analysis of pollen mother cell meiosis in *Arabidopsis*. Orcein or carmine-based stains which have been used in earlier studies of *Arabidopsis* meiotic cytology (1, 8) are adequate for providing an overview of the later (condensed) stages of meiosis, from diakinesis onwards. The more extended and diffuse chromosomes of early prophase I (leptotene, zygotene, pachytene) require a more intense stain. Klasterska and Ramel (27) applied a Giemsa C-banding protocol to this stage, but this carries the risk of chromosome distortion due to the strongly destructive action of barium hydroxide. We have applied a haematoxylin–iron alum protocol, adapted from fungal cytology (28) where similar problems of small chromosome size exist, and also staining with the DNA fluorochrome DAPI (29). These protocols are not presented as the ultimate answers to LM meiotic analysis in *Arabidopsis* and there is considerable scope to experiment with other staining techniques such as fluorochromes other than DAPI, or carmine-Giemsa (30).

Protocol 5. Haematoxylin–iron alum squash technique (28)

Equipment and reagents

- Dissecting microscope
- Compound microscope
- Microscope slides with frosted ends
- No. 1, 18 × 18 mm coverslips
- Fine mounted needles
- Watchmakers' forceps (No. 5)
- Flat-ended brass rod, 2 mm diameter
- Blotting paper
- Spirit burner
- 50 mm Petri dish
- Freshly made Carnoy's fixative: 9 parts absolute ethanol, 3 parts chloroform, 1 part glacial acetic acid
- Small piece of mm^2 graph paper
- 70% (v/v) ethanol
- Haematoxylin–iron alum: 2% (w/v) haematoxylin in 50% (v/v) propionic acid (solution A); 0.5% (w/v) iron alum in 50% (v/v) propionic acid (solution B). Mix equal quantities of solutions A and B and leave overnight before use. Can be used for one week.
- 4 M HCl
- Saturated solution of ferric chloride in 50% (v/v) acetic acid
- Distilled water

A. Growth of plants

1. To maximize inflorescence number and flower bud size, grow plants in a short day and low temperature regime (8 h day, 8–10 °C) to promote vegetative growth.

5: Classical and molecular cytogenetics of Arabidopsis

2. After three to five months transfer the plants to a long day and higher temperature regime to promote flowering.[a]

B. *Fixing flower buds*
1. Detach immature flower buds in the appropriate size range[b] and immerse in Carnoy's fixative in glass vials.
2. Label the vials by inserting pencil-written slips of card in the vials.
3. Allow the buds to fix at room temperature for several hours or overnight, with occasional agitation to encourage penetration of the fixative.
4. Transfer to a freezer (–20 °C) for storage.

C. *Making chromosome preparations*
1. Transfer the fixed buds to 70% ethanol.
2. Place the fixed buds in a small Petri dish (50 mm diameter) and arrange the buds in an ascending size order using a dissecting microscope. Slip a small piece of mm^2 graph paper under the Petri dish as a guide to bud size.
3. Transfer individual buds to 4 M HCl for 3–4 min.[c]
4. Rinse the buds in distilled water for 3–4 min, then transfer to a clean glass slide.
5. Add a drop of haematoxylin–iron alum mixture and tap the buds with a brass rod to release pollen mother cells into the stain. Leave for 3–4 min to absorb the stain.
6. Place a coverslip over the stain plus cell mixture and check staining and the meiotic stage under a compound microscope.
7. Warm the slide over a spirit burner for a few seconds (do not allow to boil) and check staining under the microscope.
8. When staining intensity is satisfactory, squash the preparation using firm vertical thumb pressure applied to the area of the coverslip under a double layer of blotting paper.
9. Continue warming the slide over the spirit burner to improve differentiation of the chromosomes and the cytoplasm.
10. Seal the coverslip edges with nail varnish or other sealant and store temporary preparations in a freezer (–20 °C) when not being used.

[a] Usable flower buds can also be obtained from plants grown more conventionally in long day and high temperature conditions.
[b] Meiosis is generally found in flower buds in the size range 300–500 μm, but bud size is not always a reliable guide to meiotic stage and should be determined empirically in each case.
[c] Acid hydrolysis applied before staining has the effect of reducing cytoplasmic staining and improving differentiation of chromosomes and cytoplasm. However, the duration of hydrolysis is critical and if prolonged causes under-staining of chromosomes and nucleolus.

Protocol 6. Air dried spread and DAPI stained meiotic preparations

Equipment and reagents

- Dissecting microscope
- Compound phase-contrast microscope
- Hot plate or temperature regulated heating block
- Oven or incubator
- Humid chamber
- Microscope slides with frosted ends
- Black solid watch glass
- Pasteur pipette
- Fine mounted needles
- Watchmakers' forceps (No. 5)
- 50 mm Petri dish
- Small piece of mm² graph paper
- Hair drier
- Compound microscope with epifluorescence
- No. 1, 22 × 22 mm coverslips
- Blotting paper
- Citrate buffer: 10 mM Na citrate pH 4.5
- Enzyme mixture containing 0.3% (w/v) pectolyase (Sigma), 0.3% (w/v) cellulase (Sigma), and 0.3% (w/v) cytohelicase (Sigma)
- Ice-cold 100% ethanol:glacial acetic acid (3:1)
- 60% (v/v) acetic acid
- DAPI (4',6-diamidino-2-phenylindole) staining solution: 1 μg/ml in Vectashield antifade mounting medium

A. *Growth of plants*

1. As for *Protocol 5*.

B. *Fixing flowerbuds*

1. As for *Protocol 5*. Ethanol:acetic acid (3:1) fixative may be substituted for Carnoy's fixative.

C. *Making chromosome preparations*

1. Transfer fixed buds through two changes of citrate buffer (5 min each).[a]

2. Incubate the buds in enzyme mixture at 37°C in a humid chamber for 1–2 h.

3. When sufficiently digested, buds may be transferred to citrate buffer and maintained at 4°C.[b]

4. Place the buds in a 50 mm Petri dish and arrange the buds in ascending size order using a dissecting microscope. Slip a small piece of mm² graph paper under the Petri dish as a guide to bud size.

5. Transfer a single enzyme digested bud to a clean slide and remove excess buffer.

6. Add a small drop (about 3 μl) of 60% (v/v) acetic acid and macerate the bud with a pair of fine needles in a minimum amount of acetic acid, taking care to avoid the material drying out.

7. Add a further 20 μl of 60% (v/v) acetic acid to the slide and stir the droplet gently with a needle on a hot plate (45°C) for 1 min.

5: Classical and molecular cytogenetics of Arabidopsis

8. Flood the slide with ice-cold 3:1 ethanol:acetic acid fixative, initially by adding drops of fixative all around the acetic acid droplet.
9. When the fixative and acetic acid have thoroughly mixed, tilt the slide and then add a few drops more fixative.
10. Drain the slide, shake to remove drops, and air dry using a hair drier.
11. Check slides for meiotic stages by phase-contrast microscopy.

D. *DAPI staining*

1. Place a drop (5 μl) of DAPI/Vectashield on the chromosome preparation.
2. Place a 22 × 22 mm coverslip over the preparation and wait 1–10 min for staining.[c]
3. Press slide between sheets of filter paper to remove excess mounting medium, and observe under epifluorescence microscope using an appropriate filter set for UV illumination.
4. Photograph suitable preparations, using either colour transparency film (Fujichrome 400) or colour print film (Fujicolor 400) (see Section 3.1). Image processing is often useful to enhance images (see Section 3.1).
5. Store slides horizontally at 4 °C.

[a] Steps 1–3 are carried out in a black solid watch glass under a dissecting microscope, adding and removing solutions with a Pasteur pipette.
[b] The correct degree of digestion is best determined empirically by the success or otherwise of subsequent spreading; under-digested buds will give under-spread chromosome preparations, and vice versa.
[c] The more elaborate DAPI staining procedure described in *Protocol 2* may also be applied.

5. Meiotic chromosome analysis by electron microscopy

Many critically important events of meiosis including chromosome synapsis and several steps in the recombination process occur during early prophase I. Successive observers have commented that prophase I stages in *Arabidopsis* are weakly stained, diffuse, or 'ghost-like' (8). Following DAPI staining it is possible, with care and selection of well spread nuclei, to trace and measure pachytene bivalents of *Arabidopsis* (7). However, this procedure is not easy; the yield of analysable nuclei is low and many nuclei are too clumped or tangled to allow unambiguous analysis.

At the ultrastructural level, prophase I chromosomes develop dense axial cores which come together during synapsis to form a characteristic tripartite structure, the synaptonemal complex (SC). Surface spreading techniques have been developed for making whole-mount two-dimensional preparations of entire prophase I nuclei in which the complete complement of SCs (and/or axes) can be visualized and analysed (31). SC karyotyping has been widely exploited to investigate chromosome number and morphology in lower eukaryotes such as fungi and protozoa in which conventional light microscopical cytology is difficult (31), and Albini (6) adapted the surface spreading technique to produce an SC karyotype for *Arabidopsis*. This technique has considerable potential for high resolution physical mapping when combined with *in situ* hybridization (7, 17).

Protocol 7. Synaptonemal complex surface spreading (6)

Equipment and reagents

- Dissecting microscope
- Compound microscope
- Transmission electron microscope
- Cavity glass slides
- Flat glass slides
- Fine mounted needles
- Watchmakers' forceps (No. 5)
- Flat-ended brass rod 2 mm diameter
- Slide warming plate
- Moisture chamber, consisting of a plastic box with a tight-fitting lid, lined with wet paper towels, and containing supports for slides
- 50°C oven or incubator
- Sharp scalpel blade
- Indelible felt-tip pen
- Nylon gauze (e.g. Nybolt 3XXX-300, Swiss Silk Bolting Cloth Mfg Co Ltd.) cut into 22 × 50 mm pieces
- EM grids
- Tissues
- Plastic backed filter paper (Benchcote)
- Vacuum coating unit
- Digestion medium: 0.05 g snail gut enzyme (Cytohelicase, LKB), 0.125 g polyvinyl pyrrolidone, 0.19 g sucrose in 12.5 ml sterile distilled water
- Spreading medium: 2% (v/v) Lipsol (LIP Equipment and Services Ltd.) detergent in distilled water
- 0.75% (w/v) Optilux plastic (Falcon plastic Petri dishes) dissolved in chloroform
- Freshly made fixative: 4% (w/v) paraformaldehyde and 1.5% sucrose in distilled water, adjusted to pH 8.2 with borate buffer
- Distilled water
- 50% (w/v) silver nitrate solution in distilled water

A. *Growth of plants*

1. As for *Protocol 1*.

B. *Coating slides with plastic*

1. Clean slides by polishing with tissue.
2. Dip the slides one at a time, slowly and steadily, into the Optilux solution and allow to air dry.
3. Make coated slides hydrophilic by glow discharging them in a vacuum coating unit at 0.1 torr for 1–2 min.

C. *Surface spreading*

1. Screen buds of approximately the correct size (300 μm) by light microscopy of acetocarmine or lactopropionic orcein squashes to

confirm that they contain pollen mother cells at prophase I of meiosis.

2. Dissect the remaining anthers from about ten buds and place them in the depression of a cavity slide containing 40 µl of digestion medium.

3. Tap the anthers with the brass rod to release pollen mother cells into the digestion medium and transfer the contents of the cavity slide onto a plastic coated slide.

4. After 4 min digestion, add 20 µl of spreading medium to the cell suspension.

5. After 2 min exposure to the spreading solution, add 60 µl of fixative to the slide and allow the whole mixture to dry down onto the slide on a hot plate set at 25–30 °C.

6. Let the slides stand at room temperature overnight and then wash thoroughly in distilled water and air dry.

7. Stain the slide by placing two drops of silver nitrate solution on the slide and covering with a small piece of nylon gauze (22 × 50 mm) pre-soaked in silver nitrate solution. Place the slides in a moisture chamber and incubate for 2 h at 60 °C.

8. Wash the slides in tap-water to remove the nylon gauze and excess stain, and allow to air dry.

9. Scan the slides by light microscopy and identify the locations of well spread and stained nuclei with an indelible felt-tip pen.

10. To transfer the spread nuclei to grids, score the plastic around the area of the slide containing one or more spread nuclei and float onto a clean water surface. Place EM grids on the marked sites and pick up both plastic film and grids on a piece of plastic backed filter paper.

11. Scan the grids by electron microscopy and photograph suitable spread nuclei.

References

1. Steinitz-Sears, L. M. (1963). *Genetics*, **48**, 483.
2. Laibach, F. (1907). *Beih. Botan. Zentralbl.*, **22**, 191.
3. Winge, O. (1925). *La Cellule*, **35**, 305.
4. Jaretzky, R. (1928). *Jahrb. Wiss. Botan.*, **68**, 1.
5. Ambros, P. and Schweizer, D. (1976). *Arabidopsis Inf. Serv.*, **13**, 167.
6. Albini, S. M. (1994). *Plant J.*, **5**, 665.
7. Fransz, P., Armstrong, S., Alonso-Blanco, C., Fischer, T. C., Torres-Ruiz, R. A., and Jones, G. (1998). *Plant J.*, **13**, 867.
8. Steinitz-Sears, L. M. and Lee-Chen, S. (1970). *Can. J. Genet. Cytol.*, **12**, 217.
9. Koornneef, M., van Eden, J., Hanhort, C. J., Stam, P., Branksma, F. J., and Feenstra, W. J. (1983). *J. Hered.*, **74**, 265.

10. Schweizer, D., Ambros, P., Grundler, P., and Varja, F. (1987). *Arabidopsis Inf. Serv.*, **25**, 27.
11. Maluszynska, J. and Heslop-Harrison, J. S. (1991). *Plant J.*, **1**, 159.
12. Brandes, A., Thompson, H., Dean, C., and Heslop-Harrison, J. S. (1997). *Chromosome Res.*, **5**, 238.
13. Murata, M., Heslop-Harrison, J. S., and Motoyoshi, F. (1997). *Plant J.*, **12**, 31.
14. Maluszynska, J. and Heslop-Harrison, J. S. (1993). *Ann. Bot.*, **71**, 479.
15. Kamm, A., Galasso, I., Schmidt, T., and Heslop-Harrison, J. S. (1995). *Plant Mol. Biol.*, **27**, 853.
16. Trask, B. J. (1991). *Trends Genet.*, **7**, 149.
17. Albini, S. M. and Schwarzacher, T. (1992). *Genome*, **35**, 551.
18. Fransz, P., Alonso-Blanco, C., Liharska, T. B., Peeters, A. J. M., Zabel, P., and De Jong, J. H. (1996). *Plant J.*, **9**, 421.
19. Bouharmont, J. (1969). *Chromosomes Today*, **2**, 197.
20. Maluszynska, J., Maluszynski, M., Rebes, G., and Wietrzyk, E. (1990). *4th Int. Conf. Arabidopsis Res*. Vienna, University of Vienna.
21. Albini, S. M., Jones, G. H., and Parker, J. S. (1992). *Agric. Food Res. Counc./Plant Mol. Biol. Arabidopsis Newslett.*, **9**, 5.
22. Ross, K. J., Fransz, P., Armstrong, S. J., Vizir, I., Mulligan, B., Franklin, F. C. H., et al. (1997). *Chromosome Res.*, **5**, 551.
23. Peirson, B., Bowling, S. E., and Makaroff, C. A. (1997). *Plant J.*, **11**, 659.
24. Martinez-Zapater, J. M., Estelle, M. A., and Somerville, R. C. (1986). *Mol. Gen. Genet.*, **204**, 417.
25. Murata, M. and Motoyoshi, F. (1995). *Chromosoma*, **104**, 39.
26. Schwarzacher, T., Leitch, A. R., and Heslop-Harrison, J. S. (1994). In *Plant cell biology: a practical approach* (ed. N. Harris and K. J. Oparka), pp. 127–55. Oxford University Press, Oxford.
27. Klasterksa, I. and Ramel, C. (1980). *Arabidopsis Inf. Serv.*, **17**, 1.
28. Henderson, S. A. and Lu, B. C. (1968). *Stain Technol.*, **43**, 233.
29. Ross, K. J., Fransz, P., and Jones, G. H. (1996). *Chromosome Res.*, **4**, 507.
30. De Jong, J. H. (1978). *Stain Technol.*, **53**, 169.
31. von Wettstein, D., Rasmussen, S. W., and Holm, P. B. (1984). *Annu. Rev. Genet.*, **18**, 331.

6

Tissue culture, transformation, and transient gene expression in *Arabidopsis*

KEITH LINDSEY and WENBIN WEI

1. Introduction

Tissue culture and transformation technologies have revolutionized studies on plant biology. They allow investigations, in a way previously only dreamed of, of the relationships between, plant morphology on the one hand and gene expression, metabolism, and cell differentiation on the other. Disorganized growth, such as occurs during callus formation, can be experimentally induced and the developmental, biochemical, and genetic consequences studied (1). Furthermore, and of fundamental significance, disorganized tissues can be induced to reorganize and to regenerate intact organs and whole plants. This observation has been crucial in the development of the concept of the totipotency of plant cells, and has been equally fundamental to the establishment of reproducible techniques for the genetic engineering of plants. Thus, it is possible to introduce foreign genes into cultured plant cells, protoplasts, or tissues, and regenerate from them stably transformed individuals that will transmit the introduced transgenes to their progeny. It is beyond the scope of this chapter to discuss the commercial or agronomic implications of plant genetic engineering, but applications that are of a more basic scientific nature, and to which *Arabidopsis thaliana* lends itself ideally as an experimental organism, can briefly be considered here.

There are broadly five types of experiment that require the use of transgenic techniques in addressing questions of the molecular biology of plants.

(a) The spatial expression pattern of a given, previously isolated, gene. It is possible to carry out this type of study by the *in situ* localization of the transcript in tissue sections (e.g. 2), but many workers prefer to construct promoter–reporter gene fusions (such as the *uid*A [*gus*A] reporter gene) (3), and use a combination of qualitative (histochemical) and quantitative (fluorimetric) methods to determine the relative activity of the isolated gene promoter in different organs and tissues.

(b) To investigate gene function by modifying experimentally the level of expression of a given gene. This includes first, overexpression studies, in which a relatively strong promoter, such as the CaMV 35S RNA gene promoter, is used to replace the native gene promoter; and secondly, down-regulation studies, in which the expression of the gene is experimentally reduced, either by expressing an antisense version of the gene in question or by sense co-suppression (4). A variation of this experimental strategy is to engineer the ectopic expression of the gene, driving its transcription in the wrong cell type using an alternative promoter.

(c) To utilize stable transformation for insertional mutagenesis. The aim is to disrupt gene function by the random insertion of foreign DNA, be it T-DNA or transposable elements, which therefore tag the mutation and so facilitate the subsequent isolation of the mutant gene. T-DNA insertional mutagenesis requires an efficient transformation procedure, so that relatively large populations of transgenics can be produced for screening purposes (5). While native transposons do exist in *Arabidopsis* (e.g. 6), a large amount of effort has gone into introducing, and improving the efficiency of transposition of, heterologous (most commonly maize) transposons (e.g. 7–10); the introduction of the transposon systems is by means of T-DNA-mediated transformation. Insertional mutagenesis in *Arabidopsis* is covered in detail in Chapter 7.

(d) The use of transgenic techniques for genetic cell ablation. The rationale here is to transform plants with a gene encoding a cell autonomous cytotoxin, such as the RNase Barnase (11) or the translation inhibitor diphtheria toxin (12, 13), under the transcriptional control of a cell type-specific promoter. This allows one to address questions of the role of a particular cell type during development, by specifically inactivating it and analysing the developmental consequences.

(e) The use of genetic complementation and/or shotgun cloning. This can be used to identify genes, by transforming mutants with an expression library, in order to isolate functionally complementing clones (14); or to confirm the identity of a gene, cloned following a programme of mutagenesis, by transforming and complementing the respective mutant with the wild-type allele (e.g. 15, 16); or by rescuing dominant selectable markers on large, tagged genomic fragments (17).

One feature of the genetic transformation of plants that becomes a problem if the aim of an experiment is to analyse elements, within a gene promoter, that may modulate either the level or tissue specificity of expression of that gene, is the phenomenon of position effects. The problem relates to the fact that both the level and spatial pattern of expression of a transgene are influenced, to a greater or lesser extent, by the site of insertion within the host genome. For example, the fortuitous insertion of a root gene promoter–*gus* fusion in the vicinity of a native gene promoter containing a strong pollen-

6: Tissue culture and transformation

specific enhancer might up-regulate the level of expression of the foreign promoter if it is relatively weak, and may cause the transgene to be expressed ectopically in pollen (18). The extent of methylation of the surrounding DNA may also influence the methylation state, and hence activity, of the introduced gene (19, 20). It is therefore difficult to assess the relative activities of promoter deletions in stable transformation experiments, since each insertion event will be different and a given transgene construct will, in ten independent transformations, be subject to the influence of ten different native enhancer (or silencer) elements. Indeed, this phenomenon is exploited in transgenic techniques designed to identify tissue-specific genes and enhancer elements (enhancer and promoter trapping) (18, 21–23).

An alternative strategy to carry out a functional analysis of gene promoters is therefore to avoid stable integration altogether, and to carry out transient expression studies. The aim of this approach is to introduce a promoter–reporter gene fusion into protoplasts or cells, and to assay the activity of the reporter enzyme within a relatively short period of time (typically one to two days post-transfection). Expression of the gene fusion is not dependent on integration into the host genome, and the vast majority of the introduced transgene copies are expected to be expressed extrachromosomally, as free plasmid. Therefore, it is assumed that the level of expression of a particular gene fusion will be independent of any influence of native gene sequences, allowing a direct comparison of the relative activities of a range of gene constructs. Other factors do, however, influence the level of transient expression of a gene, including the metabolic state of the transfected cells, their tissue origin, the quantity and conformation of the introduced DNA, and the delivery conditions (be they chemical, such as by polyethylene glycol; electrical (electroporation); or physical, such as by microprojectile bombardment or injection). These factors are discussed in detail elsewhere (e.g. 24, 25).

Arabidopsis thaliana has a number of features that have led to its adoption as a model species for classical and molecular genetic studies (26), not least of which is the fact that it is amenable to both stable transformation and transient expression studies. There are currently four general approaches to the stable transformation of *Arabidopsis*, each involving inoculation with *Agrobacterium tumefaciens*:

(a) The inoculation of explanted tissues, including cotyledons and leaf tissue (27) and roots (28).
(b) The inoculation of intact seeds (29).
(c) The injection of agrobacteria into intact plant shoot meristems (30).
(d) The vacuum infiltration of whole plants with a liquid suspension of agrobacteria (31).

Arguably the most demonstrably reproducible of these approaches, and the method we routinely use, is root tissue inoculation. Vacuum infiltration is now also becoming quite widely used. The seed inoculation system devised by Feldmann and Marks (29) is clearly very successful in the hands of the inventors of the technique, and has been used to great advantage in the generation of insertional mutants, but has proved to be a poor traveller. To our knowledge, no other laboratory has managed to get the method working efficiently, for reasons that are difficult to define.

The root explant inoculation system, which can be very efficient, has therefore been used widely. While regeneration of shoots from leaf, cotyledon, and root explants is equally efficient, a potential advantage of using roots is that cells of this organ in *Arabidopsis* are predominantly diploid, while aerial tissues are more commonly polyploid (32). Certainly, we have found those regenerants from root explants we have analysed invariably to be diploid (33), an important consideration if the aim of the work is to generate insertional mutants.

2. Stable transformation by *Agrobacterium tumefaciens*

The method for stable transformation of *Arabidopsis* described in *Protocol 1* is a modification of the method described originally by Valvekens *et al.* (28), and in our hands and in the hands of colleagues in other institutions, has proved to be of improved efficiency compared to the original. The principal variations are:

- the inclusion of silver ions in some of the culture media
- the immobilization of the root explants in solidified medium (33)

With these slight alterations, we have found that the frequency of recovery of transformants is increased between 10- and 100-fold, with up to 250 transformed shoots being produced per plate.

Protocol 1. Transformation of *Arabidopsis thaliana* by inoculation of root explants with *Agrobacterium tumefaciens*[a]

Equipment and reagents

- Micropore® gas-permeable tape (3M)
- Nylon or stainless steel mesh (100 μm pore diameter)
- Sterile filter paper
- Gas-permeable transparent SunCap® film (C6920, Sigma)
- Sterile perlite (autoclave at 121°C, 15 min)
- Aracon® tubes (Beta Developments, Gent)
- Sterile 9 cm Petri dishes (e.g. Falcon 3003)
- Glass culture tubes (e.g. C5916, Sigma)

- Polypot containers (e.g. SC020, Northern Media Supply Ltd., UK)
- Transfer/Pasteur pipettes
- Bleach solution: 5% (v/v) commercial bleach plus 0.05% (v/v) Tween 20
- 70% (v/v) ethanol
- 1.25 mg/ml silver thiosulfate stock solution: add dropwise a solution of 2.5 mg/ml silver nitrate to an equal volume of a solution of 14.6 mg/ml sodium thiosulfate pentahydrate (filter sterilize)

6: Tissue culture and transformation

- Germination medium I (GMI): half-strength Murashige and Skoog medium (34) (Sigma), 10 g/litre sucrose (pH 5.8 with 1 M KOH), 8 g/litre agar (autoclave at 121°C, 15 min); then add 5 mg/litre silver thiosulfate solution (filter sterilized) before plating out
- Stock solution of 4 mg/ml 2,4-dichlorophenoxyacetic acid (2,4-D) in dimethyl sulfoxide (DMSO)
- Stock solution of 0.4 mg/ml kinetin in DMSO
- Callus inducing medium (CIM): Gamborg's B5 medium (35) (Sigma), 0.5 g/litre MES, 20 g/litre glucose pH 5.8, 8 g/litre agar (autoclave at 121°C, 15 min); then add 5 mg/litre silver thiosulfate solution (filter sterilized), 0.5 mg/litre 2,4-D, 0.05 mg/litre kinetin
- Stock of rifampicin 20 mg/ml in methanol
- Stock solution of 200 mg/ml vancomycin (filter sterilized)
- Stock solution of 50 mg/ml kanamycin sulfate (filter sterilized)
- Stock solution of 20 mg/ml 2-isopentenyladenine (2-iP) in DMSO
- Stock solution of 1.2 mg/ml indole-3-acetic acid (IAA) in DMSO
- Shoot inducing medium (SIM): Gamborg's B5 medium (Sigma), 0.5 g/litre MES, 20 g/litre glucose pH 5.8, 8 g/litre agar (autoclave at 121°C, 15 min); then add 850 mg/litre vancomycin, 35 mg/litre kanamycin sulfate, 5 mg/litre 2-iP, 0.15 mg/litre IAA
- Shoot overlay medium (SOM): as SIM, but replacing the agar with 8 g/litre low melting point agarose (SeaPlaque, FMC) (autoclave at 121°C, 15 min)
- Shoot elongation medium (SEM): half-strength Murashige and Skoog Medium (Sigma), 10 g/litre sucrose pH 5.8, 8 g/litre agar (autoclave at 121°C, 15 min)
- Compost (36): Irish moss peat (Joseph Bently & Sons), John Innes potting compost No. 3, horticultural potting grit (Joseph Bently & Sons), coarse vermiculite (Vermiperl Medium Grade, Silverperl Ltd.), mixed in a ratio of 2:2:2:1 (autoclave at 121°C, 30 min)
- LB (Luria-Bertani) medium: 10 g/litre Bacto tryptone, 5 g/litre Bacto yeast extract, 10 g/litre NaCl pH 7.5 (autoclave at 121°C, 15 min)
- *Agrobacterium* culture medium: sterile LB medium, 50 mg/litre kanamycin sulfate (or other appropriate selective agent), 100 mg/litre rifampicin (or other appropriate selective agent)
- *Agrobacterium* culture dilution medium: Gamborg's B5 salts solution (Sigma), 20 g/litre glucose, 0.5 g/litre MES pH 5.7 (autoclave at 121°C, 15 min)
- Sterile Gamborg's B5 salt solution (autoclave at 121°C, 15 min)
- Sterile double distilled water (autoclave at 121°C, 15 min)

A. Preparation of roots for inoculation

1. Vernalize *Arabidopsis* seeds at 4°C for one week, to maximize germination frequency.

2. Surface sterilize the seeds as follows.

    (a) Immerse in 70% (v/v) ethanol for 30 sec.

    (b) Remove the ethanol with a Pasteur pipette.

    (c) Immerse the seeds in bleach solution for 15–20 min.

3. Remove bleach solution with a transfer pipette and wash the seeds six times in sterile double distilled water.

4. Suspend the seeds in a drop of sterile water and, using a pipette, transfer them to the surface of GMI agar plates. Spread out the seeds evenly (approx. 40 per 9 cm Petri dish) and seal the plates with Micropore® tape.

5. Germinate the seeds by incubation for two days in dim light (approx. 22°C), and then for a further three to four weeks in a standard light regime for cultures (at least 50 μmol/m^2/s) to allow root growth.

Protocol 1. *Continued*

B. *Root explantation and inoculation with Agrobacterium*

1. When the seedlings are approx. three weeks old and have produced abundant (but not yet greening) roots,[b] gently remove them from the agar with forceps and excise the root systems.
2. Place the intact root systems on the surface of agar plates containing CIM, and incubate at approx. 20–22 °C for three days.
3. Transfer the roots to a sterile Petri dish and cut into segments of approx. 0.5 cm in length.
4. Grow up a liquid suspension of *Agrobacterium* cells at 29 °C on a rotary shaker (200 r.p.m.) for 48 h, then dilute with sterile *Agrobacterium* dilution medium, to a final optical density of 0.1 at 600 nm.
5. Suspend the root explants in 20 ml of *Agrobacterium* culture medium and incubate at room temperature for 2 min. Drain off the bacterial suspension by placing the roots on a nylon or stainless steel mesh (100 μm pore diameter). Remove excess liquid by blotting the roots on sterile filter paper.
6. Transfer the roots to CIM agar plates, and culture for 48–72 h.
7. Wash the roots (over a nylon/steel mesh) with sterile *Agrobacterium* culture dilution medium to remove excess bacterial growth,[c] and blot the roots dry on sterile filter paper.

C. *Selection and regeneration of transformed plantlets*

1. Suspend the root explants in molten SOM, at a temperature of about 30 °C. Use the roots of approx. 40 seedlings (approx. 0.1 g tissue, corresponding to approx. 300 individual root segments) per 10 ml of medium.
2. Pour 10 ml of the root/SOM onto the surface of agar plates (9 cm) containing 40 ml of SIM. Ensure that the root explants are evenly dispersed. Seal the plates with gas-permeable Micropore® tape,[d] and incubate at approx. 25 °C in fluorescent light of at least 50 μmol/m²/s. From approx. three weeks, green (putatively transformed) calluses develop on the root explants, and kanamycin-resistant shoots will regenerate over a period of several weeks.
3. When the shoots have expanded leaves, transfer to SEM (20 ml) in either 60 ml polypots sealed with gas-permeable transparent SunCap® film, or glass culture tubes sealed with cotton wool bungs or with SunCap® film.[e] Under these conditions, at least 80% of the transformed shoots are expected to flower and set seed.[f,g]

D. *Bulking up of T2 plants*

1. Fill Petri dishes with a layer of sterile perlite, and add sterile Gamborg's B5 salt solution until the perlite has absorbed the liquid to its full capacity, but no more.

6: Tissue culture and transformation

2. Plate out and vernalize the seeds (approx. 20 per plate) from T1 plants and culture for two days in dim light, and subsequently under full illumination, at approx. 22°C.[h]
3. When the seedlings are approx. 1-2 cm tall, transfer to sterile, moist compost. Grow for one week in high humidity in a micropropagator (approx. 20°C), and gradually reduce the local humidity by increasing ventilation over the following week.
4. When the seedlings are 3-5 cm high, cover the seedlings with Aracon® tubes, and collect seed over the following few months.[i]

[a] According to the method of Clarke et al. (33) with permission.
[b] Roots which are beginning to green are too old to respond to the regeneration conditions.
[c] Following both inoculation with *Agrobacterium* and washing with *Agrobacterium* culture dilution medium, remove as much excess liquid as possible from the roots. This is to prevent an overgrowth of the roots by the bacteria.
[d] Micropore® tape allows gas exchange, preventing the accumulation of moisture inside the Petri dishes.
[e] Both SunCap film and cotton wool bungs allow gas exchange, which is assumed to prevent both ethylene accumulation (which may inhibit shoot expansion) and to reduce humidity in the culture vessels (which prevents anther dehiscence and consequently seed set).
[f] Seed set is also sensitive to the presence of kanamycin in the shoot elongation medium, and so antibiotics should be omitted at this stage.
[g] Do not put more than two or three shoots in one polypot, as this acts to increase humidity and prevent seed set.
[h] Primary transformants (T1 plants) may set only relatively few seeds *in vitro*, typically about 100 seeds. Germinability of T1 seeds (T2 seedlings) is also commonly poor (a typical viability is 70%, but this is variable between transgenic lines). We observe a higher frequency of T2 seedling survival if seeds are germinated on perlite rather than directly in compost or on agar medium.
[i] Aracon® tubes allow individual transgenic lines to be grown up separately in close proximity, without the danger of cross-fertilization or of seed loss.

3. Transient gene expression in *Arabidopsis* protoplasts

A number of protocols have now been published that describe the transfection of *Arabidopsis* protoplasts with DNA (for transient gene expression studies see refs 37–40). The most commonly used method involves the use of polyethylene glycol as the transfection agent, an approach that has been utilized in both the transient and stable transformation of a number of species. While it is possible to introduce DNA directly into intact cultured plant cells and tissues by, for example, electroporation (40–42), the plant cell wall represents a barrier to DNA uptake, and the use of protoplasts avoids this problem.

Protoplast isolation typically involves the removal of the cell wall with fungal enzymes (cellulases, pectinases, hemicellulases), in the presence of an osmoticum (sorbitol, mannitol, or sucrose) which prevents protoplast bursting (a role normally performed by the cell wall), and calcium ions, which help stabilize the plasma membrane. The quality and yield of protoplasts depends on a number of factors (43), including:

- the age and type of the starting material (young, healthy tissues such as leaves or cotyledons are usually best)
- the concentration of wall-digesting enzymes
- duration of incubation in wall-digesting enzymes
- the concentration of osmoticum
- the use of pre-plasmolysis treatments

Protoplasts are extremely delicate structures, and the isolation and transfection procedures must be carried out with appropriate care.

Protocol 2 describes the isolation of protoplasts from the aerial parts of *Arabidopsis* seedlings, and utilizes the cell wall-degrading enzyme cocktail described by Damm and Willmitzer (44). Essentially, the plant tissue is digested with a cellulase and a pectinase in the presence of mannitol and calcium, and the protoplasts are purified by floatation on a dense 'cushion' of sucrose. The transfection process involves the use of polyethylene glycol (PEG). Protoplasts are incubated with a mixture of PEG and DNA (both plasmid, encoding a reporter gene of interest, and 'carrier' DNA, which increases the level of transient expression of the reporter). The PEG causes the DNA to adhere to the protoplasts, and a subsequent washing step induces uptake of the DNA/PEG agglomerates. If the PEG and/or protoplast concentrations are too high, a high frequency of protoplast fusion may occur. The final PEG concentration used in *Protocol 2* is 20%, which causes only a low frequency (< 5%) of fusion events. Transfected protoplasts are then cultured for one to two days, to allow transcription and translation of the unintegrated reporter gene.

Protocol 2. Isolation of *Arabidopsis* leaf protoplasts, and their transfection by polyethylene glycol[a]

Equipment and reagents

- Haemocytometer
- Benchtop centrifuge with swing-out rotors
- Nescofilm
- Sterile 50 μm and 100 μm stainless steel or nylon sieves
- Sterile 5 cm and 9 cm Petri dishes
- Sterile transfer/Pasteur pipettes
- Sterile 15 ml centrifuge tubes, sterile beakers
- 25 °C incubator
- Pre-plasmolysis solution: 0.5 M mannitol (autoclave at 121 °C, 15 min)

- Germination medium II (GMII): half-strength Murashige and Skoog medium (Sigma), 10 g/litre sucrose (pH 5.8 with 1 M KOH), 8 g/litre agar (autoclave at 121 °C, 15 min)
- Enzyme solution: 1% (w/v) Cellulase Onozuka R-10, 0.25% (w/v) Macerozyme R-10, 8 mM $CaCl_2.2H_2O$, 0.5 M mannitol pH 5.5 (filter sterilize)
- W5 wash solution: 154 mM NaCl, 125 mM $CaCl_2.2H_2O$, 5 mM KCl, 5 mM glucose, 0.5 M mannitol pH 5.8 (autoclave at 121 °C, 15 min)

6: Tissue culture and transformation

- 21% (w/v) sucrose (autoclave at 121°C, 15 min)
- Mannitol/Mg solution: 15 mM $MgCl_2$, 0.1% morpholinoethanesulfonic acid (MES), 0.4 M mannitol pH 5.6 (autoclave at 121°C, 15 min)
- Plasmid DNA, 1 mg/ml
- Sheared carrier DNA (herring sperm or salmon sperm), 5 mg/ml
- PEG solution: 40% (w/v) PEG 6000 in a solution of 0.4 M mannitol, 0.1 M $Ca(NO_3)_2.4H_2O$ pH 8.0 (pH takes 2–3 h to stabilize)
- Protoplast culture medium: Murashige and Skoog medium (Sigma), 0.4 M glucose, 0.4 M mannitol, 1 mg/litre 2,4-D (see *Protocol 1*), 0.15 mg/litre kinetin (see *Protocol 1*) pH 5.8 (autoclave at 121°C, 15 min)

A. Preparation of aseptic seedlings

1. Surface sterilize seeds, vernalize, and germinate according to *Protocol 1*, except plate the seeds onto GMII medium.

B. Isolation of protoplasts

1. When seedlings are three to four weeks old, collect 4–6 g fresh weight (FW) of the aerial parts.

2. Incubate 1.0–1.5 g FW tissue in approx. 15–20 ml of pre-plasmolysis solution in a 9 cm Petri dish for 1 h at 25°C. Do this for four Petri dishes' worth of plant material.

3. Remove the pre-plasmolysis solution with a Pasteur pipette and add 25–30 ml of enzyme solution per Petri dish.

4. Seal the dishes with Nescofilm and incubate in the dark, overnight at 25°C, without shaking.

5. Using a sterile Pasteur pipette, gently suck up and down the digested material, to release the protoplasts into the medium.

6. Using a sterile Pasteur pipette to transfer the solution, filter the digest through a sterile 100 μm sieve, followed by a 50 μm sieve placed over a sterile beaker, to remove undigested plant material. Then transfer the filtrate to sterile 15 ml centrifuge tubes.

7. Pellet the protoplasts by centrifugation at 60 *g* for 5 min, and then *gently* resuspend the protoplasts in 10 ml of W5 wash solution, using a Pasteur pipette.

8. Repeat step 7.

9. Pellet the protoplasts again, and, for each tube, resuspend in 2.5 ml of mannitol/Mg solution.

10. Using a Pasteur pipette load 2.5 ml of the protoplast suspension onto the surface of a 5 ml 'cushion' of 21% (w/v) sucrose in a centrifuge tube.[b]

11. Centrifuge at 60 *g* for 10 min.[c]

12. Using a Pasteur pipette, carefully remove the viable protoplasts, seen as a green band on the surface of the sucrose cushion, and transfer them to a fresh centrifuge tube.

Protocol 2. *Continued*

13. Count the protoplasts using a haemocytometer and adjust the density to 3×10^6/ml in mannitol/Mg solution.

C. *PEG-mediated transfection*

1. Transfer 10^6 protoplasts (in 0.3 ml) to a 15 ml centrifuge tube.
2. Add 50 μg plasmid DNA plus 100 μg sheared carrier DNA.
3. Add 0.3 ml of PEG solution, mix *very gently* by rotating the tube, and leave at room temperature for 30 min.
4. Dilute very slowly with 10 ml of W5 solution: add one drop at a time over a period of about 20 min.[d]
5. Pellet the protoplasts by centrifugation at 60 *g* for 5 min.
6. Resuspend the protoplasts in 9 ml of protoplast culture medium.
7. Culture 3×10^6 protoplasts (i.e. 3 ml of protoplast culture) in 5 cm Petri dishes in the dark at 25°C for 24–48 h.
8. Assay reporter gene activity in extracts of 10^6 protoplasts.

[a] Adapted from Wei and Lindsey (37).
[b] Take care not to disturb the interface between the two layers.
[c] During centrifugation, the viable protoplasts will float on the surface of the sucrose cushion, at the interface between the two layers, while cell debris will be pelleted.
[d] It is important to add the W5 solution very slowly, since protoplasts treated by PEG are particularly susceptible to shear forces. It is the addition of W5, rather than the PEG itself, which induces DNA uptake.

3.1 Reporter gene enzyme assays

Transient gene expression assays utilize various reporter genes, the translation products of which satisfy three important criteria:

(a) Assays are quantitative, sensitive, and simple to perform.
(b) Endogenous levels of the protein in plant tissue are low or, ideally, absent.
(c) The protein is stable in plant cells.

Typically the proteins are enzymes, commonly, but not exclusively, of bacterial origin (see *Table 1*).

Two reporter genes are currently used more commonly than any other. These are β-glucuronidase (GUS), encoded by *gus*A (*uid*A) (3), and firefly luciferase (LUC), encoded by *luc* (45).

3.1.1 GUS fluorimetry

The *gus*A gene is from *Escherichia coli*, and the coding region is commercially available in vectors that allow the insertion of diverse transcriptional regu-

6: Tissue culture and transformation

Table 1. Screenable reporter genes and encoded proteins for use in plants (see also ref. 25)

Gene	Gene product
nos	Nopaline synthase
ocs	Octopine synthase
mas	Mannopine synthase
ags	Agropine synthase
acs	Agrocinopine synthase
cat	Chloramphenicol acetyltransferase
uidA	β-Glucuronidase
lacZ	β-Galactosidase
nptII	Neomycin phosphotransferase
luc	Firefly luciferase
luxA, luxB	Bacterial luciferase
bar	Phosphinothricin acetyltransferase
gfp	Green fluorescent protein

latory sequences upstream and downstream of the coding region, as both transcriptional and translational fusions (e.g. from Clontech). In stably transformed plants, or even in plant tissues showing transient expression of the *gus*A gene following microprojectile bombardment (e.g. 46), the enzyme activity can be localized *in situ* by histochemistry, using a number of different substrates (e.g. from Biosynth), but most commonly 5-bromo-4-chloro-3-indolyl glucuronide (X-gluc). The most sensitive quantitative assay for the enzyme is a fluorimetric method (see *Protocol 3*), in which the substrate 4-methylumbelliferyl glucuronide (MUG) is cleaved by the enzyme to produce 4-methylumbelliferone (MU), which is fluorescent at an excitation wavelength of 365 nm, plus glucuronic acid. The product MU can be assayed at very low (fM) concentrations, using a fluorimeter; we use a Perkin-Elmer LS-50 luminescence spectrometer, which has a facility for a 96-well microtitre plate reader.

Protocol 3. GUS enzyme assay[a]

Equipment and reagents

- 37°C water-bath
- Fluorimeter
- Microcentrifuge
- Spectrophotometer and cuvettes
- Sterile Eppendorf tubes
- Plastic tissue grinders
- 100 mM sodium phosphate buffer pH 7.0: make up 100 ml stocks of each of 0.2 M Na$_2$HPO$_4$ and 0.2 M NaH$_2$PO$_4$. To make a 0.2 M stock phosphate buffer, add 30.5 ml of 0.2 M Na$_2$HPO$_4$ to 19.5 ml of 0.2 M NaH$_2$PO$_4$. Dilute 1:1 with water to make 100 mM phosphate buffer.
- Pipettes and pipette tips
- GUS buffer: 100 mM sodium phosphate buffer, 10 mM 2-mercaptoethanol, 0.1% (v/v) Triton X-100, 1 mM EDTA
- Substrate: 5 mM 4-methylumbelliferone glucuronide (MUG) in GUS buffer (store in the dark at –20°C)
- Acid washed sand (Sigma)
- Bradford's reagent (Bio-Rad)
- Sterile distilled water
- W5 wash solution (see *Protocol 2*)
- 0.2 M Na$_2$CO$_3$
- Bovine serum albumin (BSA; Sigma)

Protocol 3. *Continued*

A. *Crude extraction of GUS enzyme*

1. Using a Pasteur pipette, gently transfer the protoplasts from the culture dishes to 15 ml centrifuge tubes, and pellet at 60 *g* for 5 min.[b] Discard the supernatant.
2. Wash the protoplasts[c] by resuspending in 5 ml of W5 wash solution, and pellet by centrifugation at 60 *g* for 5 min. Discard the supernatant.
3. Resuspend the protoplasts in GUS buffer, at a density of 10^6 protoplasts in 200 µl buffer, and transfer to an Eppendorf tube.
4. Grind 10^6 protoplasts in 200 µl of GUS buffer in the Eppendorf tube with a small amount of acid washed sand, storing the samples on ice.
5. Vortex for 10 sec.
6. Pellet the cell debris in a microcentrifuge at 10 000 *g* for 5 min.
7. Transfer a 100 µl aliquot of the supernatant to a fresh Eppendorf tube containing 0.7 ml of GUS buffer. Save the remainder for a protein assay (see part C).
8. Equilibrate at 37 °C for 5 min.

B. *GUS enzyme assay*

1. Before beginning the assay, set up five Eppendorf tubes per sample tested, each containing 0.9 ml of 0.2 M Na_2CO_3.[d]
2. Add 200 µl of 5 mM MUG as substrate (i.e. to a final concentration of 1 mM), to the protoplasts in GUS buffer to start the reaction. Incubate at 37 °C.
3. Sample 100 µl aliquots at 0, 10, 20, 40, and 60 min, adding directly to 0.9 ml of 0.2 M Na_2CO_3.
4. Measure the fluorescence at an excitation wavelength of 365 nm and an emission wavelength of 455 nm.
5. Express the data as either mol MU produced/min/10^6 protoplasts, or mol MU produced/min/µg protein. A standard curve should be constructed by measuring the fluorescence of serial dilutions of the reaction product, 4-methylumbelliferone (MU). For protein determination, see part C.

C. *Protein determination*

1. Remove 5 µl of the crude cell extract (part A, step 7) and add it directly to a 1 ml cuvette containing 200 µl of Bradford's reagent plus 795 µl of sterile distilled water. Mix well.
2. Incubate at room temperature for 10 min.[e]

3. Read the absorbance, against a blank (800 µl water plus 200 µl Bradford's reagent), at 595 nm. BSA can be used to construct a calibration curve.

[a] According to the method of Jefferson et al. (3), with minor modifications.
[b] Protoplasts, or other plant material to be analysed, can be frozen prior to the assay and stored at −80°C without any loss in activity. Do not, however, freeze samples in the GUS buffer as this does result in significant loss of activity.
[c] It is important to wash the protoplasts thoroughly, since cell wall-digesting enzymes may have GUS-like activity which may interfere with the assay.
[d] Na_2CO_3 stops the reaction and also enhances the fluorescence.
[e] The samples can be left for up to an hour but no longer, because after this time the protein precipitates in the GUS buffer.

3.1.2 GUS histochemistry

The histochemical localization of GUS enzyme activity relies on the cleavage of X-gluc, which generates an insoluble blue precipitate in GUS-positive cells in a two-phase reaction: a colourless indoxyl intermediate is oxidized to form the blue product. The intermediate may leak between cells, so potentially giving artefactual cell localization results, but this problem can be alleviated by:

(a) Carrying out the reaction in a neutral pH buffer.
(b) Including in the buffer the oxidant potassium ferricyanide and/or potassium ferrocyanide, which speeds-up the precipitation reaction (47, 48). Potassium ferricyanide alone is usually adequate for this purpose (49).

Protocol 4 describes a method for GUS histochemistry in *Arabidopsis* tissues. Intact *Arabidopsis* seedlings (up to approx. two weeks old), and roots, flowers, and leaves of mature plants can be stained whole in Eppendorf tubes or in the wells of microtitre plates, without the need for vacuum infiltration.

Protocol 4. GUS histochemistry[a]

Equipment and reagents

- 37°C water-bath or incubator
- Eppendorf tubes, 96-well microtitre plates
- Nescofilm
- X-gluc stock: 20 mM solution in *N,N*-dimethylformamide (stored in 100 µl aliquots at −20°C)
- 95% (v/v) ethanol
- 100 mM phosphate buffer pH 7.0 (see *Protocol 3*)
- Substrate solution: 1 mM X-gluc in 100 mM phosphate buffer pH 7.0, 10 mM EDTA, 1 mM potassium ferricyanide, 0.1% (v/v) Triton X-100

Method

1. Place the tissue in an Eppendorf tube or the well of a microtitre dish[b] and cover with 100–200 µl of substrate solution.
2. Incubate at 37°C overnight (or for shorter periods if appropriate).

Protocol 4. *Continued*

3. If necessary soak the tissue in 95% (v/v) ethanol to clear the tissue of chlorophyll; the blue precipitate is stable in ethanol.

[a] According to the method of Jefferson *et al.* (3), with modifications by Stomp (50).
[b] If the incubation is being carried out in a well plate make sure that the sample is well sealed to prevent it drying out.

3.1.3 Luciferase (LUC) assay

The luciferase enzyme (LUC) from firefly is encoded by the *luc* gene (51) which, like GUS, can be used as a reporter gene in both transient expression and stable transformation studies; plasmids containing *luc* are commercially available (e.g. from Promega). The assay is highly sensitive; plants have essentially no background luciferase activity, providing a high signal-to-noise ratio. The quantitative assay procedure is carried out in cell extracts, although it is possible to visualize LUC activity in intact plants (45). *Protocol 5* describes the *in vitro* assay for LUC activity.

The LUC enzyme catalyses the oxidation of the substrate luciferin, a heterocyclic carboxylic acid, in the presence of ATP, resulting in the production of a flash of light. The intensity of the flash of light is directly proportional to the enzyme activity over several orders of magnitude, and the light emissions are measured by a luminometer. Cell extracts are placed in the reaction chamber of the luminometer, the substrates are injected directly, and the light emitted is quantified using a photon counter and a photomultiplier tube, for instance a Clinilumat luminometer (Berthold Instruments UK Ltd.).

Protocol 5. LUC enzyme assay[a]

Equipment and reagents

- Eppendorf tubes, Pasteur pipettes, plastic tissue grinders
- Luminometer, cuvettes
- Microcentrifuge
- Extraction buffer: 0.1 M KHPO$_4$ pH 7.5, 1 mM DTT (DTT is added immediately prior to use)
- Acid washed sand (Sigma)
- Luciferin solution: 0.1 M D-luciferin (Sigma) in 0.1 M KHPO$_4$ pH 7.5 (make up immediately before use)
- ATP buffer: 50 mM Hepes pH 7.5, 20 mM MgCl$_2$, 10 mM ATP (make up immediately before use)
- W5 wash solution (see *Protocol 3*)

Method

1. Using a Pasteur pipette, gently transfer the protoplasts from culture dishes to 15 ml centrifuge tubes, and pellet at 60 *g* for 5 min.

2. Discard the supernatant and wash the protoplasts by resuspending in 5 ml of W5 wash solution. Pellet by centrifugation at 60 *g* for 5 min. Discard the supernatant.

3. Resuspend the protoplasts in extraction buffer, at a density of 10^6 protoplasts in 500 μl buffer, and transfer to an Eppendorf tube.
4. Grind 10^6 protoplasts (or 10–100 mg fresh weight tissue, if assaying stably transformed tissues) per 500 μl buffer in the Eppendorf tube with a small amount of acid washed sand, storing the samples on ice.
5. Centrifuge the samples at 10 000 g for 2 min.
6. Transfer the supernatant to fresh tubes on ice.[b]
7. Place the ATP buffer in one of the substrate chambers of the luminometer.
8. Place the luciferin solution in the second substrate chamber.
9. Dilute the cell extracts by mixing 50 μl of extract with 450 μl of ice-cold extraction buffer.[c]
10. Add 25 μl of the diluted extract to a sample cuvette.
11. Start the machine running. The substrates for the reaction are injected automatically into the cuvette.
12. Total light emission (arbitrary light units) is measured over a period of 10 sec.
13. Repeat the assay using a fresh cuvette for each sample.
14. Express the data as either arbitrary light units produced/10 sec/10^6 protoplasts, or arbitrary light units produced/10 sec/μg protein (see *Protocol 3C*).

[a] Based on the method of Ow *et al.* (45) with minor modifications.
[b] The crude protein extracts can be frozen at −70 °C with no loss of enzymatic activity.
[c] This dilution step can be omitted if the activity in the sample is found to be very low.

Acknowledgements

We gratefully acknowledge financial support for our work on *Arabidopsis* from BBSRC, EC, The Gatsby Charitable Foundation, Shell International Renewables, and Zeneca Agrochemicals.

References

1. Lindsey, K. and Yeoman, M. M. (1985). In *Cell culture and somatic cell genetics of plants* (ed. I. K. Vasil), Vol. 2, pp. 61–101. Academic Press, London.
2. Koltunow, A. M., Treuttner, J., Cox, K. H., Wallroth, M., and Goldberg, R. B. (1990). *Plant Cell*, **2**, 1201.
3. Jefferson, R. A., Kavanagh, T. A., and Bevan, M. W. (1987). *EMBO J.*, **6**, 3901.
4. Flavell, R. B. (1994). *Proc. Natl. Acad. Sci. USA*, **91**, 3490.
5. Choe, S. and Feldmann, K. A. (1998). In *Transgenic plant research* (ed. K. Lindsey) pp. 57–73. Harwood Academic, Reading.

6. Tsay, Y.-F., Frank, M. J., Page, T., Dean, C., and Crawford, N. M. (1993). *Science*, **260**, 342.
7. Bancroft, I., Bhatt, A. M., Sjodin, C., Scofield, S., Jones, J. D. G., and Dean, C. (1992). *Mol. Gen. Genet.*, **233**, 449.
8. Bancroft, I., Jones, J. D. G., and Dean, C. (1993). *Plant Cell*, **5**, 631.
9. Keller, J., Lim, E., James, D. W. Jr., and Dooner, H. (1992). *Genetics*, **131**, 449.
10. Swinburne, J., Balcells, L., Scofield, S. R., Jones, J. D. G., and Coupland, G. (1992). *Plant Cell*, **4**, 583.
11. Mariani, C., De Beuckeleer, M., Truettner, J., Leemans, J., and Goldberg, R. B. (1990). *Nature*, **347**, 737.
12. Bellen, H. J., D'Evelyn, D., Harvey, M., and Elledge, S. J. (1992). *Development*, **114**, 787.
13. Worrall, D., Patel, S., Lindsey, K., and Twell, D. (1996). *Plant Sci.*, **113**, 59.
14. Minet, M., Dufour, M. E., and Lacroute, F. (1992). *Plant J.*, **2**, 417.
15. Yanofsky, M. F., Ma, H., Bowman, J. L., Drews, G. N., Feldmann, K. A., and Meyerowitz, E. M. (1990). *Nature*, **346**, 35.
16. Koncz, C., Mayerhofer, R., Koncz-Kalman, Z., Nawrath, C., Reiss, B., Redei, G. P., *et al.* (1990). *EMBO J.*, **9**, 1337.
17. Gallois, P., Lindsey, K., Malone, R., Kreis, M., and Jones, M. G. K. (1992). *Nucleic Acids Res.*, **20**, 3977.
18. Topping, J. F., Wei, W., and Lindsey, K. (1991). *Development*, **112**, 1009.
19. Pröls, F. and Meyer, P. (1992). *Plant J.*, **2**, 465.
20. Meyer, P. (1998). In *Transgenic plant research* (ed. K. Lindsey) pp. 263–75. Harwood Academic, Reading.
21. Koncz, C., Martini, N., Mayerhofer, R., Koncz-Kalman, Z., Körber, H., Rédei, G. P., *et al.* (1989). *Proc. Natl. Acad. Sci. USA*, **86**, 8467.
22. Kertbundit, S., De Greve, H., DeBoeck, F., van Montagu, M., and Hernalsteens, J.-P. (1991). *Proc. Natl. Acad. Sci. USA*, **88**, 5212.
23. Lindsey, K., Wei, W., Clarke, M. C., McArdle, M. F., Rooke, L. M., and Topping, J. F. (1993). *Transgenic Res.*, **2**, 33.
24. Bates, G. W., Piastuch, W., Riggs, C. D., and Rabussay, D. (1988). *Plant Cell Tissue Organ Cult.*, **12**, 213.
25. Lindsey, K. and Jones, M. G. K. (1990). In *Plant cell line selection: procedures and applications* (ed. P. J. Dix), pp. 317–39. VCH Verlagsgesellschaft, Weinheim.
26. Meyerowitz, E. M. (1989). *Cell*, **56**, 263.
27. Schmidt, R. and Willmitzer, L. (1988). *Plant Cell Rep.*, **7**, 583.
28. Valvekens, D., van Montagu, M., and van Lijsebettens, M. (1988). *Proc. Natl. Acad. Sci. USA*, **85**, 5536.
29. Feldmann, K. A. and Marks, M. D. (1987). *Mol. Gen. Genet.*, **208**, 1.
30. Chang, S. S., Park, S. K., Kim, B. C., Kang, B. J., Kim, D. U., and Nam, H. G. (1994). *Plant J.*, **5**, 551.
31. Bechtold, N., Ellis, J., and Pelletier, G. (1993). *C. R. Acad. Sci. Paris, Sci. de la Vie*, **316**, 1194.
32. Galbraith, D. W., Harkins, K. R., and Knapp, S. (1991). *Plant Physiol.*, **96**, 985.
33. Clarke, M. C., Wei, W., and Lindsey, K. (1992). *Plant Mol. Biol. Rep.*, **10**, 178.
34. Murashige, T. and Skoog, F. (1962). *Physiol. Plant.*, **15**, 473.
35. Gamborg, O. L., Miller, R. A., and Ojima, K. (1968). *Exp. Cell Res.*, **50**, 151.
36. Russell, J., Fuller, J., Wilson, Z., and Mulligan, B. (1991). In *Arabidopsis: the com-*

6: Tissue culture and transformation

pleat guide (ed. D. Flanders and C. Dean). AFRC PMB *Arabidopsis* Programme. http://genome-www.stanford.edu/Arabidopsis/comguide.html

37. Wei, W. and Lindsey, K. (1991). In *Arabidopsis: the compleat guide* (ed. D. Flanders and C. Dean). AFRC PMB *Arabidopsis* Programme. http://genome-www.stanford.edu/Arabidopsis/comguide.html
38. Doelling, J. H. and Pikaard, C. S. (1993). *Plant Cell Rep.*, **12**, 241.
39. Hoffman, A., Halfter, U., and Morris, P.-C. (1994). *Plant Cell Tissue Organ Cult.*, **36**, 53.
40. Lindsey, K. and Jones, M. G. K. (1987). *Plant Mol. Biol.*, **10**, 43.
41. Dekeyser, R. A., Claes, B., De Rycke, R. M. U., Habets, M. E., Van Montagu, M., and Caplan, A. B. (1990). *Plant Cell*, **2**, 591.
42. D'Halluin, K., Bonne, E., Bossut, E., De Beuckeleer, M., and Leemans, J. (1992). *Plant Cell*, **4**, 1495.
43. Potrykus, I. and Shillito, R. D. (1986). In *Methods in enzymology*, (eds. A. Weissbach, and H. Weissbach) Vol. 118, p. 549. Academic Press.
44. Damm, B., Schmidt, R., and Willmitzer, L. (1989). *Mol. Gen. Genet.*, **217**, 6.
45. Ow, D. W., Wood, K. V., DeLuca, M., De Wet, J. R., Helinski, D. R., and Howell, S. H. (1986). *Science*, **234**, 856.
46. Seki, M., Komeda, Y., Iida, A., Yamada, Y., and Morikawa, H. (1991). *Plant Mol. Biol.*, **17**, 259.
47. Mascarenhas, J. P. and Hamilton, D. A. (1992). *Plant J.*, **2**, 405.
48. Wilkinson, J. E., Twell, D., and Lindsey, K. (1994). *Plant Sci.*, **97**, 61.
50. Stomp, A.-M. (1990). In *Editorial Comments*, p. 5. United State Biochemical, Cleveland.
51. de Wet, J. R., Wood, K. V., Helinski, D. R., and DeLuca, M. (1985). *Proc. Natl. Acad. Sci. USA*, **82**, 7870.

7

Transposon and T-DNA mutagenesis

MARK G. M. AARTS, CSABA KONCZ and ANDY PEREIRA

1. Introduction

Insertional mutagenesis with a known mobile DNA insert can generate mutations that are marked by a molecular tag. When the insertion disrupts a gene, causing a mutant phenotype, the tagged mutant gene can easily be cloned using the DNA insert as a molecular probe. DNA sequences of the cloned mutant gene can then be used to isolate the corresponding wild-type allele. This procedure, known as gene tagging, facilitates the analysis of both mutant and gene.

In higher organisms the gene tagging technique was first used in *Drosophila*, with well studied transposable elements as tags (1). Although transposable elements were long since known to be present in plants, the active use of plant transposons, for gene tagging purposes, first required their molecular characterization. Both maize and snapdragon transposons subsequently became standard tools for gene isolation (2).

The success of transposon tagging encouraged the development of heterologous tagging systems, using maize transposons in species lacking well characterized endogenous transposons, such as tobacco and *Arabidopsis thaliana* (3). At the same time the T-DNA, a unique insertion element transferred by *Agrobacterium tumefaciens* into plants, appeared to be equally suited for use as a molecular tag. Mutants caused by T-DNA insertion were found after generating sufficient numbers of T-DNA transformants and they were used for the isolation of genes in *Arabidopsis* (4–6; see also Chapter 6).

In this chapter various transposon and T-DNA tagging systems for *Arabidopsis* are described including strategies for their use in mutant analysis and gene isolation. *Arabidopsis* has a special advantage over other species for gene tagging, apart from being a model organism, as it is relatively easy to grow large populations in a very small area.

2. Transposon tagging

2.1 Endogenous transposable elements

Transposable elements (7) have been discovered and studied in detail by Barbara McClintock, who attributed the genetic instability of certain maize traits to genetic elements which are able to change their location (i.e. transpose) within the genome. Elements which transpose on their own are called autonomous elements, in contrast to non-autonomous elements (often deletion derivatives of autonomous elements), which require the presence of the autonomous elements for mobility. *Activator* (*Ac*), an autonomous element discovered by McClintock, is capable of activating a family of non-autonomous *Dissociation* (*Ds*) elements. *Suppressor-mutator* (*Spm*), another autonomous element, can activate non-autonomous *defective Spm* (*dSpm*) elements. Parallel to the discovery of *Spm* an autonomous element called *Enhancer* (*En*) was described, and found to activate non-autonomous *Inhibitor* (*I*) elements (8). After isolation and DNA sequencing of both the *En* and *Spm* elements from maize (9, 10) they turned out to be virtually identical, none the less both names are still used.

Ac and *En/Spm* both encode transposase proteins and contain short terminal inverted repeats. Additional subterminal regions with short inverted and direct repeats are required for transposition. *Ac* and *En/Spm* belong to different transposon families, which cannot activate each other. Soon after molecular characterization of *Ac* and *En/Spm* their extraordinary use in gene isolation was demonstrated by the cloning of transposon tagged genes encoding various steps of the anthocyanin biosynthesis pathway in maize. The mutation frequencies at target loci observed with *Ac* and *En/Spm* in maize are about $1-10 \times 10^{-6}$ (2).

Although ubiquitous in plants, transposable elements from only a few species other than maize (*Antirrhinum majus*, *Petunia hybrida*) are characterized in sufficient detail to allow their use in gene tagging (11, 12). A poor knowledge of endogenous transposons is mainly the reason why maize transposons were exploited to design heterologous transposon tagging systems. Experiments with heterologous systems were initially carried out with *Ac* and *En/Spm* in tobacco (reviewed in ref. 3), followed a little later in *Arabidopsis*. Whether the recently characterized active endogenous transposon of *Arabidopsis* (13) will replace the use of available heterologous transposon tagging systems remains to be seen.

2.2 Transposon tagging systems in *Arabidopsis*

Transposon tagging in heterologous species can be employed using either one or two element systems. In a one element system, an autonomous transposable element is used as a mutagen. In a two element system, a non-autonomous transposable element is used, activated *in cis* or *trans* by the expression of a transposase, e.g. from a stable derivative of an autonomous element (14). In either case, the transposon is cloned into a selectable marker gene (e.g. for antibiotic resistance), blocking its expression (*Figure 1*). Excision of the element restores the activity of the excision marker gene, which can be followed with a selective agent (e.g. an antibiotic) in a phenotypic excision

7: Transposon and T-DNA mutagenesis

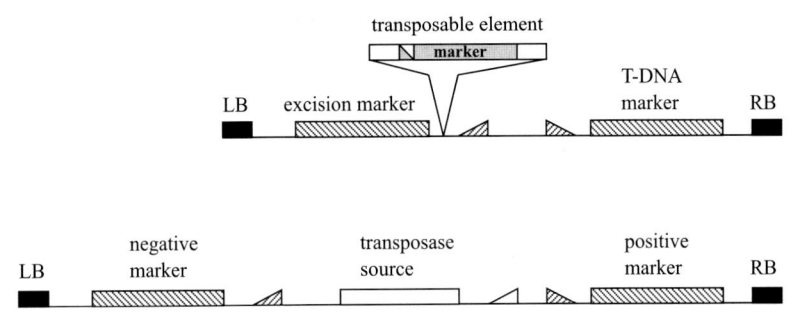

Figure 1. Basic design of a heterologous two element transposon tagging system. A marked T-DNA (*top*), contains a marked non-autonomous transposable element, inserted in an excision marker gene. Transposition of the transposable element occurs in the presence (*in cis* or *in trans*) of a T-DNA containing a transposase source with accompanying positive and negative selection markers. Triangles represent promoter sequences, LB and RB are left and right T-DNA borders respectively.

assay (15). In *Arabidopsis* both one and two element systems have been developed from the maize *Ac-Ds* and *En/Spm-I/dSpm* elements.

2.2.1 *Ac-Ds* systems

The unaltered *Ac* element from maize is not very active in *Arabidopsis*. The germinal excision frequency (the fraction of seedlings in which excision has occurred among the total number of seedlings in the progeny of a plant with *Ac*) of 0.2–0.5% is insufficient for efficient gene tagging (16, 17). Deletion of a methylation-sensitive CpG-rich *Nae*I fragment from the 5' end of *Ac* increased the germinal excision frequency to a level suitable for gene tagging (18).

More efficient transposon tagging systems were designed based on two element systems. *Ds* elements, carrying a selectable marker, were mobilized by stable transposase sources yielding different germinal excision frequencies depending on the strength and timing of the promoter fused to the transposase gene (18–22). Frequencies of over 30% were achieved using an *Ac* transposase construct driven by the CaMV 35S promoter (20–22). This high frequency of excision was not as advantageous as expected. Over-expression of *Ac* transposase in tobacco inhibited late transposition of *Ds* elements (23), thus a predominantly early transposition yielded only a few different *Ds* inserts in the progeny (24).

2.2.2 Tagging with *Ac-Ds*

Many *Ac-Ds* two element systems in *Arabidopsis* have been published as summarized in *Table 1*. Although only tested in a few cases, the *Ac* transposase producing lines and the *Ds* lines can easily be combined from different systems, provided the selectable markers are compatible. A tagging strategy

Table 1. Ac-Ds two element systems

Transposase construct[a]	Stable Ac transposase sources		Ds elements				
	Positive selection	Negative selection[b]	T-DNA selection	Ds excision selection	Ds element selection	Ref.
---	---	---	---	---	---	---
wtAc3'Δ	Kanamycin	X-Gluc	Kanamycin	Streptomycin	None	18
ΔNaeIsAc	Kanamycin	X-Gluc				18
ΔNaeIsAc	Kanamycin	NAM				18
35S-Ac	Hygromycin	None	Hygromycin	Kanamycin	35S readout	19
			None	Kanamycin	Methotrexate	19
pAc-Ac	Kanamycin	None	Kanamycin	Streptomycin	None	20, 24
ocs-Ac	Kanamycin	X-Gluc	Kanamycin	Streptomycin	Hygromycin, 35S readout	20, 24
35S-Ac	Kanamycin	None				20, 24
35S-Ac	Methotrexate	X-Gluc	Hygromycin	Kanamycin	Chlorsulfuron	22
rbcS-Ac	Methotrexate	X-Gluc				22
chs-Ac	Methotrexate	X-Gluc				22
35S-Ac	Kanamycin	X-Gluc	Kanamycin	Chlorsulfuron	hygromycin, promoter or enhancer trap	21
35S-ΔNaeIAc	Kanamycin	X-Gluc				21
wtAc3'Δ	Hygromycin	None	Hygromycin	Kanamycin	BASTA®	54
35S-Ac	Kanamycin	NAM	Kanamycin, NAM	Selection for transposition	Kanamycin, enhancer/gene trap	45

[a] Key to the constructs:
wtAc3'Δ: wild-type Ac element with 3' terminal deletion.
ΔNaeIsAc: wild-type Ac element with 5' NaeI fragment deletion.
35S-Ac: CaMV 35S promoter fused to Ac transposase gene.
pAc-Ac: Ac promoter fused to Ac transposase gene.
ocs-Ac: octopine synthase promoter fused to Ac transposase gene.
rbcS-Ac: ribulose 1,5-biphosphate carboxylase small subunit promoter fused to Ac transposase gene.
chs-Ac: chalcone synthase promoter fused to Ac transposase gene.
35S-ΔNaeIAc: CaMV 35S promoter fused to Ac transposase gene with NaeI fragment deletion.
[b] X-Gluc: p-nitrophenyl β-D-glucuronide. NAM: α-naphthalene-acetamide.

7: Transposon and T-DNA mutagenesis

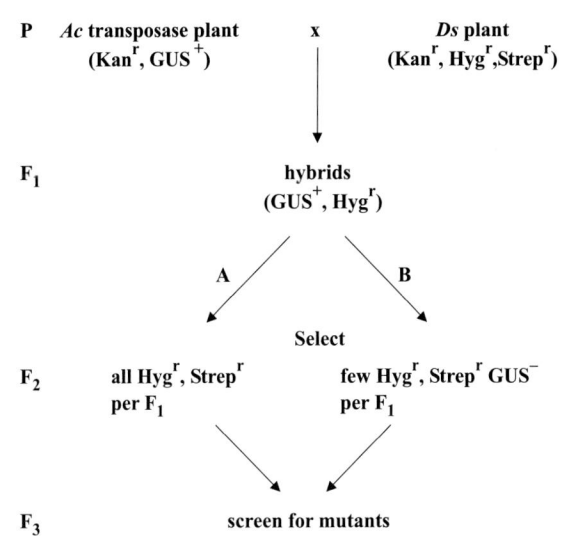

```
P    Ac transposase plant         x           Ds plant
     (Kan^r, GUS^+)                            (Kan^r, Hyg^r, Strep^r)
                                  │
                                  ▼
F₁                            hybrids
                           (GUS^+, Hyg^r)

                          A  /      \  B
                            Select
F₂       all Hyg^r, Strep^r      few Hyg^r, Strep^r GUS^−
         per F₁                  per F₁

F₃                       screen for mutants
```

Figure 2. Strategy for tagging with two element *Ac-Ds* transposon systems in *Arabidopsis thaliana*. Strategy A is used with the ΔNaeIsAc transposase source. Strategy B is used with the CaMV 35S-*Ac* transposase source. The parental plants are both homozygous for the T-DNAs containing the *Ac* transposase source marked with β-glucuronidase (GUS⁺), or the *Ds* marked with hygromycin resistance (hyg^r), inserted in the streptomycin resistance (strep^r) marker for excision. T-DNAs are marked with kanamycin resistance (kan^r).

used for the isolation of two tagged mutants (25, 26) is demonstrative for the application of two element *Ac-Ds* systems. Briefly, plants homozygous for an *Ac* transposase source marked by β-glucuronidase (GUS⁺) and kanamycin resistant (kan^r) markers were crossed with plants homozygous for a *Ds* element, expressing hygromycin phosphotransferase (HPT), inserted within a streptomycin resistance excision reporter gene located on the T-DNA (*Figure 2*). F₃ progenies were screened for mutations.

The *DRL1* (*d*eformed *r*oots and *l*eaves) locus tagged by *Ds* was identified in F₃ progeny derived from a population of 487 hygromycin and streptomycin resistant (hyg^r/strep^r) F₂ plants using the ΔNaeIsAc transposase source (25) (*Figure 2A*). This transposase source gives rise to about 5% germinal excision, and predominantly yields independent *Ds* inserts (18). Therefore all hyg^r/strep^r F₂ plants are used to collect F₃ seed for mutant screening.

The *alb3* (albino) mutant (26) was isolated using a slightly different strategy (*Figure 2B*). As the CaMV 35S-*Ac* transposase source often yields just one independent *Ds* insertion per F₂ family (24), only one or a few hyg^r/strep^r F₂ progeny was collected from each F₁ plant to produce F₃ progeny for mutant screening. To obtain lines with stable *Ds* insertions, GUS⁻ F₂ plants lacking the *Ac* transposase were chosen. The tagged *alb3* mutant was found in the progeny of 201 (hyg^r/strep^r/GUS⁻) F₂ plants from 1678 F₁ individuals.

Unstable *alb3* mutants, resulting from somatic *Ds* excision, were found in progeny of GUS⁺ (= *Ac* transposase⁺) F₂ siblings of the *alb3* segregating line.

Marking *Ds* elements by a selectable marker is essential to obtain many plants that have inherited a *Ds* after excision. With the 35S-*Ac* transposase source, 90% of the hyg^r/strep^r F₂ plants contained a transposed *Ds* (24), whereas only 50% with the Δ*NaeIsAc* transposase source (27). In the other hyg^r/strep^r plants, *Ds* excised from only one of two SPT::*Ds* T-DNA alleles, without subsequent reinsertion.

2.2.3 *En/Spm-I/dSpm* systems

Based on experiments with tobacco using a wild-type *En/Spm* element of maize (28) a one element *En/Spm* system was developed for *Arabidopsis*, that surprisingly gave a much higher transposition frequency as compared to tobacco or potato or to *Ac* in *Arabidopsis* (29). The germinal excision frequency averaged 7.5% and remained constant over a number of generations. Excisions of *En/Spm* occurred often independently from each other, with a high reinsertion frequency. A two element system in which a non-autonomous *dSpm* element was activated by two transposase genes, controlled by CaMV 35S promoters, has not been analysed in detail yet (29).

At the CPRO-DLO an '*in cis En-I* two element system' was developed (30, 31), which harbours both the *En* transposase source and a non-autonomous *I* element inserted in a HPT marked T-DNA. The expression of the two transposase genes is controlled by a CaMV 35S promoter fused to a truncated immobile *En* element. This system combines the advantages of the one and two element systems, having both continuous transposition and the ability to stabilize *I* elements. One transgenic plant, carrying two loci with multiple T-DNA inserts, was allowed to self-fertilize to generate different populations, carrying transposed *I* elements at many different positions. Instead of the germinal excision frequency, the frequency of independent transposition was assessed by DNA hybridization, as the fraction of unique novel inserts compared to all *I* element inserts in the progeny of a certain parental plant. This frequency varied from 5–10% in progeny of plants carrying over five *I* element copies, to almost 30% in progeny of single copy *I* element plants.

2.2.4 Tagging with *I* elements

In principle every generation of *I* element-carrying lines obtained after selfing can be screened for mutants. Outcrossing a mutant with wild-type Landsberg *erecta* (L*er*) for one or two generations will segregate out the transposase source, help to reduce the number of inserts, and yield stable mutants for the isolation of *I* element tagged genes.

The *En-I* was used for tagging and isolation of the *MS2* (male sterility) gene (30), and others including *CER1* (eceriferum) (32), *LFY* (leafy flowers) (M. V. Byzova, unpublished results), *AP1* (apetala), *AB13* (abscisic acid-insensitive),

LEC1 (leafy cotyledon) (M. Koornneef, K. Léon-Kloosterziel, and A. J. M. Peters, unpublished results), *GL2*, *ANL2* (anthocyaninless) (51), *LAD* (late anther dehiscence) (M. A., unpublished results), *WIL* (wilting) (A. P., unpublished results), and *SAP* (sterile apetala). All of these mutants showed either somatic or germinal reversion in the presence of transposase.

2.3 Which system to use?
2.3.1 Random tagging
The use of *Ac-Ds* systems for random gene tagging is labour-intensive, because it requires an *in vitro* selection for excision and reinsertion. Assuming that 2–4% of all *Ds* inserts give rise to mutations with visible phenotypes (25, 26), screening 2500–5000 F_3 families carrying transposed *Ds* is expected to yield about 100 tagged mutants. This may seem a lot, however a saturation mutagenesis in *Arabidopsis* may require the generation and screening of over 100 000 inserts in F_3 as was estimated by Feldmann (4).

The *En-I* system seems to be more adapted for random tagging. With an independent transposition frequency of 10–30%, it is possible to generate a large population of different inserts in only a few generations. For example, starting with a single M_0 plant, which harbours a hemizygous *En* transposase source and ten different *I* element inserts, at least 10% of all *I* elements are expected to reside at a new location in the next generation. There will thus be on average 1000 different hemizygous inserts in 1000 M_1 plants. M_2 seeds of these 1000 M_1 plants can be harvested in bulk. Sowing 12 000 bulked seeds gives a 95% probability of recovering all new homozygous inserts (assuming equal seed set and viability). Theoretically, 100 000 different inserts can be obtained from 10 000 M_5 plants, when starting with 100 M_0 plants (as unrelated as possible) containing ten *I* elements at different positions. When M_0 plants hemizygous for the *En* transposase source are used, on average 37.5% of the M_2 plants will have stable homozygous *I* elements, reducing the chance of losing mutants that are not distinguishable due to a high excision frequency. Lines containing transposed *I* elements will become available through the seed stock centres.

2.3.2 Targeted tagging
An advantage of two element *Ac-Ds* systems lies especially in targeted gene tagging. Targeted tagging involves transposon mutagenesis of an already mapped locus. It has been employed in maize, using either linked or unlinked transposable elements (33, 34), and recently also in *Arabidopsis* (35). As with wild-type *Ac* in maize (36) most *Ds* elements transpose to positions genetically linked to their original genomic location (27). Choosing *Ds* elements closely linked to a target locus will increase the chance of tagging that locus when compared to using unlinked *Ds* inserts. Some of the *Ds* T-DNA inserts, available from the seed stock centres, are already genetically mapped (27, 37)

I/dSpm donor plants   x   target mutant plants

$$\left(\frac{ms}{ms} ; \frac{A \; I/dSpm}{A \; I/dSpm} ; \frac{T}{-}\right) \qquad \left(\frac{MS}{MS} ; \frac{a}{a}\right)$$

20,000 F_1 progeny

$$\left(\text{most } \frac{MS}{Ms} ; \frac{A \; I/dSPM}{a} ; \frac{T}{-} \text{ or } -\right)$$

Screen for target mutant plants

$$\left(\frac{Ms}{ms} ; \frac{a::I/dSpm}{a} ; \frac{T}{-} \text{ or } -\right)$$

Figure 3. Strategy for targeted tagging of locus A, with a linked *I* element. *I* element donor plants are homozygous for a male sterile mutation (ms) and the *I* element linked to the *A* allele, and they are hemizygous for the *En* transposase (T). The target mutant plants are homozygous for the mutant *a* allele. Only F_1 plants with an *I* element insertion in the *A* allele (*a::I*) will show a mutant phenotype, that can be stable or unstable, depending on the presence or absence of the *En* transposase source (T or –).

and the appropriate ones can be chosen to start a targeted transposon tagging approach. When no *Ds* element is close (within 5–10 cM) to a target gene, such an insertion can be generated by mobilizing *Ds* elements from the closest mapped T-DNA carrying *Ds*. The new inserts can then be mapped relative to the T-DNA, using the T-DNA encoded excision marker and another marker encoded by *Ds* (27) as well as relative to other markers using RILs or YACs (as described in Chapters 3 and 8 respectively).

The high transposition frequency of the *En-I* also offers good possibilities for targeted gene tagging. Linked transposition of *I* elements was observed (31) although at a lower frequency than for *Ac-Ds*. An efficient strategy of targeted gene tagging is the use of male sterility to generate large F_1 populations (*Figure 3*). In this case a plant homozygous, for example, for an EMS-induced mutation at the target locus (A) is crossed with a nuclear male sterile plant (available from the seeds stock centres) (38), which is hemizygous for *En* transposase source and homozygous for an *I* element insert which is closely linked (< 5–10 cM) to the target A. In a large F_1 population heterozygous for the target A, homozygous mutants resulting from *I* element transposition into the target locus are screened or selected for. Using a transposon donor plant with one linked *I* element and a transposition frequency of 10%, a F_1 population of 20 000 plants is expected to carry 2000 new *I* element inserts, enriched for insertion near or at the target locus. As about 4000–5000 F_1 seeds can be obtained per male sterile plant (Aarts, unpublished results), five male sterile and five target mutant plants are sufficient to produce this

7: *Transposon and T-DNA mutagenesis*

required population. To obtain as many independent new insertion sites as possible, it is preferable to use more transposon donor plants (e.g. 20–50) with less crosses per plant.

2.4 Genetic and molecular analysis of a putatively transposon tagged mutant

A stepwise analysis (see *Protocol 1*) is required as a general strategy for cloning a transposon tagged gene. Initially an analysis of segregation of the elements and the mutation using revertants is necessary, prior to embarking on cloning (see *Protocol 1*). This protocol is applicable for the analysis of recessive mutations, which do not affect fertility. For rare dominant transposon-induced mutants, and infertile or inviable mutants (such as embryo lethals), these procedures will have to be adjusted.

For the analysis it is very important to use as few generations as possible. In each generation there is a chance of transposon excision, which in the worst case may generate a secondary transposon insertion closely linked to an empty, but mutated, target site. This may seriously complicate the genetic analysis. The observation of an unstable phenotype caused by excision, either as wild-type somatic sectors in a mutant background, or as wild-type germinal revertants in a mutant progeny, indicates a transposon-induced mutation.

Inverted PCR (IPCR) (39) is a reliable way to isolate DNA sequences flanking a transposon insert (*Protocol 2*). Plants with homozygous as well as hemizygous inserts can be used for IPCR. IPCR-derived flanking DNA fragments are used as probes for Southern DNA hybridization analysis of wild-type, mutant, and revertant plants to confirm the successful cloning of fragments from the tagged gene.

Protocol 1. Strategy for analysis of a putatively transposon tagged mutant

Equipment and reagents

- Liquid nitrogen
- −80 °C freezer
- Electrophoresis equipment and reagents (53)
- *Arabidopsis* plant material: wild-type line; line actively expressing transposase
- DNA isolation reagents
- Reagents for Southern hybridization (53)

Method

1. Harvest leaf material from the putatively transposon tagged mutant for DNA isolation (see *Protocol 2* or Chapters 3 and 8 for specific protocols). Cross (see Chapter 4 for conducting crosses in *Arabidopsis*) the mutant with wild-type (e.g. Landsberg *erecta*; cross 1) and with a transposase line if the mutant was not known to express transposase (cross 2).

Protocol 1. *Continued*

2. Follow segregation of the phenotype in the progeny and screen for presence of the transposase locus (e.g. by selecting for antibiotic resistance) (see Chapter 1 for growing *Arabidopsis* under selective conditions). If the mutant contains the transposase locus, screen 100–1000 progeny for wild-type looking revertants to test the stability of the mutant phenotype.

3. If the mutant does not contain a transposase locus, screen siblings to find a family expressing transposase. When found, screen the progeny as in step 2. Alternatively screen the F_2 from cross 2 for families with transposase. When found, screen progeny as in step 2. Only proceed when revertants are found.

4. Identify a transposon insertion co-segregating with the mutation. Preferably use a population without transposase (e.g. F_2 from cross 1), segregating 3:1 for wild-type:mutant. Alternatively, use a population with transposase, such as the F_2 from cross 2, or revertant and mutant progeny of the original mutant when the mutant contained the transposase locus. Perform a Southern blot analysis on about 50 plants for an *I* element tagged mutant (multiple inserts), or on about 10 plants (half mutant, half wild-type revertant) for a *Ds* tagged mutant (single or a few inserts). Load equal amounts of DNA per lane to distinguish between homozygous and hemizygous inserts. Make use of other populations if no co-segregating transposon can be identified.

5. Isolate genomic DNA flanking the co-segregating transposable element by IPCR (*Protocol 2*) using preferably DNA template from a plant lacking the transposase locus and carrying less than five copies of transposon inserts (especially for *I* element-induced mutants). If no plants with less than five inserts was found, use a backcross of the mutant with wild-type (F_1 cross 1 × wild-type) to reduce the transposon copy number.

6. Confirm the cloning of genomic DNA flanking the transposon by:

    (a) Hybridizing the IPCR probe to a Southern blot containing DNA from mutant and revertant plants, to reveal homozygous inserts in mutant and hemizygous or no inserts in revertant plants.

    (b) Analysis of the insertion site. Determine the sequence of DNA fragments carrying genomic DNA flanking the transposon insert using the IPCR fragments as template. Design PCR primers for the amplification of the insertional target site from wild-type DNA. PCR amplify the target site sequences from wild-type, revertant, and mutant alleles without inserts. Clone the PCR products and determine their DNA sequence. All revertants should have at least one allele with (near) wild-type DNA sequence. All mutants should

> have only alleles featuring frameshifts, aberrant termination, or amino acid exchanges.
> 7. Isolate genomic and cDNA clones from appropriate λ phage libraries using the IPCR products as probes, and determine their DNA sequence.

Homozygous mutant plants may not always contain a transposon insert in both mutant alleles. Occasionally, one of the inserts may transpose and leave an excision footprint behind, thus generating a stable mutant allele. When transposon insertions occur in coding regions, this is rather a rule than exception. Upon insertion, both *Ds* and *I* element generate a target site duplication of 8 bp and 3 bp respectively. Excision of these elements often deletes or duplicates a few base pairs leading to a frameshift or generating a stop codon. An adequate proof for cloning the correct gene can therefore be obtained by correlating the sequence of excision alleles with the plant phenotype. Revertant plants should have at least one allele for encoding a wild-type-like protein, whereas both alleles of a mutant should display an aberrant reading frame.

> **Protocol 2.** Isolation of DNA probes flanking *I* element inserts in *Arabidopsis thaliana* by IPCR
>
> *Equipment and reagents*
> - Eppendorf tubes
> - Liquid nitrogen
> - Eppendorf-shaped grinders
> - 14°C, 37°C, 65°C incubator
> - Thermocycler
> - DNA extraction buffer: 0.3 M NaCl, 50 mM Tris pH 7.5, 20 mM EDTA, 2% (w/v) sarkosyl, 0.5% (w/v) SDS, 5 M urea, 5% (v/v) phenol (equilibrated) (52); the first five ingredients are mixed as a 2 × stock solution, and urea and phenol are added before use
> - Phenol (saturated):chloroform (1:1)
> - 100%, 70% (v/v) ethanol
> - Isopropanol
> - Primers:
> - TE: 10 mM Tris–HCl, 1 mM EDTA pH 8
> - DNase-free RNase A (10 mg/ml stock)
> - Enzymes: *Hin*fl, DNA polymerase I Klenow fragment, *Sal*l
> - Spermidine (Sigma): prepare a 100 mM stock, store at −20°C
> - 2.5 mM dNTPs (2.5 mM each dNTP)
> - 0.3 M sodium acetate pH 5.5
> - T4 DNA ligase
> - 0.45 M NaCl
> - *Taq* DNA polymerase
> - 10 × PCR buffer
> - 10 × TBE: 108 g Tris base, 55 g boric acid, 9.3 g Na$_2$EDTA per litre
> - 1.2% (w/v) agarose (electrophoresis grade) in 0.5 × TBE
>
> 1st PCR ILJ1 (120 ng/μl) GAA TTT AGG GAT CCA TTC ATA AGA GTG T
> IRJ1 (120 ng/μl) TTG TGT CGA CAT GGA GGC TTC CCA TCC GGG GA
> 2nd PCR ILJ2 (120 ng/μl) ATT AAA AGC CTC GAG TTC ATC GGG A
> IRJ2 (120 ng/μl) AGG TAG TCG ACT GAT GTG CGC GC
> 3rd PCR ITIR (105 ng/μl) GAC ACT CCT TAG ATC TTT TCT TGT AGT G
>
> **A.** *Isolation of DNA from a single plant (52)[a] (especially for I elements)*
> 1. Harvest 100–150 mg of leaf or preferably inflorescence tissue per plant in an Eppendorf tube. Freeze in liquid N$_2$.

Protocol 2. *Continued*

2. Grind the tissue to a fine powder in the tube. Add 150 μl of DNA extraction buffer and grind once more. Add an additional 300 μl of extraction buffer and mix. Leave samples at room temperature until 18 or 24 samples are prepared.

3. Phenol:chloroform extract (450 μl) the samples. Precipitate the DNA with 0.7 vol. isopropanol. Keep tubes at room temperature for 5 min, then centrifuge for 5 min in an Eppendorf centrifuge at full speed. Wash the DNA pellet with 70% (v/v) ethanol and briefly dry them.

4. Dissolve the DNA pellets in 100 μl of TE containing 10 μg/ml RNase. DNA samples may be stored at 4°C for a few months or at −20°C.

B. *Preparation of DNA template for IPCR of I elements*

1. Digest 300 ng of DNA with 20 U *Hin*fI in 100 μl of 1 × *Hin*fI buffer containing 1 mM spermidine (3 h at 37°C).

2. Add 1 μl of 2.5 mM dNTPs and 1 U of DNA polymerase I Klenow fragment. Incubate for 15 min at room temperature. Phenol:chloroform extract and precipitate the DNA with 0.1 vol. of 0.3 M NaAc and 1 vol. of isopropanol for 20 min at −20°C (53).

3. Centrifuge the DNA for 20 min, wash the pellet in 70% (v/v) ethanol, and air dry. Resuspend in 99 μl of 1 × ligation buffer (53). Add 2.5 U of T4 DNA ligase and self-ligate the DNA fragments overnight at 14°C.

4. Inactivate the ligase by heating the sample at 65°C for 10 min. Use half of the DNA for step 5.

5. Add 25 μl of 0.45 M NaCl, 10 U of *Sal*I, and incubate for 3 h at 37°C.

6. NaAc/isopropanol precipitate both non-treated and *Sal*I digested DNA samples (see step 4), wash the pellet with 70% (v/v) ethanol, and dry. Resuspend the DNA in 30 μl of sterile, distilled H_2O.

C. *Inverse PCR for I elements*

1. Transfer the DNA template into a PCR tube and add 4 μl of 10 × PCR buffer, 2 μl primer ILJ1, 2 μl primer IRJ1 (both at 120 ng/μl), and 2 μl dNTPs (2.5 mM each). Prepare 10 μl of 1 × PCR buffer with 2.5 U of *Taq* DNA polymerase.

2. PCR reaction:

    (a) 5 min at 95°C (hot start).

    (b) Add 10 μl *Taq* DNA polymerase solution.

    (c) Set 25 cycles of PCR: 1 min 95°C, 1 min 55°C, 3 min 72°C.

    (d) Elongate for 5 min at 72°C.

3. Transfer 2 μl aliquot to a new PCR tube. Add 38 μl of 1 × PCR buffer,

7: Transposon and T-DNA mutagenesis

containing 2 µl primer ILJ2, 2 µl primer IRJ2 (both at 120 ng/µl), and 2 µl dNTPs (2.5 mM each).

4. Second PCR for 25 cycles using the conditions described in step 2.

D. Cloning of IPCR fragments for I elements

1. Size separate IPCR fragments on a 1.2% (w/v) TBE–agarose gel. Cut out the DNA bands from the gel, elute, and clone in an appropriate PCR cloning vector (53).

2. To obtain probes with very little *I* element sequence, use the (cloned) IPCR fragments for a third PCR with primer ITIR hybridizing to both terminal inverted repeats (TIR) of the *I* element.

3. Use 25 ng of linearized plasmid in a 50 µl PCR reaction (see part C), containing 2 µl of ITIR primer (at 105 ng/µl), but with annealing at 50°C instead of 55°C.

4. Clone PCR fragments as described in part D, step 1.

[a] Protocol adapted for *Arabidopsis* by Robert Whittier (personal communication) and used at CPRO-DLO for fast DNA analysis (Southern, PCR) of single plants.

Protocol 3. Isolation of DNA probes flanking *Ds* element inserts in *Arabidopsis thaliana* by IPCR

Equipment and reagents

- As *Protocol 2* (except for the following)
- Restriction enzymes: *Sau*3A or *Bst*Y1
- Primers (20 µM):
  - *Ds* 5′ 1st PCR: A3: ATA CGA TAA CGG TCG GTA CGG G
  - D74: GGA TAT ACA AAA CGG TAA ACG GAA ACG
  - 2nd PCR: D73: TTT CCC ATC CTA CTT TCA TCC CTG
  - E4: CAA AAC GGT AAA CGG AAA CGG AAA CGG TAG
  - *Ds* 3′ 1st PCR: B39: TTT CGT TTC CGT CCC GCA AGT TAA ATA
  - B38: GGA TAT ACC GGT AAC GAA AAC GAA CGG
  - 2nd PCR: D71: CCG TTA CCG ACC GTT TTC ATC CCT A
  - D75: ACG AAC GGG ATA AAT ACG GTA ATC
- 60°C incubator
- Carrier yeast tRNA (10 mg/ml)

A. Isolation of DNA from a single plant (52)

1. Follow *Protocol 2*, part A.

B. Preparation of DNA template for IPCR of Ds elements

1. Digest 1 µg of DNA with 25 U *Sau*3A or *Bst*Y1 at 60°C for 2–3 h in a 60 µl reaction volume. Remove 5 µl before adding the enzyme and compare with a 5 µl aliquot after the reaction on an agarose gel.

2. Ethanol precipitate the digested DNA. Centrifuge and wash with 70% (v/v) ethanol, air dry the pellet, and resuspend in 40 µl H$_2$O.

Protocol 3. *Continued*

3. Use 10 μl for self-ligation in 400 μl volume (2.5 U T4 DNA ligase, 14°C, overnight).

4. Add 2 μl of carrier tRNA (10 mg/ml) and 40 μl of 3 M NaAc pH 5.5. Phenol:chloroform extract and precipitate the DNA by adding 1 ml of cold 100% (v/v) ethanol to the supernatant, at −70°C for at least 30 min.

5. Centrifuge the DNA (10 min). Wash the pellets with 500 μl of 70% (v/v) ethanol, dry, and resuspend in 10 μl H$_2$O. Compare a 2 μl aliquot with a sample of digested, but unligated DNA by electrophoresis. Estimate DNA concentration by fluorimetry or A$_{260}$/A$_{280}$.

C. *IPCR for Ds elements*

1. Use 5 μl (50 ng) of DNA template in a 100 μl PCR reaction mix, containing 1 × PCR buffer, 5 μl of each first set primer (20 μM), 8 μl of 2.5 mM dNTP mix, and 2.5 U *Taq* DNA polymerase.

2. PCR reaction:

    (a) 5 min 94°C.

    (b) 35 cycles of 30 sec 94°C, 30 sec 55°C, and 3 min 72°C.

3. Use 5 μl for a further PCR reaction with a second set of primers, using conditions as in step 2.

4. Check 10 μl on a 2% (w/v) agarose minigel.

D. *Cloning of IPCR fragments for Ds elements*

1. Size separate IPCR fragments on a 1.2% (w/v) TBE–agarose gel. Cut out the DNA bands from the gel, elute, and clone in an appropriate PCR cloning vector (53).

2.5 Further applications of transposon tagging

2.5.1 Promoter or enhancer trapping with transposable elements

Less than 4% of *Ds* insertions were found to yield a visible mutant phenotype in *Arabidopsis* (25, 26). To benefit from the other 96%, transposons can be equipped with a reporter gene that is activated when the transposon is integrated in the vicinity of a transcriptional regulatory region, such as a promoter or enhancer sequence. In *Drosophila melanogaster*, P-elements containing a promoterless *lacZ* gene fused to the weak P-element transposase promoter are successfully used to detect transcribed genomic regions by transposon insertions (40). In *Nicotiana* and *Arabidopsis* this technique was

7: Transposon and T-DNA mutagenesis

first exploited using T-DNA insertions carrying promoterless reporter genes fused to the right T-DNA border (41–44, and discussed later).

Fedoroff and Smith (21) demonstrated the use of an *Ac-Ds* based promoter and enhancer trapping system in *Arabidopsis*. They constructed *Ds* elements which carried a promoterless *uidA* gene, linked either to a minimal promoter (the –46 to +6 region of the CaMV 35S core promoter) or directly to the 5' terminus of *Ds*. Combination of a *Ds-uidA* element with a CaMV 35S-*Ac* transposase source in plants resulted in β-glucuronidase (GUS) expressing sectors after transposition. Recently a similar, but more advanced enhancer and gene trapping system has been described, in which plants are selected with *Ds* elements transposed to loci unlinked to the T-DNA donor locus (45, 46). Currently populations of plants containing independent transpositions are being built up to assay for GUS expression.

2.5.2 Insertion trapping by PCR

A novel technique, which may be widely used in *Arabidopsis*, exploits the abundance of transposons for identification of insertions in specific genes. Originally developed for *Drosophila melanogaster* (47) this PCR-based approach was taken to screen a library of P-elements in a fly population using two primers, one specific for terminal sequence of the P-element, the other derived from a target gene in which P-element insertions are desired. The resolution of PCR screening in *Drosophila* permitted the detection of one individual with the right insert in a population of 1000 flies. Analogously, with the aid of the Tc1 transposon in *Caenorhabditis elegans*, Zwaal *et al.* (48) have found 23 inserts for 16 different genes in a library of 960 worm cultures, pooled in a $10 \times 10 \times 10$ three-dimensional matrix of 30 pools.

In plants this system has been applied to isolate mutants from *Petunia hybrida* using the dTph1 transposon (49, 76) and recently also in *Arabidopsis* using populations of plants containing T-DNA (50) or *En-I* transposable element inserts (82, 83). A prerequisite for this type of insertion–trapping, is a very high frequency of independent transpositions. For random inserts in *Arabidopsis*, only *En-I* systems seem to approach such high frequencies. Currently three such populations have been constructed. At the MPI für Züchtungsforschung, the AMAZE population of 8000 lines carrying 48000 independent *En* insertions has been generated (83). At CPRO-DLO a population of 5000 plants carrying 45000 independent *I* element insertions has been prepared (Speulman *et al.*, in prep.) and at the Sainsbury lab of the John Innes Institute the SLAT population of about 48000 lines has been grown in pools of 50, of which approximately 80% carry independent insertion events (84). Assuming a random distribution of inserts, the combined populations will on average contain an insert every 1 kb. With these transposon mutagenized populations it will not only be possible to obtain mutants for previously isolated genes by reverse genetics, but also to analyse mutations displaying very subtle phenotypes.

3. T-DNA tagging

3.1 The use of T-DNA as insertional mutagen

T-DNA tagging is based on a unique DNA transfer system of *Agrobacterium tumefaciens*, a soil borne plant pathogen, that causes grown galls mostly on dicotyledonous plants. *Agrobacteria* are capable of transferring a segment of their Ti or Ri (tumour-, or root-inducing) plasmids into plant cells. The transferred DNA (termed T-DNA) is flanked by 25 bp direct imperfect border repeats, and is stably integrated into the plant nuclear genome. Genes carried by the T-DNA are expressed in plants and encode functions for the synthesis of plant growth factors and specific metabolites (called opines), that can be used as sole carbon source by *Agrobacterium*. Deletion of oncogenes located between the 25 bp boundaries of the T-DNA does not affect the process of T-DNA transfer into plants, which is primarily regulated by Ti and Ri plasmid encoded virulence (*vir*) gene functions expressed in bacteria. Therefore, any foreign DNA can simply be transferred into plants by the help of T-DNA based vectors provided that the virulence gene functions are supplied *in cis* or *trans* in *Agrobacterium* (55). A wide variety of T-DNA based gene transfer vectors, as well as methods for transformation and regeneration of transgenic fertile plants, are now available for dicotyledonous, and few monocotyledonous species (see Chapter 6).

Molecular analysis of the T-DNA integration process revealed that the T-DNA is randomly integrated by illegitimate recombination into plant genomic loci that are potentially transcribed (43, 56, 57). It was thus predictable that, if T-DNA insertions occur frequently in genes, T-DNA induced insertional mutations causing visible phenotypes should likely arise when larger populations of T-DNA transformed plants are generated. The T-DNA was first used as a molecular tag to identify and isolate gene fusions in *Nicotiana* species and subsequently, when *Arabidopsis* became the model plant for molecular genetic research, a number of groups started generating large populations of T-DNA transformed *Arabidopsis* lines (4–6, 58).

3.2 Random tagging

As the T-DNA integration process is apparently not sequence-specific, T-DNA tagging is especially suited for random mutagenesis. In comparison to transposon tagging, an advantage of T-DNA tagging is that the T-DNA inserts are stable. Once integrated, the inserts remain at their original position, although recombination between multiple inserts may occur. By simply selfing transformants, large numbers of transformed seed can be obtained and distributed, so that many laboratories can screen the same T-DNA insertion library at different places over the world. With an efficient transformation method it is possible to generate a saturated population of T-DNA inserts. Every new transformant adds to the collection of T-DNA inserts, so that eventually the whole genome of *Arabidopsis* will be covered with insertions.

Feldmann (4) estimated that a population containing 105 000 randomly distributed T-DNA tags should be sufficient to saturate the *Arabidopsis* genome, to achieve a 95% probability of an average resolution of 2 kb between the inserts. Several laboratories using various transformation procedures contribute now to the approach of saturation mutagenesis. The available transformants can either be screened just for the segregation of mutants, or additionally for gene fusion insertions, depending on whether the T-DNA vector contains a reporter gene for detecting gene fusion insertions. In addition, T-DNA insertions in known genes (i.e. coding for expressed sequence tags, ESTs) can simply be identified by PCR aided screening of DNA pools from T-DNA insertional mutant lines (50, 76), and novel genes can efficiently be identified by random sequencing plant DNA fragments flanking the ends of T-DNA tags isolated by plasmid rescue or PCR amplification.

3.3 Available populations of T-DNA transformants

Table 2 summarizes the published populations of T-DNA transformants, four of which are available from the seed stock centres (38, 59) (see also Chapters 1 and 2). A very large population of transformants was obtained from the seed transformation experiments of Feldmann (4). Over 8000 transformants were generated in the Wassilewskija ecotype of *Arabidopsis*. On average 1.4 insertions are present in these transgenic lines, often represented by several T-DNA copies inserted into a single locus as direct or inverted repeats (4). Progeny from 4900 transformed lines are available from the seed stock centres as pools from 20 different transformed lines (38). Of the 8000 seed transformants screened, 15–26% segregated an offspring displaying visible

Table 2. Populations of T-DNA transformants

T-DNA selection	Gene fusion detection	Reporter gene	Bacterial *ori*	Population size	Ref.
Hygromycin	Transcription	*aph(3′)II*	Yes	>3000	62
Hygromycin	Translation	*aph(3′)II*	Yes		62
Kanamycin	–	–	Yes	>8000	4
Kanamycin	Transcription	*uidA*	No	171	41
Kanamycin	Translation	*uidA*	No	191	41
Kanamycin	Transcription	*uidA*	No	430	66
Kanamycin–hygromycin	–	–	No	>1500	77
Kanamycin or basta	Transcription–translation	*uidA*	No	>4000	65
Kanamycin	Transcription	*uidA*	No	12 000	78
Basta	35S readout	–	Yes	8550	79
Kanamycin	Transcription	GAL4-GFP	No	>100	80

mutant phenotypes (4). The first gene that was isolated from this insertional mutant collection was *GL1*, a gene involved in trichome formation (60), but many more T-DNA tagged genes isolated from this population demonstrated the success of T-DNA tagging (5).

Another large population of over 3000 transformants was made in the Columbia ecotype by tissue culture transformation (61). A number of these transformed lines are also available from NASC (see Chapter 1) (38). From a small subset of 450 transformants, a *pale* mutant was identified, shown to be T-DNA tagged, and the corresponding *CH42* gene was isolated (62).

A problem thought to be associated with transformation by tissue culture methods is the generation of somaclonal mutants that have nothing to do with T-DNA insertion. However, it was found that non-tagged mutants were also frequently produced by other transformation methods, such as the seed transformation (4). In a screen of 1340 tissue culture transformed lines, 25.07% showed a mutant phenotype. Interestingly, the mutation frequency and the mutation spectrum reported are similar for the seed and tissue culture transformants (4, 6). The high mutation frequency seems very promising for gene tagging, but it has to be noted that many of the observed mutations did not co-segregate with a T-DNA insert. Castle *et al.* (63) performed an extensive characterization of 178 embryonic mutants derived from the seed transformants. They found that only 36% of the 115 mutants examined were actually tagged by T-DNA. Among tissue culture-derived transformants, Van Lijsebettens *et al.* (64) reported that only one out of seven mutants was T-DNA tagged. Koncz *et al.* (6) estimated that the proportion of T-DNA tagged mutants with an observable phenotype in their collection is 10–30%, stressing the importance of careful genetic linkage analysis before going into the process of gene cloning.

Recently, a very simple whole plant transformation procedure was published by Bechtold *et al.* (65) offering a practical possibility for high density gene tagging. Accordingly, these authors plan to generate a saturated T-DNA insertional mutant collection which they think will be reached by 50 000 to 100 000 independent insertions. A large number of these transformants are already available from the seed stock centres and many are expected to follow as an extended, valuable source of T-DNA tagged mutants.

3.4 Promoter/enhancer trapping

Many of the T-DNA transformed populations summarized in *Table 2* were generated using T-DNAs carrying a reporter gene for the detection of transcriptional or translational plant gene fusions. A major goal of using such promoter or enhancer trapping systems is either:

(a) To identify T-DNA insertions in coding regions using a selection or screening for the expression of translational fusions between plant genes and reporter genes.

(b) To detect T-DNA inserts in the vicinity of transcriptional regulatory elements that control gene expression spatially or temporally in response to developmental, hormonal, or environmental stimuli.

Gene fusions thus allow detection of gene mutations, without screening for a particular mutant phenotype, as well as permitting the analysis of gene expression in heterozygotes when insertional inactivation of a gene results in lethality. Koncz et al. (43), used a promoterless *aph(3')II* gene fused to the right T-DNA border in two variants, one with its own ATG start codon (and in-frame stop codons upstream), and another without ATG (and no in-frame stop codons upstream). Over 30% of the transformed plants tested expressed the APH(3')II kanamycin phosphotransferase reporter enzyme in different tissues. Similar experiments with the *uidA* reporter gene (for β-glucuronidase; GUS) resulted in the detection of 54% transcriptional and 1.6% translational fusions showing GUS activity in any tissue (44).

Transgenic lines expressing reporter gene fusions can be used to characterize promoters and their upstream regulatory sequences, as well as to isolate the genes corresponding to these sequences. To detect upstream regulatory sequences, Topping *et al.* (66) used a minimal TATA box promoter driven *uidA* gene fused to the T-DNA border. By assaying for *uidA* expression in siliques of 430 T-DNA transformants, they found 74 families displaying GUS activity. From one out of three transgenics showing embryo-specific GUS expression, they have isolated the genomic boundaries of the T-DNA insert and using these as probes, cloned the corresponding wild-type genomic and cDNA sequences. None of these lines with embryo-specific GUS expression resulted in aberrant phenotypes in homozygous offspring. Goddijn *et al.* (67) screened a similar T-DNA tagged population for down- and up-regulation of GUS expression in the syncytial cell during infection of *Arabidopsis* with nematodes. Insertional mutants found with the desired GUS expression will be used to identify regulatory sequences influenced by syncytial cell development. A large population of T-DNA tagged lines by Bechtold *et al.* (65) also contains a promoterless *uidA* gene for the detection of transcriptional and translational gene fusions, increasing its value for gene tagging experiments. Recently insertional mutagenesis and promoter trapping has been reviewed by Topping and Lindsey (68). For a practical overview of the generation of T-DNA induced reporter gene fusions in plants, the vectors to use for transformation and the cloning of regulatory sequences see Koncz *et al.* (69).

3.5 Analysis of T-DNA mutants and cloning a tagged gene

As described for the analysis of transposon-induced mutants, a stepwise protocol is given for the analysis of T-DNA induced mutations (*Protocol 4*) applicable for the collections reported in *Table 2*. The strategy is based on the assumption that the mutation is recessive, gives a clear phenotype, the homozygous mutant is fertile, and the T-DNA carries a dominant selection marker, such as antibiotic and/or herbicide resistance.

A problem encountered when screening a population of T-DNA transformants may be the occurrence of untagged mutations. It is therefore essential to rigorously confirm genetic linkage between the mutation and the T-DNA insert before attempting the isolation of the tagged locus. Because the T-DNA carries a dominant marker, rather large F_2/M_2 and F_3/M_3 populations have to be used to attempt the separation of untagged mutations from potentially closely linked T-DNA inserts. Finding no mutant without T-DNA among 1000 mutants means linkage within about 3.2 cM. *Protocol 4* describes the screening of mutants for presence of T-DNA. When working with a single T-DNA insert, alternatively the progeny of a wild-type plant (i.e. an M_2 family) carrying the T-DNA can be screened for mutants after selfing (i.e. this wild-type M_2 family is expected to be hemizygous for the T-DNA tagged locus). If within the T-DNA transformed progeny a wild-type family is found, not segregating mutant phenotype, the mutation is not T-DNA tagged.

When a different ecotype than the one used for transformation, is crossed with the mutant for making a segregating F_2 (provided the mutant phenotype is expressed in a different genetic background), ARMS (70) or CAPS (71) markers (see Chapter 3) can provide help to map the mutant to a chromosome arm.

Isolation of plant DNA fragments flanking the T-DNA by plasmid rescue (as described in *Protocol 5*) is only possible when the T-DNA contains an *E. coli* plasmid replication origin (*ori*). Otherwise the T-DNA insert junctions can be isolated by IPCR (see *Protocols 2* and *6*) using T-DNA-specific primers (69, 72). When the complementation of the mutant by transformation is complicated, for example when fertility is affected, different EMS or radiation-induced alleles can be sequenced and compared with the T-DNA locus (73).

Protocol 4. Analysis of putative T-DNA tagged mutants

Equipment and reagents
- Growth conditions and antibiotic selection for T-DNA
- Wild-type ecotype for crossing

Method

1. Make a segregating population by crossing the mutant with a wild-type ecotype. Select a F_1 plant containing the T-DNA marker, to produce around 4000 F_2 seeds.

2. Sow at least 100 F_2 plants on selective medium to estimate the number of T-DNA loci present in the mutant. If the segregation resistant: sensitive is significantly higher than 3:1, then cross several mutant families with the wild-type to produce several F_2 populations, at least

7: Transposon and T-DNA mutagenesis

one of which is segregating 3:1 for resistance:sensitivity. Self individual F_2 plants and screen F_3 families.

3. Sow 2000–4000 F_2 seeds of a properly segregating population in soil. Test for linkage by treating one or two leaves of 12 mutant plants with selective agent for T-DNA (determine the necessary amount first on wild-type plants).

4. When all 12 mutants are resistant, it is likely that the T-DNA is linked to the mutation (95% probability). Treat the rest of the mutant plants (about 500–1000) with the selective agent. Herbicide treatment is done by spraying, antibiotic treatment is safer to apply on leaves or, especially for hygromycin selection, by sowing progeny of mutants on antibiotic-containing medium (see Chapter 1). The mutation is not T-DNA tagged if a mutant is found without the T-DNA insert. Where possible score the mutant phenotype in Petri plates, alternatively germinate seeds on agar and transfer all mutants to antibiotic- or herbicide-containing media providing a selection for the T-DNA encoded dominant marker.

5. Isolate the T-DNA and flanking genomic DNA by plasmid rescue (*Protocol 5*) or IPCR (*Protocol 6*). Confirm the rescue and map transcript(s) using the flanking DNA fragments as probes for hybridizing Southern and Northern blots carrying DNA and RNA samples, respectively, prepared from wild-type, heterozygous, and homozygous F_2 or F_3 families.

6. Isolate genomic and cDNA clones from wild-type *Arabidopsis* libraries and perform their molecular characterization. At the same time accomplish the genetic or physical (YAC) linkage mapping of the mutation, using either classical or molecular methods as described in ref. 61 and elsewhere in this book.

7. Confirm the genetic linkage data by transformation of the T-DNA mutant with full-length genomic DNA and/or cDNA cloned in expression vectors, to demonstrate complementation. When available, use an EMS- or radiation-induced allele for transformation, to avoid silencing problems caused by multiple T-DNA inserts in the genome. Alternatively, characterize a number of mutant alleles (see text).

Protocol 5. Isolation of T-DNA flanking genomic DNA by plasmid rescue (69)

Equipment and reagents

- Reagents from *Protocol 2*A
- Equipment and reagents for Southern analysis (53)
- Bio-Rad Gene Pulser and cuvettes
- Agarose 0.8% (w/v) (electrophoresis grade) in 0.5 × TBE

Protocol 5. *Continued*

- 10 × TBE: 108 g Tris base, 55 g boric acid, 9.3 g Na$_2$EDTA per litre
- Phenol (saturated):chloroform (1:1)
- 3 M sodium acetate pH 5.5
- Isopropanol
- T4 DNA ligase
- 70% (v/v) ethanol
- LB medium (per litre): 10 g Bacto tryptone, 5 g Bacto yeast extract, 10 g NaCl
- Overnight *E. coli* culture (e.g. MC1061)
- 1 mM Hepes pH 7.0
- 1 mM Hepes pH 7.0, 10% glycerol
- SOC: 2% (w/v) Bacto tryptone, 0.5% (w/v) Bacto yeast extract, 10 mM NaCl, 2.5 mM KCl, 10 mM MgCl$_2$, 10 mM MgSO$_4$, 20 mM glucose
- Ampicillin, 100 mg/ml stock (filter sterilized)

A. *Preparation of DNA samples for electroporation*

1. Isolate DNA from mutant plants (see *Protocol 2A*, or ref. 74).

2. Determine by Southern analysis (53), which restriction enzyme is most suitable to use for plasmid rescue. The size of flanking DNA to be rescued should be optimally in the range of 1–4 kb.

3. Digest 5 μg of DNA (but as little as 100 ng can be used) with 25 U of the appropriate enzyme in 100 μl for 2 h. Check the digestion (5 μl) on an agarose gel.

4. Phenol:chloroform extract and precipitate the DNA with NaAc/isopropanol (10 min −20 °C), then centrifuge for 10 min at full speed.

5. Resuspend in sterile, distilled H$_2$O (20–50 μg/ml), add ligation buffer (53), and self-ligate overnight with 2.5 U T4 DNA ligase at 14 °C.

6. Phenol:chloroform extract, NaAc/isopropanol precipitate the DNA, wash twice with excess 70% (v/v) ethanol, dry, and resuspend in sterile, distilled H$_2$O at 10–100 μg/ml.

B. *Preparation of E. coli cells*

1. Inoculate 200 ml of LB with 2 ml of overnight *E. coli* (e.g. MC1061) culture. Grow for 2 h until OD$_{550}$ reaches 0.5, and centrifuge the cells (fixed-angle rotor) at 4 °C with 16 000 r.c.f. for 10 min.

2. Resuspend the cells in 100 ml of 1 mM Hepes pH 7.0 at 0 °C, recentrifuge, and resuspend in 50 ml of 1 mM Hepes. Recentrifuge and resuspend in 5 ml of 1 mM Hepes, 10% (v/v) glycerol.

3. Transfer the cells to Eppendorf tubes and pellet them at 4 °C (10 min, 2100 r.c.f.). Resuspend in 400 μl of 1 mM Hepes, 10% (v/v) glycerol.

4. Mix 40 μl of the cells with 10 μl of DNA in a pre-cooled cuvette for electroporation, e.g. a Bio-Rad Gene Pulser (25 μF, 2.5 kV, 200 A, and 4.8 msec). Immediately after electroporation add 1 ml of SOC (75) and incubate cells for 1 h with shaking at 37 °C.

5. Centrifuge cells briefly (20 sec) at full speed, resuspend in SOC, and plate aliquots on LB with ampicillin (100 μg/ml).

7: Transposon and T-DNA mutagenesis

Protocol 6. Amplification and direct sequencing of T-DNA tagged plant DNA fragments by long-range IPCR (LA-IPCR)

Equipment and reagents

- Thermocycler
- Automated sequencer
- Equipment and reagents for CsCl banding (74–76)
- Qiagen DNA purification tip (optional)
- ABI Prism Dye Terminator Cycle Sequencing Kit (PE Applied Biosystems)
- Restriction enzyme that cleaves within T-DNA insert
- T4 DNA ligase
- Primers for nested PCR
- Elongase (Gibco BRL)
- 0.8% (w/v) agarose (electrophoresis grade) in 0.5 × TBE
- 10 × TBE: 108 g Tris base, 55 g boric acid, 9.3 g Na$_2$EDTA per litre
- Phenol (saturated):chloroform
- Isopropanol
- 0.3 M sodium acetate pH 6.0

Method

1. Purify high quality plant DNA by CsCl banding or on a miniscale with or without CTAB precipitation as described (see *Protocol 2*, and refs 74–76). Digest it with a restriction endonuclease which cleaves within the T-DNA insert.

2. Self-ligate the digested DNA (0.5 μg) as described in *Protocols 2, 3,* and *5*.

3. Optional: digest the ligated DNA with a restriction endonuclease which does not cleave within the T-DNA, but cleaves the plant DNA fragment flanking the left or right T-DNA end.

4. Design two sets of nested PCR primers:

    (a) One pair facing the T-DNA end (left or right).

    (b) Another pair facing the restriction endonuclease cleavage site within the T-DNA which was used in step 1.

5. Use half of the DNA in elongase PCR (BRL) or LA-PCR (Takara Shuzo Co.) for long-range amplification of large plant DNA fragments with a primer set facing the T-DNA end (left or right) and the endonuclease cleavage site within the T-DNA (step 1). Assemble the reaction mixes as recommended by the suppliers. Denature the template at 95°C for 2 min, and perform 35 cycles of amplification (94°C for 30 sec, 65°C for 30 sec, 68°C for 8 min), followed by elongation at 68°C for 10 min.

6. Use one-tenth of the PCR mix to detect the product on an agarose gel, then gel isolate the PCR amplified fragments from the rest of the PCR mix.

7. If the yield is about 0.5 μg or more, purify each amplified DNA fragment with phenol:chloroform extraction and isopropanol precipitation (0.54 vol. isopropanol, 0.1 vol. 3 M Na acetate pH 6.0), or alternatively on a Qiagen tip. Use directly as a template for sequencing with an ABI

Protocol 6. *Continued*

Prism Dye Terminator Cycle Sequencing Kit and automatic DNA sequencer.

8. If the yield of PCR amplified plant DNA is low, dilute the PCR mix 500-fold and perform a repeated PCR reaction with the pair of nested primers located closer to the T-DNA end and the restriction endonuclease cleavage site used for IPCR in step 1.

3.6 Further applications of T-DNA tagging

At present T-DNA tagging is usually exploited to isolate genes from tagged mutants, by either screening for 'loss-of-function' type of mutations, or active 'promoter/enhancer trap' type T-DNA insertions, existing in the T-DNA transformed stock collections. None the less, T-DNA vectors designed for alternative applications are also available, and used for generation of new pools of transformants. One of these new applications, the activation tagging approach, is based on T-DNA constructs which carry transcriptional enhancer sequences linked to their termini. Upon integration, these transcriptional regulatory sequences may activate the expression of plant genes located in the vicinity of T-DNA inserts in a constitutive or developmentally regulated fashion (79). Many alternative designs using different promoters in combination with genes encoding transcription factors, suicide gene products, or different signalling factors are similarly expected to find application in studies of regulatory pathways and developmental processes in the future.

References

1. Bingham, P. M., Levis, R., and Rubin, G. M. (1981). *Cell*, **25**, 793.
2. Walbot, V. (1992). *Annu. Rev. Plant Physiol. Plant Mol. Biol.*, **43**, 49.
3. Haring, M. A., Rommens, C. M. T., Nijkamp, H. J. J., and Hille, J. (1991). *Plant Mol. Biol.*, **16**, 449.
4. Feldmann, K. J. (1991). *Plant J.*, **1**, 71.
5. Forsthoefel, N. R., Wu, Y., Schultz, B., Bennett, M. J., and Feldmann, K. A. (1992). *Aust. J. Plant Physiol.*, **19**, 353.
6. Koncz, C., Németh, K., Rédei, G. P., and Schell, J. (1992). *Plant Mol. Biol.*, **20**, 963.
7. Fedoroff, N. (1989). In *Mobile genetic elements* (ed. J. Shapiro), pp. 1–63. Academic Press, New York.
8. Peterson, P. A. (1970). *Theor. Appl. Genet.*, **40**, 367.
9. Pereira, A., Cuypers, H., Gierl, A., Schwarz-Sommer, Zs., and Saedler, H. (1986). *EMBO J.*, **5**, 835.
10. Masson, P., Surosky, R., Kingsbury, J., and Fedoroff, N. V. (1987). *Genetics*, **117**, 117.

11. Coen, E. S., Robbins, T. P., Almeida, J., Hudson, A., and Carpenter, R. (1989). In *Mobile DNA* (ed. D. E. Berg and M. M. Howe), pp. 413–36. Am. Soc. Microbiol., Washington DC.
12. Gerats, A. G. M., Huits, H., Vrijlandt, E., Maraña, C., Souer, E., and Beld, M. (1990). *Plant Cell*, **2**, 1121.
13. Tsay, Y.-F., Frank, M. J., Page, T., Dean, C., and Crawford, N. M. (1993). *Science*, **260**, 342.
14. Coupland, G., Plum, C., Chatterjee, S., Post, A., and Starlinger, P. (1989). *Proc. Natl. Acad. Sci. USA*, **86**, 9385.
15. Baker, B., Coupland, G., Fedoroff, N., Starlinger, P., and Schell, J. (1987). *EMBO J.*, **6**, 1547.
16. Schmidt, R. and Willmitzer, L. (1989). *Mol. Gen. Genet.*, **220**, 17.
17. Dean, C., Sjodin, C., Page, T., Jones, J., and Lister, C. (1992). *Plant J.*, **2**, 79.
18. Bancroft, I., Bhatt, A. M., Sjodin, C., Scofield, S., Jones, J. D. G., and Dean, C. (1992). *Mol. Gen. Genet.*, **233**, 449.
19. Grevelding, C., Becker, D., Kunze, R., Von Menges, A., Fantes, V., Schell, J., *et al.* (1992). *Proc. Natl. Acad. Sci. USA*, **89**, 6085.
20. Swinburne, J., Balcells, L., Scofield, S., Jones, J. D. G., and Coupland, G. (1992). *Plant Cell*, **4**, 583.
21. Fedoroff, N. V. and Smith, D. L. (1993). *Plant J.*, **3**, 273.
22. Honma, M. A., Baker, B. J., and Waddel, C. S. (1993). *Proc. Natl. Acad. Sci. USA*, **90**, 6242.
23. Scofield, S. R., English, J. J., and Jones, J. D. G. (1993). *Cell*, **75**, 507.
24. Long, D., Swinburne, J., Martin, M., Wilson, K., Sundberg, E., Lee, K., *et al.* (1993). *Mol. Gen. Genet.*, **241**, 627.
25. Bancroft, I., Jones, J. D. G., and Dean, C. (1993). *Plant Cell*, **5**, 631.
26. Long, D., Martin, M., Sundberg, E., Swinburne, J., Puangsomlee, P., and Coupland, G. (1993). *Proc. Natl. Acad. Sci. USA*, **90**, 10370.
27. Bancroft, I. and Dean, C. (1993). *Genetics*, **134**, 1221.
28. Pereira, A. and Saedler, H. (1989). *EMBO J.*, **8**, 1315.
29. Cardon, G. H., Frey, M., Saedler, H., and Gierl, A. (1993). *Plant J.*, **3**, 773.
30. Aarts, M. G. M., Dirkse, W., Stiekema, W. J., and Pereira, A. (1993). *Nature*, **363**, 715.
31. Aarts, M. G. M., Corzaan, P., Stiekema, W. J., and Pereira, A. (1995). *Mol. Gen. Genet.*, **247**, 555.
32. Aarts, M. G. M., Keijzer, C. J., Stiekema, W. J., and Pereira, A. (1995). *Plant Cell*, **7**, 2115.
33. Nelson, O. E. and Klein, A. S. (1984). *Genetics*, **106**, 779.
34. Schmidt, R. J., Burr, F. A., and Burr, B. (1987). *Science*, **238**, 960.
35. James, D. W. Jr., Lim, E., Keller, J., Plooy, I., Ralston, E., and Dooner, H. K. (1995). *Plant Cell*, **7**, 309.
36. Greenblatt, I. M. (1984). *Genetics*, **108**, 471.
37. Coupland, G. (1992). In *Methods in Arabidopsis research* (ed. C. Koncz, N.-H. Chua, and J. Schell), pp. 290–309. World Scientific, Singapore.
38. Anderson, M. and Mulligan, B. (1994). *Seed list, 1994*. Nottingham Arabidopsis Stock Centre. Nottingham, UK (http://nasc.nott.ac.uk).
39. Ochman, H., Gerber, A. S., and Hartl, D. J. (1988). *Genetics*, **120**, 621.
40. Bellen, H. J., Wilson, C., and Gehring, W. J. (1990). *BioEssays*, **12**, 199.

41. André, D., Colau, D., Schell, J., Van Montagu, M., and Hernalsteens, J.-P. (1986). *Mol. Gen. Genet.*, **204**, 512.
42. Teeri, T. H., Herrera-Estralla, L., Depicker, A., Van Montagu, M., and Palva, T. (1986). *EMBO J.*, **5**, 1755.
43. Koncz, C., Martini, N., Mayerhofer, R., Koncz-Kalman, Zs., Körber, H., Redei, G. P., et al. (1989). *Proc. Natl. Acad. Sci. USA*, **86**, 8467.
44. Kertbundit, S., De Greve, H., Deboeck, F., Van Montagu, M., and Hernalsteens, J.-P. (1991). *Proc. Natl. Acad. Sci. USA*, **88**, 5212.
45. Sundaresan, V., Springer, P., Volpe, T., Haward, S., Jones, J. D. G., Dean, C., et al. (1995). *Genes Dev.*, **9**, 1797.
46. Springer, P. S., McCombie, W. R., Sundaresan, V., and Martienssen, R. A. (1995). *Science*, **268**, 877.
47. O'Hare, K. (1990). *Trends Genet.*, **6**, 202.
48. Zwaal, R., Broeks, A., van Meurs, J., Groenen, J. T. M., and Plasterk, R. H. A. (1993). *Proc. Natl. Acad. Sci. USA*, **90**, 7431.
49. Koes, R., Souer, E., van Houwelingen, A., Mur, L., Spelt, C., Quattrocchio, F., et al. (1995). *Proc. Natl. Acad. Sci. USA*, **92**, 8149.
50. McKinney, E. C., Ali, N., Traut, A., Feldmann, K. A., Belostotsky, D. A., McDowell, J. M., et al. (1995). *Plant J.*, **8**, 613.
51. Kubo H., Peeters, A. J. M., Aarts, M. G. M., Pereira, A., and Koornneef, M. (1999). *Plant Cell*, **11**, 1217.
52. Shure, M., Wessler, S., and Fedoroff, N. (1983). *Cell*, **35**, 225.
53. Sambrook, J., Fritsch, E. F., and Maniatis, T. (ed.) (1989). *Molecular cloning: a laboratory manual* (2nd edn). Cold Spring Harbor Laboratory Press, NY.
54. Altmann, T., Schmidt, R., and Willmitzer, L. (1992). *Theor. Appl. Genet.*, **84**, 371.
55. Zambryski, P., Tempe, J., and Schell, J. (1989). *Cell*, **56**, 193.
56. Mayerhofer, R., Koncz-Kalman, Z., Nawrath, C., Bakkeren, G., Crameri, A., Angelis, K., et al. (1991). *EMBO J.*, **10**, 607.
57. Gheysen, G., Villarroel, R., and Van Montagu, M. (1991). *Genes Dev.*, **5**, 287.
58. Valvekens, D., Van Montagu, M., and Van Lijsebettens, M. (1988). *Proc. Natl. Acad. Sci. USA*, **85**, 5536.
59. Anderson, M. A. (1995). *Weeds World*, **2**, 37.
60. Herman, P. L. and Marks, M. S. (1989). *Plant Cell*, **1**, 1051.
61. Koncz, C., Schell, J., and Rédei, G. P. (1992). In *Methods in Arabidopsis research* (ed. C. Koncz, N.-H. Chua, and J. Schell), pp. 224–73. World Scientific, Singapore.
62. Koncz, C., Mayerhofer, R., Koncz-Kalman, Zs., Nawrath, C., Reiss, B., Rédei, G. P., et al. (1990). *EMBO J.*, **9**, 1337.
63. Castle, L. A., Errampalli, D., Atherton, T. L., Franzmann, L. H., Yoon, E. S., and Meinke, D. W. (1993). *Mol. Gen. Genet.*, **241**, 504.
64. Van Lijsebettens, M., Vanderhaeghen, R., and Van Montagu, M. (1991). *Theor. Appl. Genet.*, **81**, 277.
65. Bechtold, N., Ellis, J., and Pelletier, G. (1993). *C. R. Acad. Sci. Paris, Sci. de la Vie/Life Sci.*, **316**, 1194.
66. Topping, J. F., Agyeman, F., Henricot, B., and Lindsey, K. (1994). *Plant J.*, **5**, 895.
67. Goddijn, O. J. M., Lindsey, K., van der Lee, F. M., Klap, J. C., and Sijmons, P. C. (1993). *Plant J.*, **4**, 863.
68. Topping, J. F. and Lindsey, K. (1995). *Transgenic Res.*, **4**, 291.
69. Koncz, C., Martini, N., Szabados, L., Hrouda, M., Bachmair, A., and Schell, J.

7: Transposon and T-DNA mutagenesis

(1994). In *Plant molecular biology manual* (ed. S. B. Gelvin and R. A. Schilperoort), pp. B2, 1–22. Kluwer Academic Publishers, Dordrecht.
70. Fabri, C. O. and Schäffner, A. R. (1994). *Plant J.*, **5**, 149.
71. Konieczny, A. and Ausuble, F. M. (1993). *Plant J.*, **4**, 403.
72. Lindsey, K., Wei, W., Clarke, M. C., McArdle, H. F., Rooke, L. M., and Topping, J. F. (1993). *Transgenic Res.*, **2**, 33.
73. Shevell, D. E., Leu, W.-M., Gillmore, C. S., Xia, G., Feldmann, K. A., and Chua, N.-H. (1994). *Cell*, **77**, 1051.
74. Dellaporta, S. L., Wood, J., and Hicks, J. B. (1983). *Plant. Mol. Biol. Rep.*, **1**, 19.
75. Zabarowsky, E. R. and Winberg, G. (1990). *Nucleic Acids Res.*, **18**, 5912.
76. Krysan, P. J., Young, J. C., Tax, F., and Sussman, M. R. (1996). *Proc. Natl. Acad. Sci. USA*, **93**, 8145.
77. Sangwan, R. S., Bourgeois, Y., and Sangwan-Norreel, B. S. (1991). *Mol. Gen. Genet.*, **230**, 475.
78. Jack, T. http://www.dartmouth.edu/~tjack/
79. Weigel, D. http://www.salk.edu/LABS/pbio-w/index.html
80. Haseloff, J. http://www.plantsci.cam.ac.uk/Plantsci/Groups/JimHaseloff.html
81. Byzova, M. V., Franken, J., Aarts, M. G. M., de Almeida-Engler, J., Engler, G., Mariani, C. *et al.* (1999). *Genes Dev.*, **13**, 1002.
82. Wisman, E., Hartmann, U., Sagasser, M., Baumann, E., Palme, K., Hahlbrock, K., Saedler, H. and Weisshaar, B. (1998). *Proc. Natl. Acad. Sci. USA*, **95**, 12432.
83. Baumann, E., Lewald, J., Saedler, H., Schulz, B. and Wisman, E. (1998). *Theor. Appl. Genet.*, **97**, 729.
84. Jones, J. D. G. http://www.jic.bbsrc.ac.uk

8

Map-based cloning in *Arabidopsis*

JOANNA PUTTERILL and GEORGE COUPLAND

1. Introduction

Novel plant genes are often first identified as mutations that disrupt or modify plant processes. However, in most species it is still extremely difficult to isolate the affected genes if nothing is known of their sequence or the biochemical properties of the proteins they encode. Insertional mutagenesis using transposons or T-DNA has proved a successful method of isolating such genes in maize, *Antirrhinum*, *Arabidopsis*, and *Petunia* (1–5), as long as appropriate alleles caused by insertion of one of these elements are available or can be identified (see also Chapter 7). However, the isolation of such alleles is always laborious and is often the rate-limiting step in this approach. More recently map-based cloning has proved to be a powerful alternative to insertional mutagenesis in tomato (6) and *Arabidopsis* (see below). This involves isolating the gene after first accurately locating a mutation on the chromosomes relative to previously isolated DNA markers. The approach therefore overcomes the major difficulty of T-DNA or transposon tagging, in that a rare allele caused by insertion of one of these elements is not required.

Arabidopsis is well suited to the isolation of genes by map-based cloning. It contains a small plant genome of approximately 100 megabases, so that the genetic distances between markers and the gene of interest tend to represent relatively short regions of DNA (7–10). *Arabidopsis* also contains a small proportion of repeated DNA sequences, which reduces the possibility that the progressive steps involved in chromosome walking are blocked by the presence of repeats (11, 12). The number of DNA markers mapped on the *Arabidopsis* chromosomes is close to 800 and continues to increase (13–15; see also Chapter 3). Other essential materials have already been constructed and are usually made freely available, often through the *Arabidopsis* stock centres. These include good lambda, cosmid, BAC (bacterial artificial chromosome), and YAC (yeast artificial chromosome) libraries of *Arabidopsis* genomic DNA, and a variety of cDNA libraries (see also Chapter 2). Several examples are published in which these materials have been used to isolate *Arabidopsis* genes (e.g. 10, 16–23). In this chapter we will summarize the strategies and methods commonly used to proceed from an unmapped mutation to the identification of the gene by map-based cloning.

2. Locating the mutation of interest relative to DNA markers

2.1 Determining an approximate map position

The position of the mutation on the *Arabidopsis* genetic map is determined by detecting genetic linkage between the mutation and markers of known position. Broadly, two types of markers are available: phenotypic markers caused by mutations in other genes, and DNA markers that reveal polymorphisms between the DNA of different ecotypes of *Arabidopsis* that can be detected in a variety of ways.

To test for linkage between phenotypic markers and the mutation of interest, the new mutant would usually be crossed to a multiply marked line harbouring several mutations. One line is available in which all five chromosomes are multiply marked (W100; 24) (see Chapter 4), or alternatively lines in which each chromosome is marked separately can be used (see Chapter 1 and *Arabidopsis* stock centre catalogues for details). The F1 progeny created in the crosses are self-fertilized, and linkage between the new mutation and a mutation in the marked line is detected in the F2 generation because fewer than one-sixteenth (assuming both mutations are recessive) of plants are homozygous for both mutations and therefore show both phenotypes. The advantage of this approach is that no molecular biology is involved, and it can be efficient (e.g. 25). The disadvantages are that the line marked on all chromosomes can be difficult to grow and therefore multiple crosses, and the scoring of multiple F2 families is usually required. Furthermore, in some cases the presence of mutations derived from the marker line makes it difficult to observe the phenotype of the mutation of interest or vice versa.

The first step in detecting linkage between DNA markers and a mutation is to cross the mutant to another ecotype whose DNA is known to be polymorphic to that of the mutant. Usually it is advantageous for the two parents in this cross to be Landsberg *erecta* and Columbia, so that if the mutation is isolated in Landsberg *erecta* then the line is crossed to Columbia or vice versa, because most available markers have been tested between these ecotypes and the polymorphisms are well documented. The F1 progeny of this cross are self-fertilized and the F2 generation scored for mutants. For recessive mutations, often only those F2 plants that show the mutant phenotype are used for mapping because their genotype with respect to the mutation is unambiguous while phenotypically wild-type plants require to be scored in the F3 generation to determine whether they are heterozygous for the mutation. As few as 30 of these F2 mutant plants is sufficient to identify an approximate map position (e.g. 26). DNA is extracted either from individual F2 plants (*Protocol 1*), or from pools of F3 plants derived from single F2 individuals (see Chapter 3, *Protocol 3*) (51). The DNA is then tested with markers well distributed around the chromosomes, to identify a polymorphism that is homozygous in a

8: Map-based cloning in Arabidopsis

high proportion of the tested F2 individuals and is characteristic of the parental mutant line.

Protocol 1. Isolation of *Arabidopsis* DNA from single plants by miniprep[a]

Equipment and reagents
- 65°C water-bath
- 37°C oven
- Eppendorf tubes
- Eppendorf-shaped pestle
- Liquid nitrogen
- Acid washed sand (optional)
- T5E: 50 mM Tris pH 8.0, 10 mM EDTA pH 8.0
- 70% (v/v) ethanol

- Extraction buffer: 100 mM Tris–HCl pH 8.0, 50 mM EDTA pH 8.0, 500 mM NaCl, 1.25% (w/v) SDS, 8.3 mM NaOH, 0.38% (w/v) Na bisulfite
- Isopropanol
- TE: 10 mM Tris–HCl pH 8.0, 1 mM EDTA
- 5 M potassium acetate
- 7.4 M ammonium acetate; filter sterilized
- 3 M sodium acetate pH 5.2

Method
1. Place 0.15 g of leaf tissue in an Eppendorf tube and freeze in liquid nitrogen. The material can then be frozen at −70°C and used when convenient.
2. Fill tubes with liquid nitrogen and grind the leaf material in the Eppendorf tube, using an Eppendorf-shaped pestle. Adding some acid washed sand can facilitate the grinding process.
3. Add 0.7 ml of extraction buffer that is pre-heated to 65°C and mix thoroughly using a pipette tip. Place the tube in a 65°C water-bath for 10 min.
4. Add 0.22 ml of 5 M potassium acetate, mix well, and put on ice for 20–40 min.
5. Pellet by microcentrifugation at 4°C for 5 min.
6. Pour the supernatant through a 1 ml (blue) Gilson tip containing a small plug of tissue paper to act as a filter. Then add 0.7 vol. of isopropanol and mix well.
7. Centrifuge at 14 000 g for 5 min, pour out the supernatant, and rinse the pellet twice with 70% (v/v) ethanol. Drain on a paper towel for 1 min.
8. Add 300 μl of T5E, vortex for 2 sec, place at 65°C for 5 min, and vortex for 2 sec. The pellet should have dissolved.
9. Add 150 μl of filter sterilized 7.4 M ammonium acetate, vortex for 2 sec, and centrifuge in a microcentrifuge at 14 000 g for 3 min.
10. Transfer the supernatant into a fresh Eppendorf tube, add 330 μl of isopropanol, and mix well.
11. Centrifuge for 3 min at room temperature, pour out the supernatant,

Protocol 1. *Continued*

  rinse the pellet twice with 70% (v/v) ethanol, and drain onto a paper towel for 2 min.

12. Add 100 µl of T5E, vortex for 2 sec, heat to 65°C for 5 min, and then vortex for 2 sec to dissolve the pellet.

13. Add 10 µl of 3 M sodium acetate pH 5.2 and 75 µl of isopropanol. Mix well.

14. Centrifuge for 3 min, pour off the supernatant, rinse the pellet twice with 70% (v/v) ethanol, and drain onto a paper towel for 2 min.

15. Dry the pellet in a 37°C oven for 15 min, add 25 µl of TE pH 8.0, and let the pellet redissolve overnight at 4°C.

16. Vortex the DNA solution for 2 sec, heat to 65°C for 5 min, and vortex for a further 2 sec. Store at 4°C.

[a] Modified by Doug Dahlbeck (University of California, Berkeley) from ref. 51. Provides enough DNA for Southern and PCR analysis of single plants.

A variety of methods are available to detect DNA polymorphisms that have been located on the genetic map. PCR-based methods are the most powerful because small amounts of DNA isolated from single F2 plants can be tested with several different markers. Two types of PCR-based polymorphic markers are widely used: cleaved amplified polymorphic sequences (CAPS) and simple sequence length polymorphisms (SSLPs). CAPS are described in detail elsewhere in this book (see Chapter 3), and 18 have been published along with their locations on the genetic map (27). Details of new CAPS markers are available online from the *Arabidopsis* database, AtDB (see Chapter 10). SSLPs rely on the observation that the number of simple sequence repeats present in tandem at particular loci varies between ecotypes, therefore fragments differing in length are amplified from the different ecotypes by using unique primers flanking the array of repeats. 30 SSLPs have been described along with their map positions and precise details of how to perform the amplifications (28). Which of these types of marker is used is a matter of personal preference; we prefer to use SSLPs as long as the polymorphism can be easily detected on an agarose gel, and use CAPS in cases where the amplified fragment can be readily cleaved with a cheap, easily available enzyme. By using a combination of these two approaches markers distributed across all chromosomes are available. Other useful PCR-based methods, RAPDS and AFLPs, are described in the next section.

The other types of DNA marker that are widely used to obtain an approximate map position are classical restriction fragment length polymorphism (RFLP) markers. Over 500 of these are located on the chromosomes and are available from the stock centres or from the individuals who mapped them initially (13–15). RFLPs are DNA fragments that are used as hybridization probes to Southern blots of ecotype DNA cleaved with particular enzymes. Restriction site differences between the lines results in detection of DNA fragments of differing size which can be used as genetic markers. An advantage of this system is that there is a much higher density of markers available than there is for the PCR-based markers, so there is a higher probability of detecting close linkage. Until recently it was necessary to cleave the plant DNA with different enzymes to reveal polymorphisms at different loci, and the large amounts of DNA that were therefore required had to be isolated from bulked F3 plants rather than single F2 individuals. This was overcome by the development of an *Arabidopsis* RFLP mapping set (ARMS) which is 13 markers distributed on all five chromosomes, that all detect a polymorphism with *Eco*RI (26). These markers can be used in two pools of probes, and hybridized against a single Southern blot of plant DNA cleaved with *Eco*RI.

2.2 Identifying a short genetic interval containing the mutation as a prelude to isolating the gene

Prior to proceeding to isolate the gene of interest using the methods described in the next section, the gene must be located as accurately as possible relative to the available markers, or new markers more closely linked to the mutation must be identified. Accurate positioning of the gene requires as high a density of markers as possible and as high a density of cross-overs in the vicinity of the gene as can be identified.

The most straightforward way to increase the number of cross-overs close to the gene, is simply to increase the number of F2 plants analysed. So while analysing 30 F2 individuals homozygous for a recessive mutation can be sufficient to provide an approximate map location, analysing 200 should provide an accurate location: a marker that detected no cross-overs in 200 such plants would have a 95% probability of being within 1.5 cM of the mutation. The limitation of this approach is therefore that a large number of DNA samples must be extracted and analysed. Once the samples are available, markers in the vicinity of the gene are tested using the approximate map position determined as described above as a guide. For example, to facilitate the isolation of the *ABI1* locus Meyer *et al.* (21) tested 1032 F2 chromosomes with a closely linked RFLP marker.

A more elegant method to identify informative recombinants is to utilize flanking phenotypic markers to enrich for cross-overs close to the mutation of interest. The approximate map position can be used as a guide as to which flanking phenotypic markers are likely to be useful. The markers can then be used in a variety of ways depending on whether the mutations are recessive or dominant, and whether they are available in the same or different ecotypes. If the mutation of interest and the flanking mutation are both recessive and present in the same ecotype, a recombinant chromosome containing both mutations must first be made. A hybrid F1 plant containing this chromosome and one of a different ecotype is then generated and self-fertilized. F2 plants homozygous for one mutation, but not for the other, are identified by screening for plants only showing one of the phenotypes. These cross-overs must then be located between the mutation of interest and the flanking marker; if the flanking marker is very close to the gene of interest then this can greatly enrich for informative recombinants (9, 20).

As well as a large number of cross-overs, a high density of markers are required to position the gene in a small enough interval to proceed with map-based cloning. If the density of RFLP markers in the region of interest is insufficient, then it might be most efficient to try to identify more markers. In general one would need a marker within 2 cM of the gene, as covering distances of more than this is likely to involve time-consuming chromosome walking experiments, although to some extent this is being alleviated by the construction of YAC libraries with larger inserts of *Arabidopsis* DNA and by the development of physical maps (see Section 3.2).

Several strategies are available to identify more markers in the region of interest. A method described by Michelmore *et al.* (29) makes use of a combination of random amplified polymorphic DNA (RAPD) markers (30) and bulked segregant analysis. RAPD markers are identified by PCR using short oligonucleotide primers, typically ten bases long. They often amplify multiple fragments in a single reaction and those that differ in size between polymorphic parents, or amplify fragments present in one parent, but absent in another, are looked for. Different markers are made by simply varying slightly the sequence of the primer, and because multiple fragments are detected in single reactions several genomic locations are tested for polymorphism in one amplification. Bulked segregant analysis allows markers in the vicinity of the gene of interest to be identified directly. The plant DNAs used in the PCR are made from a bulk of around 20 independently isolated F2 plants homozygous for the mutation of interest, and from a similar bulk of plants homozygous for the wild-type allele at the same locus. Because these bulks differ at DNA closely linked to the mutation of interest, but should not differ at loci unlinked to the mutation, any marker that is polymorphic between the two bulks is likely to be linked to the gene of interest (29). This method can be used to select markers even more closely linked to the mutation of interest by using DNA extracted from a bulk of recombinants with cross-overs closely linked to the gene of interest as one sample, and DNA isolated from randomly isolated F2 plants without cross-overs close to the gene as the other sample.

Amplified fragment length polymorphism (AFLP) (31) is a recent alternative to RAPDs for using a PCR-based approach to identify new markers closely linked to a gene of interest. In this method, DNA of both ecotypes is cleaved with two restriction enzymes, a frequent and an infrequent cutter. Two adapters, corresponding to the two enzymes used for the cleavage, are then ligated to the digested plant DNAs. Primers which anneal to the adapters

8: Map-based cloning in Arabidopsis

are then used to amplify the restriction fragments. To reduce the complexity of the mixture of amplified fragments, the primers are extended by several bases at the 3' end so that only a subset of restriction fragments that are complementary to the extension are amplified. To facilitate detection of the products, one of the primers is end-labelled by phosphorylating the 5' end. The fragments produced by the amplification are then separated on a sequencing gel, and polymorphic fragments between the two parents are identified. To identify polymorphisms linked to the mutation of interest, the AFLP technique can be applied to bulked segregant analysis as described above (e.g. 32). An advantage of AFLPs over RAPDs is that the primers and the site to which they anneal are homologous, whereas with RAPDs the primers often anneal to sequences that are not completely homologous, with the result that amplification can be more difficult to reproduce.

3. Placing the gene on the physical map

3.1 Chromosome landing

Once the gene of interest is positioned relative to molecular markers, then the next step is to isolate a segment of DNA containing the gene. Chromosome landing is an efficient way of achieving this. The procedure requires a DNA marker that co-segregates with the gene in detailed mapping experiments. This closely linked marker is then used to screen an *Arabidopsis* YAC, BAC, P1, or cosmid library (see Chapter 2). DNA fragments, derived from each end of the insert DNA in the clone (end fragments), are used as RFLP markers and positioned relative to the gene and recombinant breakpoints. In this way, a segment of DNA containing the gene is identified without the need for time-consuming chromosome walking experiments. Chromosome landing has been used successfully in positional cloning experiments in *Arabidopsis* and tomato (e.g. 6, 16). The extent of physical mapping and sequence data that is now available for *Arabidopsis* (see Chapter 10) means that, depending on the map position, identifying contiguous DNA containing the gene of interest may be possible by database searches alone.

Other approaches may also be helpful for directly positioning the gene on the physical map. Southern blots can be probed with labelled clones from the region to try and identify chromosomal changes associated with the mutation (e.g. 21). Genomic clones can also be used to identify candidate cDNA clones of the gene in libraries derived from tissues where it is likely that the gene is expressed. These cDNAs can then be used as RFLP markers to define those that co-segregate with the gene (6). Sequence databases should also be searched regularly to see if the genomic DNA sequence of the region has been determined. If so, the DNA sequence of candidate genes in the equivalent region from the mutant can be analysed, as some of the mutant alleles would be expected to carry mutations in the coding region of the gene.

3.2 Chromosome walking with YAC clones

If chromosome landing is not feasible, then chromosome walking experiments are initiated to isolate a segment of DNA that contains the gene. In this process, DNA markers that are closest to the gene are used to identify YAC clones in *Arabidopsis* YAC libraries. End fragments are then isolated from these YAC clones and used to rescreen the library. Chromosome walking, therefore, results in isolation of an increasingly long segment of contiguous DNA (DNA contig) from the region. The gene is positioned on the DNA contig by molecular mapping with new markers isolated during the walk.

Before starting new experiments to isolate genomic clones carrying DNA from the region, the *Arabidopsis thaliana* database, AtDB, should be searched to see whether there is an existing YAC, BAC, or cosmid contig in the vicinity of the gene. In addition, a request on the *Arabidopsis* News Bulletin board may yield useful, unpublished information from other groups building physical contigs around nearby genes. If no contigs are available, then the nearest flanking markers to the gene are used to screen YAC libraries for clones containing DNA from the region. A number of *Arabidopsis* YAC libraries have been constructed from various ecotypes (see *Table 1*) and copies of some of these are available from the *Arabidopsis* Biological Resource Centre at Ohio State University, or by direct request to the laboratories involved. YAC libraries are maintained as described in *Protocol 2* and are screened by yeast colony hybridization using radiolabelled RFLP markers as probes as described in *Protocol 3*.

Table 1. Large insert genomic libraries of *Arabidopsis* DNA[a]

Library	Ecotype	Average insert size	Reference
EG YAC	Columbia	150 kb	33
EW YAC	Columbia	150 kb	34
abi1 YAC	Landsberg *erecta*	< 200 kb	33
S YAC	Landsberg *erecta*	< 150 kb	33
yUP YAC	Columbia	250 kb	35
CIC YAC	Columbia	450 kb	36
TAMU BAC	Columbia	100 kb	37
IGF YAC	Columbia	100 kb	b
Mitsui P1	Columbia	80 kb	38

[a] Information about these libraries and their availability from the *Arabidopsis* stock centres can be accessed electronically via the *Arabidopsis thaliana* database; http://genome-www.stanford.edu/Arabidopsis/
[b] T. Altmann, T. Mozo, and associates; Institut für Genbiologische Forschung, Berlin, Germany.

8: Map-based cloning in Arabidopsis

Protocol 2. Maintenance of YAC libraries

Equipment and reagents

- Kiwibrew: CM medium (52)–uracil, 40 μg/ml adenine (hemisulfate salt), 20 μg/ml L-arginine (HCl), 100 μg/ml L-aspartic acid, 100 μg/ml L-glutamic acid (monosodium salt), 20 μg/ml L-histidine, 60 μg/ml L-leucine, 30 μg/ml L-lysine (mono-HCl), 20 μg/ml L-methionine, 50 μg/ml L-phenylalanine, 375 μg/ml L-serine, 200 μg/ml L-threonine, 40 μg/ml L-tryptophan, 30 μg/ml L-tyrosine, 150 μg/ml L-valine, 11% (w/v) Casamino acids (Vitamin Assay, Difco)
- 96-prong replicating device
- 96-well microtitre dishes
- Kiwibrew agar plates: Kiwibrew, 2% (w/v) Difco agar, one NaOH pellet per litre
- 50% (v/v) glycerol, autoclaved

Method

1. For long-term storage, replicate the library into 96-well microtitre dishes containing 100 μl Kiwibrew.

   (a) Use a 96-prong replicating device to replicate yeast colonies growing on agar plates to the microtitre dishes.

   (b) Sterilize the prongs of the replicator in between each replication, by dipping the tips of prongs in water, stamping the prongs onto tissue paper, then flaming them in ethanol three times and allowing to cool.

2. Wrap the dishes in Clingfilm and grow the yeast cultures at 30°C for one to two days.

3. Add glycerol to a final concentration of 20% (v/v) and freeze cells at −70°C.

4. Maintain working stocks on Kiwibrew agar plates at 4°C. Replate working stocks of the libraries onto fresh Kiwibrew plates every three months using a 96-prong replicator.

Protocol 3. Screening YAC libraries by yeast colony hybridization according to the method of Coulson *et al.* (53) and modified by Schmidt *et al.* (47)

Equipment and reagents

- 30°C incubator
- 80°C oven
- 96-prong replicator
- Autoclaved filter paper (3MM): cut 8 × 11.4 cm and 8.5 × 11.5 cm
- Glass plate
- Hybond N nylon membrane (8 × 11.4 cm pieces; Amersham)
- Kiwibrew medium (see *Protocol 2*)
- Kiwibrew agar plates (see *Protocol 2*) plus 2% (w/v) Bacto agar
- SOE: 1 M sorbitol, 20 mM EDTA, 10 mM Tris–HCl pH 8.0
- SOE/DTT: 0.8% (w/v) DTT in SOE
- SOEM: 1% (v/v) 2-mercaptoethanol in SOE
- 100% ethanol

Protocol 3. *Continued*

- 10% (w/v) SDS
- 1 mg/ml Novozyme 234 (Novo Enzyme Products)
- Denaturation solution: 0.5 M NaOH, 1.5 M NaCl
- Neutralizing solution: 0.5 M Tris–HCl pH 7.2, 1.5 M NaCl, 0.001 M EDTA
- 20 × SSPE (stock): 3.0 M NaCl, 0.2 M NaH$_2$PO$_4$, 0.02 M Na$_2$EDTA pH 7.7
- 2 × SSPE/0.1% (w/v) SDS
- 1 × SSPE/0.1% (w/v) SDS
- 0.1 × SSPE/0.1% (w/v) SDS
- 50 × Denhardt's solution: 1% (w/v) Ficoll 400, 1% (w/v) polyvinyl pyrrolidone, 1% (w/v) BSA; store at −20°C
- Pre-hybridization solution: 5 × SSPE, 0.5% (w/v) SDS, 5 × Denhardt's solution in 25 ml, add 50 μl of 10 mg/ml sonicated denatured salmon sperm DNA

A. *Preparation of yeast colony filters*

1. Replicate all the colonies from the YAC library onto master agar plates.

    (a) Use a 96-prong replicator to replicate the colonies from plate one of the YAC library to the Kiwibrew agar master plate.

    (b) Replicate the colonies from the second plate, next to the first set of colonies on the master plate.

    (c) Offset the colonies from the third plate onto the master plate, and so on, until all the colonies from 8–24 plates have been transferred onto the master plate(s).

    (d) Generate master plates carrying two different arrays of all the colonies in the YAC library.

2. Grow the colonies overnight at 30°C.

3. Take a filter copy (master filter) of the colonies on the master plate. Place a marked 8 × 11.4 cm master filter (Hybond N, Amersham) onto the yeast colonies on the master plate. Leave the master filter on the plate for a few seconds, and then layer the filter onto a new Kiwibrew agar plate, colony side up. Incubate at 30°C for 2 h.

4. Make multiple copies of the master filter.

    (a) Wash a glass plate thoroughly with water and 100% ethanol and put a dry, autoclaved 3MM filter paper (8 × 11.4 cm) onto it. Wet a labelled Hybond N filter (8 × 11.4 cm) by placing it for a few minutes on a dry agar plate. Transfer the master filter onto the 3MM paper (colony side up), and place the wet side of the new filter carefully on the colonies of the master filter. Put two sterilized 3MM filter papers and finally a glass plate on top of the stack and apply brief hand pressure. Remove the glass plate and filter papers and put the copy filter back on its plate (colony side up).

    (b) Use the master filters up to five times to produce filter copies 'slaves', then place the master back onto its agar plate to allow it to rehydrate. The master filter can be used again (at least 80–100 times in total). Exchange the 3MM paper after four or five copies have been made.

8: Map-based cloning in Arabidopsis

5. Grow the colonies on the copy filter for one to two days at 30 °C to obtain small colonies.
6. Process the yeast colony filters.
   (a) Place the filters for 5 min on 3MM filter paper soaked with SOE/DTT (colony side up).
   (b) Wipe out the growth plates, and place 3MM filter paper sheets (8.5 × 11.5 cm) in them. Soak each filter paper with 4.5 ml of SOEM containing 1 mg/ml Novozyme 234. Place each slave (colony side up) carefully on a filter paper, avoiding air bubbles. Seal the plates in plastic bags and incubate overnight at 37 °C. Ensure that no liquid drops come in contact with the colonies.
   (c) Place filters sequentially on sheets of 3MM filter papers in trays soaked with the following solutions ensuring that no air bubbles are trapped under the slave filters:
      - 10% (w/v) SDS for 5 min
      - denaturing solution for 5 min (twice)
      - neutralizing solution for 5 min (twice)
      - 2 × SSPE for 5 min
7. Dry the filters at room temperature and bake for 2 h at 80 °C. Dry filters can then be stored at room temperature.

B. Hybridization to yeast colony filters

1. Radiolabel DNA fragments, cosmid, and lambda DNA using random primer extension (54).
2. Place the filters in a small box and cover with 25 ml (per six filters) of pre-hybridization solution. Pre-hybridize the filters at 65 °C for 1 h.
3. Denature the probe, add to the filters, and incubate overnight at 65 °C for at least 12 h.
4. Wash the filters in 2 × SSPE, 0.1% (w/v) SDS at room temperature for 10 min.
5. Repeat.
6. Replace the solution with 1 × SSPE, 0.1% SDS. Incubate at 65 °C for 15 min.
7. Replace the solution with 0.1 × SSPE, 0.1% SDS. Incubate at 65 °C for 10 min.
8. Repeat step 7 if necessary, then place the filters in a plastic bag and carry out autoradiography.
9. Score the coordinates of the hybridizing YAC colonies. If low levels of background hybridization make this difficult, then wash less stringently, or hybridize ^{35}S-radioabelled pBR322 probe to the filters.

The colony hybridization results are then confirmed by streaking out the positively hybridizing YAC colonies and rescreening single colonies. These are picked and intact YAC DNA is isolated and separated from the endogenous yeast chromosomes by pulsed-field gel electrophoresis (*Protocol 4*). Analysis of the YAC clones in this way confirms that the yeast colony contains a single YAC that hybridizes to the RFLP probe and allows the size of the insert DNA in the YAC to be determined. The degree of overlap of the YACs with the *Arabidopsis* DNA inserted within the RFLP marker is then determined. This is achieved by comparing the pattern of restriction fragments in the YAC and RFLP marker DNA that hybridize to radiolabelled marker DNA. The YACs can then be arranged relative to one another and the RFLP marker, into a YAC island. This process also reveals any YAC clones identified due to cross-hybridization between the RFLP marker and unlinked DNA from another part of the genome. Isolation of YAC DNA for routine restriction enzyme cleavage is described in *Protocol 5*.

Protocol 4. Yeast chromosome preparation and fractionation by pulsed-field gel electrophoresis

Equipment and reagents

- Pulsaphor electrophoresis system (LKB) or equivalent pulsed-field gel electrophoresis equipment
- Plug moulds
- Kiwibrew (see *Protocol 2*)
- SCE: 1 M sorbitol, 0.1 M sodium citrate, 50 mM EDTA pH 7
- SCEM: SCE, 30 mM 2-mercaptoethanol
- Novozyme: 100 mg/ml Novozyme 234 (Novo Enzyme Products) in SCE; filter sterilized before use
- LMP sorbitol agarose: 1% (w/v) InCert agarose (FMC) in 1 M sorbitol
- 1 M sorbitol
- 10 × TBE: 108 g Tris base, 55 g boric acid, 9.3 g Na$_2$EDTA; make up to 1 litre with deionized water
- LMP TBE agarose: 1% InCert (w/v) agarose (FMC) in 0.5 × TBE
- Proteinase K solution: 1 mg/ml proteinase K (Boehringer Mannheim) in 100 mM EDTA pH 8, 1% (w/v) Sarkosyl
- Lambda DNA (Gibco BRL)
- TE: 10 mM Tris–HCl pH 8.0, 1 mM EDTA
- 0.5 M EDTA pH 8.0

Method

1. Inoculate 5 ml of Kiwibrew with a single yeast colony and grow at 30 °C for 24 h.
2. Pellet cells by centrifugation for 10 min at 2000 *g* and wash cells by resuspending in 1 ml of SCE.
3. Transfer cells to a microcentrifuge tube and pellet by centrifugation at 14 000 *g* for 1 min.
4. Resuspend 50 μl of packed cell volume of yeast cells in 300 μl SCEM.
5. Add 30 μl of Novozyme and incubate for 1 h at room temperature to spheroplast the cells.
6. Pellet the spheroplasts for 30 sec in a microcentrifuge at 2500 *g* and increase the volume of the spheroplasts to 50 μl with 1 M sorbitol.

8: Map-based cloning in Arabidopsis

7. Add 80 μl of 50°C molten LMP sorbitol agarose.
8. Vortex the mixture briefly and pipette into plug moulds.
9. Once set, place the plugs into microcentrifuge tubes and add 1 ml of proteinase K solution. Incubate for 4 h at 50°C. Replace with the same solution and incubate plugs at 50°C overnight.
10. Wash plugs with TE for 30 min, repeat twice more.
11. Wash plugs with 0.5 × TBE, repeat once more.
12. Generate size markers of concatamers of lambda DNA (55).
    (a) Pipette the solution of lambda DNA gently into a microcentrifuge tube and heat to 70°C for 5 min.
    (b) Add DNA to ice-cold TE to a final concentration of 0.1 or 0.2 mg/ml.
    (c) Store on ice for 5 min and then warm the solution to room temperature.
    (d) Add an equal volume of 50°C 1% (w/v) LMP molten TBE agarose and pipette the solution into plug formers.
    (e) Place plugs in 5 vol. of 0.5 M EDTA pH 8.0 and store at 4°C overnight.
    (f) Equilibrate plugs in 0.5 × TBE, by soaking in 0.5 × TBE twice for 30 min each. Store in 0.5 × TBE at 4°C.
12. Carry out PFGE using a Pulsaphor system (LKB) or equivalent pulsed-field gel equipment.
    (a) Load one-third of a plug onto a 1% (w/v) agarose gel made up in 0.5 × TBE.
    (b) Load one-sixth of a plug of lambda DNA concatamer size markers in the first and last lane of the gel.
    (c) Electrophorese in 0.5 × TBE at 170 V, 20 sec pulse time for 36 h at 4°C.
13. Visualize DNA by staining with ethidium bromide.

Protocol 5. Preparation of yeast genomic DNA for restriction enzyme digestion and YAC end fragment isolation

Reagents
- TE/SDS: 75 mM Tris pH 7.4, 40 mM EDTA pH 8.0, 0.4% (w/v) SDS
- 4 M potassium acetate pH 4.8
- 3 M sodium acetate pH 5.2
- TE: 10 mM Tris–HCl pH 8.0, 1 mM EDTA
- 100% and 70% (v/v) ethanol
- Phenol (saturated):chloroform (1:1)
- Chloroform

Protocol 5. *Continued*

Method

1. Generate yeast spheroplasts as described in *Protocol 4*, steps 1–6.
2. Add 400 µl of TE/SDS per 50 µl packed volume of spheroplasts and lyse by incubation at 65°C for 30 min.
3. Add 100 µl of cold 4 M potassium acetate and incubate on ice for 30 min.
4. Pellet the precipitated protein and debris in a microcentrifuge at 14 000 *g* for 15 min at 4°C.
5. Transfer the supernatant to a fresh tube and ethanol precipitate the DNA by the addition of 1 ml ethanol.
6. Leave on ice for 5 min.
7. Collect the precipitate by centrifugation for 5 min in a microcentrifuge at 14 000 *g*.
8. Wash the DNA pellet in 70% (v/v) ethanol and resuspend in 500 µl TE.
9. Extract twice with phenol:chloroform, once with chloroform, and ethanol precipitate the DNA by the addition of 0.1 vol. of 3 M sodium acetate and 2.5 vol. of ethanol.
10. Resuspend the yeast genomic DNA in 50–100 µl TE.

YAC end probes are then obtained from both ends of the *Arabidopsis* DNA inserted within each YAC (see *Protocols 6* and *7*) (39). The YAC end probes are first hybridized to filters containing DNA of all YACs in the appropriate contigs or islands, to confirm their arrangement within the islands, and to test the possibility that neighbouring islands might overlap to form one contig. If the islands are not joined, then the YAC end probes, from YACs at the extremities of the islands, are used as RFLP markers to confirm that they are genetically linked to the gene.

Protocol 6. Plasmid rescue of the left end of YACs adapted from a method from the Meyerowitz laboratory

Reagents

- SOC: 2% (w/v) Bacto tryptone, 0.5% (w/v) Bacto yeast extract, 10 mM NaCl, 2.5 mM KCl, 10 mM MgCl$_2$, 10 mM MgSO$_4$, 20 mM glucose
- *Xho*I or *Nde*I and *Bam*HI
- T4 DNA ligase (Gibco BRL)
- Phenol (saturated):chloroform (1:1)
- Chloroform
- 100% and 70% (v/v) ethanol
- 3 M sodium acetate pH 5.2
- *E. coli* DH5α cells
- Agar plates plus carbenicillin (100 µg/ml)

A. *Preparation of plasmid DNA*

1. Prepare yeast DNA for restriction enzyme digestion as described in *Protocol 5*.

8: Map-based cloning in Arabidopsis

2. Digest 2 µg of total yeast DNA with either *Xho*I or *Nde*I in a digestion volume of 400 µl.
3. Extract digests once with phenol:chloroform, once with chloroform, and ethanol precipitate by the addition of 40 µl of 3 M sodium acetate and 1 ml of 100% ethanol.
4. Collect the pellet by centrifugation in a microcentrifuge at 14 000 *g* for 15 min at 40°C. Wash the pellet with 70% (v/v) ethanol, dry, and resuspend in 360 µl of distilled water.
5. Circularize DNA fragments by ligation in a volume of 400 µl overnight at 16°C in the presence of 2 U of T4 DNA ligase.
6. Heat inactivate ligase for 15 min at 68°C.
7. Ethanol precipitate the DNA as described in step 3, and resuspend finally in 10 µl of distilled water.

B. *Preparation of competent cells for electroporation (DH5α)*
1. Inoculate 1 litre of L broth with 0.01 vol. of a fresh overnight culture of DH5α.
2. Grow cells at 37°C with vigorous shaking to an OD_{600} of 0.5.
3. Chill the flask on ice for 15–30 min and pellet cells by centrifugation in a cold rotor at 4000 *g* for 15 min.
4. Resuspend pellet in 1 litre of cold sterile water.
5. Repeat the centrifugation step.
6. Resuspend in 500 ml of cold sterile water.
7. Repeat the centrifugation step.
8. Resuspend in 20 ml of 10% (v/v) glycerol.
9. Repeat the centrifugation as in step 3.
10. Resuspend to a final volume of 2–3 ml in 10% (v/v) glycerol.
11. Aliquot 40 µl of cell suspension per microcentrifuge tube, freeze in liquid N_2, and store cells at −70°C.

C. *Electroporation using a Bio-Rad electroporator and analysis of transformants*
1. Pre-cool cuvettes and cuvette holder on ice.
2. Set Gene Pulser at 2.5 kV (max), 25 µF (max).
3. Set pulse controller at 200 Ω.
4. Remove competent cells (DH5α), prepared as above, from freezer and thaw on ice.
5. Pipette 40 µl of cells onto 2 µl of ligated DNA in a microcentrifuge tube. Incubate on ice for 30–60 sec.
6. Transfer cells to ice-cold cuvettes, tap cuvettes so that the cells are on the bottom, and pulse.
7. Immediately add 1 ml of SOC medium.

Protocol 6. *Continued*

8. Transfer the cell suspension to 17 × 100 mm polypropylene tubes.
9. Shake at 37°C for 1 h.
10. Transfer cells to a microcentrifuge tube, centrifuge at 2500 *g* for 1 min. Remove 800 μl of the supernatant and resuspend cells in the remaining supernatant.
11. Plate approx. half of the cells on a carbenicillin plate (100 μg/ml).
12. Prepare plasmid DNA from three or four clones for each rescue.
13. Release inserts of rescued plasmids.

    (a) Release inserts of rescued plasmids derived from EG and abi YAC library clones by *Bam*HI/*Xho*I or *Bam*HI/*Nde*I double digests.

    (b) Release inserts of rescued plasmids derived from yUP clones by *Eco*RI/*Xho*I or *Eco*RI/*Nde*I double digests.

    (c) To release inserts of rescued plasmids derived from EW or S YAC libraries (where the cloning site has been lost), try *Eco*RI/*Nde*I or *Eco*RI/*Xho*I double digests as *Eco*RI does not cleave the vector. The size of the fragment containing the vector will be greater than 5.5 kb. If other enzymes are used to release the insert, then check for vector containing fragments by Southern hybridization analysis.

    (d) Use PCR primers designed to flanking vector sequences to amplify the insert DNA in rescued plasmids as an alternative method (see *Protocol 7*).

Protocol 7. Amplification by IPCR of YAC end fragments for use as radiolabelled probes

Equipment and reagents

- Thermocycler
- LMP agarose
- Random Prime Labelling Kit (e.g. Pharmacia Ready To Go)
- 10 × PCR buffer: 100 mM Tris–HCl pH 8.3, 500 mM KCl, 20 mM MgCl$_2$, 0.1% (w/v) gelatine, autoclaved; then add Tween 20 to 0.05% (v/v) and NP-40 to 0.05% (v/v)
- dNTP mix: 5 mM of each dNTP
- *Taq* DNA polymerase
- T4 DNA ligase (Gibco BRL)
- *Alu*I, *Hae*III, *Eco*RV, *Hinc*II, *Fok*I
- Phenol (saturated):chloroform (1:1)
- 100% and 70% (v/v) ethanol
- 3 M sodium acetate pH 5.2

- IPCR primers:
  C69 5' ctgggaagtgaatggagacata 3'
  C70 5' aggagtcgcataagggagag 3'
  C71 5' agagccttcaacccagtcag 3'
  C77 5' gtgataaactaccgcattaaagc 3'
  C78 5' gcgatgctgtcggaatggac 3'
  D71 5' tcctgctcgcttcgctactt 3'
  JP1 5' aagtactctcggtagccaag 3'
  JP4 5' ttcaagctctacgccgga 3'
  JP5 5' gtgtggtcgccatgatcgcg 3'
  F2 5' acgtcggatgctcactataggatc 3'
  F3 5' gacgtggatgctcactaaagggatc 3'
  F6 5' acgtcggatgactttaatttatcacta 3'
  F7 5' acgtcggatgccgatctcaagatta 3'

8: Map-based cloning in Arabidopsis

Method

1. Prepare yeast genomic DNA as described in *Protocol 5*.
2. Digest 100 ng of yeast genomic DNA with *Alu*I, *Hae*III, *Eco*RV, or *Hinc*II.
3. Extract the digests once with phenol:chloroform, then precipitate with ethanol by the addition of 0.1 vol. of 3 M sodium acetate and 2.5 vol. of 100% ethanol.
4. Leave at −70°C for 30 min. Collect the DNA precipitate by centrifugation at 14 000 *g* for 15 min at 4°C.
5. Wash the pellet with 70% (v/v) ethanol, dry, and resuspend in 50 μl of sterile distilled water.
6. Circularize the DNA fragments by ligation in a volume of 100 μl overnight at 16°C in the presence of 2 U of T4 DNA ligase.
7. Heat kill the ligase by incubation of the ligation reaction at 68°C for 10 min.
8. Carry out IPCR reactions.
    (a) Set up IPCR in a volume of 100 μl containing 10 μl of the ligation mixture as template DNA, 10 μl of 10 × PCR buffer, 4 μl dNTP mix, 1 μl of each of the two primers (100 ng/μl), and 1 U of *Taq* DNA polymerase (0.3 μl of 5 U/μl *Taq* DNA polymerase), 83.7 μl water.
    (b) Primer pairs used for IPCR are the C and D primers as previously described (40) and the JP series (from M. Hirst, IMM Molecular Genetics Group, Oxford).
    - For left end IPCR:-
    for templates originally cleaved with *Alu*I, *Eco*RV: D71 and C78.
    for templates originally cleaved with *Hae*III: JP1 and JP5.
    - For right end IPCR:-
    for templates originally cleaved with *Alu*I, *Hinc*II: C69 and C70.
    for templates originally cleaved with *Hae*III: C69 and JP4.
    (c) Overlay with mineral oil as required.
    (d) PCR using the following conditions: 3 min at 95°C, then 35 cycles consisting of 1 min at 94°C, 1 min at 56°C, and 2 min at 72°C, with a final incubation at 72°C for 10 min.
9. Check aliquots of the IPCR reaction on a 1.5% (w/v) agarose gel.
10. Reamplify 1 μl of the IPCR reaction as described in step 8a by nested PCR using the F primer series as recommended by I. Hwang (MGH, Boston).
    (a) The F primers anneal very near the cloning site and therefore reduce the amount of vector sequence present in the PCR product. In addition, they introduce a *Fok*I site very close to the destroyed cloning site of EW and S YACs.

Protocol 7. *Continued*

(b) PCR conditions: 95°C for 3 min, then 30 cycles of 1 min at 94°C, 1 min at 45°C, and 3 min at 72°C.

(c) Primers used for reamplification of left end IPCR products.
- EG, abi, and S YACs:-

for templates originally cleaved with *Alu*I:	F2 and C77.
for templates originally cleaved with *Hae*III:	F2 and JP5.
for templates originally cleaved with *Eco*RV:	F2 and C78.

- EW, yUP, and CIC YACs:-

for templates originally cleaved with *Alu*I:	F6 and C77.
for templates originally cleaved with *Hae*III:	F6 and JP5.
for templates originally cleaved with *Eco*RV:	F6 and C78.

(d) Primers used for reamplification of right end IPCR products.
- EG, abi, and S YACs:-

for templates originally cleaved with *Alu*I:	F3 and C71.
for templates originally cleaved with *Hae*III:	F3 and JP4.
for templates originally cleaved with *Hinc*II:	F3 and C70.

- EW, yUP, and CIC YACs:-

for templates originally cleaved with *Alu*I:	F7 and C77.
for templates originally cleaved with *Hae*III:	F7 and JP4.
for templates originally cleaved with *Hinc*II:	F7 and C70.

11. Purify the resulting PCR product away from vector sequences by cleaving with the enzyme originally used in the digestion together with *Bam*HI (EG and abi YACs) or *Eco*RI (yUP, CIC YACs) followed by electrophoresis on 1% (w/v) LMP agarose gels.
12. Cut out the gel strip containing the PCR product.
13. Radiolabel the YAC end probe using random priming in molten agarose.
14. To cleave off the remaining vector sequence, add *Fok*I and incubate the labelling reaction for 1 h at 37°C.
15. Process the probe and add to the hybridization reaction.

To seal the gaps between the islands, YAC end probes that map closest to the gene can be used to find new YACs in the libraries, so extending the YAC islands by chromosome walking. The chromosome walk to the gene may also be monitored directly by assessing the physical distance remaining between the islands. Megabase-sized plant DNA is isolated and cleaved with rare cutting restriction enzymes (40) (see Chapter 9). The nearest flanking DNA markers are radiolabelled and hybridized to plant DNA. An estimate of physical distance between the DNA markers can then be made that is based on the size of the smallest DNA fragment that hybridizes to both of the probes. If this analysis indicates that the physical distance between the probes

8: Map-based cloning in Arabidopsis

remains large, then it might be an advantage to try and isolate more markers that are closer to the gene, and then resume walking from these DNA markers.

Isolation of a DNA contig containing the gene can be time-consuming and complicated if a large number of chromosome walking steps are required. We found for example, that 40% of the *Arabidopsis* YAC clones analysed during our chromosome walk to the flowering time gene *CO* (9) were problematic in walking experiments, as they appeared to harbour deletions, or were chimeric between unlinked unique or repetitive sequences. Co-linearity of YAC inserts with the plant genome can be assessed directly. This can be done by cleaving YAC and plant DNA with rare cutting restriction enzymes and comparing the patterns of fragments after hybridization to gel purified, radiolabelled YAC DNA (40). The alternative is to build a contig that throughout its length contains each region cloned in at least three YACs. This redundancy helps to overcome problems caused by individual YAC clones. It is also often an advantage to screen more than one YAC library for clones carrying DNA from the region, as particular regions of DNA are not always evenly represented in the libraries (41).

While the YAC contig is being completed new RFLP markers that become available during the chromosome walking can be used to analyse recombinants with cross-overs in the vicinity of the gene, and therefore to position the gene on the YAC contig. Another advantage, therefore, of a YAC contig with redundancy, is that the YAC end probes, or the larger cosmid or lambda genomic clones they hybridize to, are a valuable source of RFLP markers for fine-mapping the gene. Mapping also confirms that the YAC end probes are reliable and the YAC is not likely to be chimeric.

4. Identification of the gene

4.1 Location of the gene by molecular complementation

Once the gene has been positioned to as small a region of the physical map as possible, then experiments can be initiated to identify the gene by molecular complementation. In the case of a recessive mutation, clones carrying wild-type DNA are introduced into the plants homozygous for the mutant allele. To identify genes affected by dominant, gain-of-function mutations, clones carrying DNA from mutant plants are introduced into wild-type and the resulting transgenics assessed for the mutant phenotype (19–21). If the mutant is difficult to transform, complementing clones can be introduced into wild-type plants and complementation is demonstrated by analysis of the progeny from a cross between these transgenics and the mutant (16).

Complementation tests are carried out by cloning DNA containing the gene into plant transformation vectors, introducing these into plants by *Agrobacterium*-mediated transfer (see also Chapter 6), and analysing the phenotype of the resulting transgenic plants. Candidate cDNAs have been fused to a constitutive plant promoter such as the CaMV 35S promoter and then used directly to complement the mutant phenotype (6, 17). Alternatively, the cDNA clones can be used to identify genomic clones which are then tested for their ability to complement the mutant phenotype.

Complementation with genomic DNA first requires isolation of cosmids containing the gene. Genomic libraries in plant transformation vectors can be constructed by either subcloning a BAC (*Protocols 8* and *9*) or a purified YAC containing the gene, or by making a genomic library from the yeast strain that contains the YAC (42) (see also Chapter 9). The alternative, is to use YAC end probes or YAC subclones to identify genomic clones in an *Arabidopsis* genomic library that has been constructed in a plant transformation vector (e.g. 43). Complementation experiments, up until recently at least, have involved time-consuming and labour-intensive tissue culture procedures. Therefore, we initiated complementation experiments to identify the *CO* gene once we had narrowed the location of the gene down to a 48 kb region of genomic DNA (22). Three overlapping cosmids spanning this region were introduced into mutant plants. Each cosmid in our contig overlapped by approximately 6 kb with the next one in the sequence allowing, at best, the position of the gene to be narrowed down to the 6 kb region of DNA. Whatever strategy is taken, the contig should have enough redundancy to try and ensure that the gene is contained on at least one of the cosmids, and not split between two of them. Also, to reduce the possibility that the gene extends beyond the end of the cosmid contig, the sites of the RFLPs used for positioning the mutation on the contig should not be located at the extreme ends of the contig.

Protocol 8. Preparation of BAC DNA

Equipment and reagents

- Miracloth
- Solution 1: 0.9% (w/v) glucose, 25 mM Tris pH 8, 10 mM EDTA (optional 5 mg/ml lysozyme)
- Solution II: freshly prepared 1% (w/v) SDS, 0.2 M NaOH
- Solution III: 3 M potassium acetate made up in 2 M acetic acid
- TE: 10 mM Tris pH 8.0, 1 mM EDTA
- LB: 1% (w/v) Bacto tryptone, 0.5% (w/v) yeast extract, 1% (w/v) NaCl pH 7.0 (containing appropriate selective antibiotics)
- 4 M lithium acetate
- 3 M sodium acetate
- 100% ethanol
- Isopropanol
- Phenol:chloroform (1:1)
- RNase A (10 mg/ml)

Method

1. Inoculate 25 ml of LB containing 12.5 µg/ml chloramphenicol (TAMU BACs) or 30 µg/ml kanamycin (IGF BACs) with a single colony and grow at 37 °C for 14–20 h.
2. Pellet cells by centrifugation for 15 min at 4000 *g*.

3. Resuspend the cells in 2 ml of solution I and leave for 5 min at room temperature.
4. Add 4 ml of solution II. Mix by gentle inversion. The solution should clear. Incubate for 5 min on ice.
5. Add 3 ml of solution III and invert gently to mix. A white precipitate should form. Freeze at −80°C for 10–15 min, then thaw at room temperature.
6. Centrifuge for 15 min at 4000 g at room temperature.
7. Gently pour the supernatant through miracloth to remove the floating debris. Collect the filtrate into a fresh tube and precipitate the DNA by adding 6 ml of ice-cold isopropanol. Freeze at −80°C for 10–15 min. Warm to room temperature then centrifuge for 20 min at 4°C at 4000 g to pellet the DNA.
8. Pour off the supernatant and invert the tubes to drain. Dry the sides of the tubes with a paper towel. Add 2 ml of TE and resuspend the DNA.
9. Add 2 ml of 4 M LiAc, mix, place on ice for at least 20 min, then centrifuge at 2000 g for 15 min.
10. Transfer the supernatant to a fresh tube, and precipitate the DNA by the addition of 0.4 ml of 3 M NaOAc and 9 ml of 100% ethanol. Mix and place on ice for at least 20 min.
11. Centrifuge at 4000 g for 15 min. Pour off the supernatant, invert to drain, dry the sides of the tubes, and resuspend in 0.4 ml of TE.
12. Transfer to a 1.5 ml microcentrifuge tube, add 40 µl of 3 M NaOAc and 0.4 ml of phenol:chloroform. Shake to mix, centrifuge at 14000 g for 2 min to resolve the phases.
13. Transfer the aqueous phase to a fresh tube and add 0.9 ml of 100% ethanol. Place on ice for 5 min, centrifuge at 14000 g for 5 min, remove supernatant, dry pellet, and resuspend in 50 µl TE with 0.5 µl of 10 mg/ml RNase.

Protocol 9. Subcloning BAC clones into cosmid vector 04541[a]

Equipment and reagents

- 65°C incubator
- Gigapack XL or Gold extract (Stratagene)
- Agarose gel electrophoresis equipment
- Low melting point agarose
- 10 × TBE (see *Protocol 4*)
- Lambda DNA/*Hind*III marker
- Agarase (New England Biolabs)
- 2 × agarase buffer: 20 mM Tris pH 7.0, 2 mM EDTA, 400 µg/ml BSA
- Phenol:chloroform (1:1)
- 3 M sodium acetate pH 5.2
- 100% ethanol
- 0.5 M EDTA pH 8.0
- Calf intestinal alkaline phosphatase (CIAP)
- *Taq*I, *Cla*I, acetylated BSA (New England Biolabs)
- TE: 10 mM Tris–HCl pH 8.0, 1 mM EDTA
- High concentration ligase (2000 U/µl) (New England Biolabs)

Protocol 9. *Continued*

Method

1. Digest 20 μg of 04541 DNA with *Cla*I.
2. Add 1 μl (10 U) of CIAP and incubate at 37°C for 30 min.
3. Add 0.5 M EDTA pH 8.0 to a final concentration of 25 mM and incubate at 65°C for 20 min.
4. Extract the DNA with phenol:chloroform and precipitate the DNA by adding 0.1 vol. of 3 M sodium acetate, and 2 vol. of 100% ethanol.
5. Dissolve the DNA in 20 μl of water.
6. Prepare 180 μl of a pre-digestion mix containing 5–10 μg of BAC DNA, 0.1 mg/ml acetylated BSA, and 1 × *Taq*I restriction buffer.
7. Prepare ten dilutions of *Taq*I in 1 × buffer on ice. Dilutions should be in the range of 1–0.001 U/μl.
8. Set up a small scale digestion reactions by adding 18 μl of the pre-digestion mix and 2 μl of each enzyme dilution to individual tubes.
9. Incubate the digestion reactions at 65°C for 30 min.
10. Cool on ice and add 2 μl of 0.5 M EDTA pH 8.0.
11. Load 10 μl of each reaction onto a 0.6% (w/v) agarose gel in 0.5 × TBE buffer. Run the gel slowly along with lambda DNA/*Hin*dIII marker, to determine which digestion produces the maximum yield of DNA of the desired size (18–25 kb).
12. Using the optimized conditions determined above, perform large scale digestion reactions with (5–10 μg of BAC DNA), keeping the same DNA concentration.
13. Analyse a small aliquot of the digestion reaction by electrophoresis to confirm that the digestion is adequate.
14. If the size of the digested BAC DNA is in the desired range, load the rest of the reaction onto a 0.8% (w/v) low melting point agarose gel.
15. Run the gel slowly and for a short distance.[b]
16. Excise a slice of gel containing the band of DNA of the appropriate size.
17. Wash the gel slice in ten volumes of TE.
18. Remove the buffer, transfer the gel slice to a 1.5 ml centrifuge tube, and incubate it at 65°C for 10 min.
19. Add to the melted gel 1 vol. of 2 × agarase buffer pre-warmed to 65°C, and mix. Incubate at 65°C for 10 min.
20. Transfer the tube to 37°C and allow to equilibrate for 10 min. Add 1 U of agarase per 150 μl of gel slice, and incubate at 37°C overnight.
21. Centrifuge briefly and incubate at 65°C for 20 min.

8: Map-based cloning in Arabidopsis

22. Equilibrate to 37 °C, add 0.5 U of agarase per 150 µl of original gel slice, and incubate at 37 °C for 4 h.
23. Heat to 65 °C and add 0.1 vol. of 3 M sodium acetate pH 5.2. Extract twice with 1 vol. of phenol:chloroform and precipitate DNA with 2.5 vol. of ethanol.
24. Dissolve the BAC DNA in TE, and determine its concentration by fluorimetry or $A_{260/280\ nm}$.
25. Set up ligation tubes adding 1 µl (1 µg) of dephosphorylated vector DNA, and 2 µl of 3 M sodium acetate pH 5.2. To that, add partially digested BAC DNA and water to a final volume of 20 µl. Use at least two different ratios of insert to vector DNA (e.g. 1:1 M ratio or 3:1 M ratio). Precipitate the DNA by adding 50 µl of 100% ethanol.
26. Dissolve the DNA pellet in 2 µl of water. Add 1 µl of 3 × ligation buffer and 0.2 µl of high concentration of T4 DNA ligase (2000 U/µl).
27. Incubate at 4 °C for three to four days.
28. Package using Stratagene Gigapack XL or Gold extract following the manufacturer's instructions. Use *E. coli* Sure (tetracycline sensitive) or XL1-Blue MR as a host bacterial strain.
29. Plate the cosmid library on selective medium (LB plus 10 µg/ml tetracycline). Incubate overnight at 37 °C.
30. Pick colonies into 384-well microtitre plates containing 50 µl of LB/tetracycline/glycerol per well. Incubate again overnight at 37 °C.
31. Make plates for colony lifts using a 384-prong replicator. Master glycerol plates can then be frozen on dry ice, and stored at −80 °C.

[a] Cosmid 04541 was modified from SLJ1711 (56) by insertion of a cos site by Clare Lister (Dean laboratory, John Innes, Norwich).
[b] Do not allow bromophenol blue dye to migrate more than 8 cm.

Agrobacterium-mediated gene transfer into *Arabidopsis* is described in Chapter 6 of this book. We found that the Valvekens procedure involving transformation of root explants was satisfactory (44). Although this method has been largely replaced by the use of vacuum infiltration transformation methods, it may still be useful for transformation of certain mutants such as those with reduced stature or fertility that can be difficult to transform by vacuum infiltration. We obtained five to ten independent transformants for each of the three cosmids and two to three clones of each transformant to allow for plant death in tissue culture or poor seed set of the primary transformants. Complementation of some mutant phenotypes have been tested in the primary transformants (17). Other mutant phenotypes are semi-dominant, or sensitive to environmental cues and affected by tissue culture conditions. In these cases the primary regenerants (T1 generation) are allowed to self-fertilize and complementation phenotypes are analysed in the T2 generation. T2 plants are plated on medium containing selective agents (e.g. kanamycin) to confirm that they carry the T-DNA. T2 families that segregate kanamycin-resistant individuals at a ratio of three kanamycin-resistant : one kanamycin-sensitive, suggesting that there is a single locus of insertion of the T-DNA, are then analysed for complementation phenotypes. Linkage tests are carried out to verify that the complementation phenotype always co-segregates with the kanamycin resistance phenotype associated with the T-DNA.

4.2 Determining the structure of the gene

The easiest way to determine the structure of the complementing gene is to isolate a cDNA clone that maps to the complementing region. The relative abundance of such cDNA clones in a library from a particular tissue may provide an indication of the most likely candidate for the gene (17). The cDNA clones that map to the region can also be used as probes in Northern or in *in situ* hybridization analysis (see Chapter 4) to identify those that are expressed in a manner that is consistent with the predicted expression pattern of the gene, or those that identify changes in mRNA transcripts in the mutant plants.

In the absence of a candidate cDNA, the DNA sequence of the complementing region is determined to try and define the coding region of the gene. The sequence is analysed for protein open reading frames (ORFs) and putative exons, and assessed for homology with other genes in the databases, at the protein and nucleotide sequence level, using analysis programs such as *BLAST* and *TFASTA* (see Chapter 10). Once a potential coding region has been identified, a number of tests can be carried out to finally prove that the correct gene has been identified. Further complementation analysis with constructs carrying the gene can be performed. Also, the DNA sequence can be determined of the equivalent region from plants that are homozygous for the mutant alleles. A copy of the mutant allele is amplified from DNA derived from mutant plants by PCR and sequenced directly. After identifying candidate ORFs for the gene in the genomic sequence, the expression of the gene can be determined by RT-PCR (45).

5. Perspectives

In most of the examples of map-based cloning cited above, a great deal of effort was expended using YAC clones to construct a physical map of the region of interest. An international collaboration has been initiated to construct a physical map based on YACs of the entire *Arabidopsis* genome (46–48), and it is likely therefore that in the near future individual scientists will not need to perform chromosome walking experiments themselves.

Furthermore, the *Arabidopsis* genome initiative plans to complete the sequence of the *Arabidopsis* genome in the year 2000. The availability of *Arabidopsis* genomic sequence (e.g. 49) enables instant identification of candidate genes in a specific region. It also allows new DNA mapping markers to be more easily designed to the region of interest. Larger numbers of markers will also be located on the *Arabidopsis* genetic maps, increasing the likelihood that one will be positioned close to the gene of interest, and the integration of these onto a single mapping population of recombinant inbreds will ensure that many of the markers are accurately mapped relative to one another (15, and Chapter 3). The development of easier and faster transformation protocols that do not require regeneration of plants from tissue explants (50) are simplifying the complementation steps.

However, the requirement for extremely accurate genetic mapping of the gene of interest will remain and the construction of large numbers of chromosomes with cross-overs in the vicinity of the gene of interest will probably become the rate-limiting step in map-based cloning strategies.

References

1. Walbot, V. (1992). *Annu. Rev. Plant Physiol. Mol. Biol.*, **43**, 49.
2. Bancroft, I., Jones, J., and Dean, C. (1993). *Plant Cell*, **5**, 631.
3. Aarts, M. G. M., Dirkse, W. G., Stiekema, W. J., and Pereira, A. (1993). *Nature*, **363**, 715.
4. Long, D., Martin, M., Sundberg, E., Swinburne, J., Puangsomlee, P., and Coupland, G. (1993). *Proc. Natl. Acad. Sci. USA*, **90**, 10370.
5. Chuck, G., Robbins, T., Nijjar, C., Ralston, E., Courtney-Gutterson, N., and Dooner, H. (1993). *Plant Cell*, **90**, 371.
6. Martin, G. B., Brommonschenkel, S. H., Chunwongse, J., Frary, A., Ganal, M. W., Spivey, R., et al. (1993). *Science*, **262**, 1432.
7. Leutweiler, L. S., Hough-Evans, B. R., and Meyerowitz, E. M. (1984). *Mol. Gen. Genet.*, **194**, 15.
8. Argumanathan, K. and Earle, E. D. (1991). *Plant Mol. Biol. Rep.*, **9**, 208.
9. Putterill, J., Robson, F., Lee, K., and Coupland, G. (1993). *Mol. Gen. Genet.*, **239**, 145.
10. Pepper, A., Delaney, T., Washburn, T., Poole, D., and Chory, J. (1994). *Cell*, **78**, 109.
11. Meyerowitz, E. (1992). In *Methods in Arabidopsis research* (ed. C. Koncz, N.-H. Chua, and J. Schell), pp. 290–309. World Scientific, Singapore.
12. Gibson, S. and Somerville, C. (1992). In *Methods in Arabidopsis research* (ed. C. Koncz, N.-H. Chua, and J. Schell), pp. 119–43. World Scientific, Singapore.
13. Chang, C., Bowman, J. L., DeJohn, A. W., Lander, E. S., and Meyerowitz, E. M. (1988). *Proc. Natl. Acad. Sci. USA*, **85**, 9856.
14. Nam, H. G., Giraudat, J., den Boer, B., Moonan, F., Loos, W., Hauge, B., et al. (1989). *Plant Cell*, **1**, 699.
15. Lister, C. and Dean, C. (1993). *Plant J.*, **4**, 745.
16. Giraudat, J., Hauge, B. M., Valon, C., Smalle, J., Parcy, F., and Goodman, H. M. (1992). *Plant Cell*, **4**, 1251.

17. Arondel, V., Lemieux, B., Hwang, I., Gibson, S., Goodman, H. M., and Somerville, C. R. (1992). *Science*, **258**, 1353.
18. Leyser, H. M. O., Lincoln, C. A., Timpte, C., Lammer, D., Turner, J., and Estelle, M. (1993). *Nature*, **364**, 161.
19. Chang, C., Kwok, S., Bleeker, A., and Meyerowitz, E. (1993). *Science*, **262**, 539.
20. Leung, J., Bouvier-Durand, M., Morris, P.-C., Guerrier, D., Chefdor, F., and Giraudat, J. (1994). *Science*, **264**, 1448.
21. Meyer, K., Leube, M. P., and Grill, E. (1994). *Science*, **264**, 1452.
22. Putterill, J., Robson, F., Lee, K., Simon, R., and Coupland, G. (1995). *Cell*, **80**, 847.
23. Li, J. and Chory, J. (1997). *Cell*, **90**, 929.
24. Koornneef, M., Hanhart, C. J., Van Loenen-Martinel, E. P., and Van der Veen, J. H. (1987). *Arabidopsis Inf. Serv.*, **23**, 46.
25. Bancroft, I. and Dean, C. (1993). *Genetics*, **134**, 1221.
26. Fabri, C. O. and Schäffner, A. R. (1994). *Plant J.*, **5**, 149.
27. Konieczny, A. and Ausubel, F. (1993). *Plant J.*, **4**, 403.
28. Bell, C. J. and Ecker, J. R. (1994). *Genomics*, **19**, 137.
29. Michelmore, R. W., Paran, I., and Kesseli, R. V. (1991). *Proc. Natl. Acad. Sci. USA*, **88**, 9828.
30. Williams, J. G. K., Kubelik, A. R., Livak, K. J., Rafalski, J. A., and Tingey, S. V. (1990). *Nucleic Acids Res.*, **18**, 6531.
31. Vos, P., Hogers, R., Bleeker, M., Reijans, M., van de Lee, T., Hornes, M., *et al.* (1995). *Nucleic Acids Res.*, **23**, 4407.
32. Thomas, C. M., Vos, P., Zabeau, M., Jones, D. A., Norcott, K. A., Chadwick, B. P., *et al.* (1995). *Plant J.*, **8**, 785.
33. Grill, E. and Somerville, C. (1991). *Mol. Gen. Genet.*, **226**, 484.
34. Ward, E. R. and Jen, J. C. (1990). *Plant Mol. Biol.*, **14**, 561.
35. Guzman, P. and Ecker, J. R. (1988). *Nucleic Acids Res.*, **16**, 11091.
36. Creusot, F., Fouilloux, E., Dron, M., Lafleuriel, J., Picard, G., Billault, A., *et al.* (1995). *Plant J.*, **8**, 763.
37. Choi, S., Creelman, R. A., Mullet, J. E., and Wing, R. A. (1995). *Plant Mol. Biol. Rep.*, **13**, 124.
38. Liu, Y. G., Mitsukawa, N., Vazquez-Tello, A., and Whittier, R. F. (1995). *Plant J.*, **7**, 351.
39. Liu, Y.-G. and Whittier, R. F. (1995). *Genomics*, **25**, 674.
40. Bancroft, I., Westphal, L., Schmidt, R., and Dean, C. (1992). *Nucleic Acids Res.*, **20**, 6201.
41. Schmidt, R., Putterill, J., West, J., Cnops, G., Robson, F., Coupland, G., *et al.* (1994). *Plant J.*, **5**, 735.
42. Leung, J. and Giraudet, J. (1998). In *Methods in molecular biology*, Vol. 82, *Arabidopsis protocols* (ed. J. Martinez-Zapater and J. Salinas), pp. 277–303. Humana Press Inc., Totowa, NJ, USA.
43. Olszewski, N. E., Martin, F. B., and Ausubel, F. M. (1988). *Nucleic Acids Res.*, **16**, 10765.
44. Valvekens, D., van Montagu, M., and Van Lijsebettens, M. (1988). *Proc. Natl. Acad. Sci. USA*, **85**, 5536.
45. Frohman, M. A., Dush, M. K., and Martin, G. R. (1988). *Proc. Natl. Acad. Sci. USA*, **85**, 8998.

8: Map-based cloning in Arabidopsis

46. Schmidt, R. and Dean, C. (1992). In *Genome analysis 4* (ed. K. E. Davies and S. M. Tilghman), pp. 71–98. Cold Spring Harbor, NY.
47. Schmidt, R., Cnops, G., Bancroft, I., and Dean, C. (1992). *Aust. J. Plant Physiol.*, **19**, 341.
48. Schmidt, R., West, J., Love, K., Lenehan, Z., Lister, C., Thompson, H., et al. (1995). *Science*, **270**, 480.
49. Bancroft, I., Bent, E., Love, K., et al. (1998). *Nature*, **391**, 485.
50. Bechtold, N., Ellis, J., and Pelletier, G. (1993). *C. R. Acad. Sci. Paris*, **316**, 1194.
51. Tai, T. and Tanksley, S. (1991). *Plant Mol. Biol. Rep.*, **8**, 297.
52. Ausubel, F. M., Brent, R., Kingston, R. E., Moore, D. D., Seidman, J. G., Smith, J. A., and Struhl, K. (ed.) (1988). In *Current protocols in molecular biology*, Vol. 2, p. 13.1.2. John Wiley and Sons.
53. Coulson, A., Waterston, R., Kiff, J., Sulston, J., and Kohara, Y. (1988). *Nature*, **335**, 184.
54. Feinberg, A. P. and Vogelstein, B. (1983). *Anal. Biochem.*, **132**, 6.
55. Bancroft, I. and Wolk, C. P. (1988). *Nucleic Acids Res.*, **16**, 7405.
56. Jones, J. D. G., Shlumukov, L., Carland, F., English, J., Scofield, S. R., Bishop, G. J., et al. (1992). *Transgenic Res.*, **1**, 285.

9

Physical mapping: YACs, BACs, cosmids, and nucleotide sequences

IAN BANCROFT

1. Introduction

Physical mapping of genomes encompasses a range of techniques aimed at the establishment of the arrangement and composition of DNA. The techniques affording the broadest view are cytogenetic, whereby whole chromosomes are observed under the microscope. These approaches are relatively difficult in *Arabidopsis*, due to the small size of its chromosomes (see Chapter 5). In contrast, the paucity of repetitive DNA (1, 2) and the small size of the *Arabidopsis* genome, *c.* 130 Mb, of which 100 Mb is low copy (3), makes the *Arabidopsis* genome an excellent subject for clone-based physical mapping. In this chapter, we discuss how to conduct analyses involving various types of clones of *Arabidopsis* genomic DNA. These techniques have contributed to the development of a detailed understanding of genome structure and have provided valuable resources for gene identification.

The most detailed description of a genome is its nucleotide sequence, which represents a complete genetic blueprint. Within a short time of writing, the complete nucleotide sequence of the genome of *Arabidopsis* will have been determined and will become a key resource for plant scientists studying processes in *Arabidopsis* and many other plants. We discuss the systematic approaches that have been used to generate *Arabidopsis* nucleotide sequences and the opportunities that the sequence data offer for mapping.

2. Genome mapping with YAC clones

Yeast artificial chromosome (YAC) vectors are capable of propagating exogenous DNA fragments of as much as 1 Mb in size as linear chromosomes in yeast (4, 5). They thus represent the vector system currently capable of maintaining the largest inserts. The size of clone inserts is a critical factor contributing to their usefulness in physical mapping experiments. The larger the insert size, the fewer clones that are needed to represent a genomic region of a given size, so YACs have been used extensively for physical mapping in many organisms, including *Arabidopsis*.

In designing experiments involving YACs, it is important to appreciate some of the limitations of the YAC cloning system. YAC libraries are prone to containing substantial proportions of chimeric clones. This was a particular problem with early libraries, when small fragments of DNA were not completely excluded by the size fractionation of genomic DNA in preparation for cloning. These small fragments became ligated to larger fragments and included in the YACs. Also, YACs do not maintain all inserts stably, repeated sequences in particular can be deleted out. Care must therefore be taken to ensure that YACs truly represent the region of the genome that they are supposed to. Chromosome walking using YAC end probes is particularly difficult due to the chimerism problem, and all walking steps and contigs need to be double-checked using overlapping clones (see Chapter 8). YAC DNA is also difficult to obtain free of contamination by endogenous yeast DNA. Consequently, control experiments (using yeast DNA) need to be conducted whenever using gel purified YACs as hybridization probes.

Several YAC libraries containing *Arabidopsis* genomic DNA have been constructed. Four of them were constructed using DNA from the Columbia ecotype and have been the main subjects for systematic genome scale mapping: yUP (6), EW (7), EG (8), and CIC (9). These libraries have average insert sizes of approximately 250 kb, 150 kb, 150 kb, and 420 kb, respectively. Of these, the CIC library has proven the most productive for physical mapping due to its large insert size and low level of chimerism. These libraries, individual clones from them, and colony filter sets are available from the *Arabidopsis* Biological Resource Centre (ABRC) and may by ordered online (see *Table 1* for URL and Chapter 2).

Near-complete YAC-based physical maps have been reported for *Arabidopsis* chromosomes 4 (10), 2 (11), and 3 (12), with extensive contigs reported for chromosome 5 (13). The most extensive YAC-based physical mapping data for all five chromosomes are available through the Internet, as listed in *Table 1*. The primary strategy for the construction of these maps was the use of RFLP markers as hybridization probes to the YAC libraries. This approach firmly anchors the YAC map to the genetic map. Following the systematic anchoring of markers, limited chromosome walking involving the generation (by various means) and hybridization of YAC end-specific probes, was conducted.

Although largely superseded as the clone type of choice for physical mapping, the extensive YAC maps of *Arabidopsis* are a valuable resource. *Arabidopsis* YACs can be used as RFLP markers and to rapidly construct long-range restriction maps of defined regions (14). They can be used to rapidly map clones by hybridization, as almost the whole genome can be represented in a set of under 350 clones (see *Table 2*). They can also be subcloned as smaller fragments into vectors such as binary cosmids (see

9: Physical mapping: YACs, BACs, cosmids, and nucleotide sequences

Table 1. Internet sources for *Arabidopsis* physical mapping tools and information (see also Chapter 10)

URL (http://)	Information available
genome.bio.upenn.edu/physical-mapping/ch1-graphics-all-legend.html	Chromosome 1 YAC map
weeds.mgh.harvard.edu/goodman/c2.html	Chromosome 2 YAC map
genome-www.stanford.edu/Arabidopsis/Chr3-INRA/	Chromosome 3 YAC map
genome-www.stanford.edu/Arabidopsis/JIC-contigs/Chr4_YACcontigs.html	Chromosome 4 YAC map
genome-www.stanford.edu/Arabidopsis/JIC-contigs/Chr5_YACcontigs.html	Chromosome 5 YAC map
www.mpimp-golm.mpg.de/101/bac.html	IGF BAC genome-wide contigs
www.kazusa.or.jp/arabi/	Chromosomes 3 and 5 integrated physical maps
synteny.nott.ac.uk/agr/agr.html	*Arabidopsis* Genome Resource (AGR), integrated genetic, physical, and sequence maps of chromosomes 4 and 5
genome.wustl.edu/gsc/arab/arabsearch.html	BAC fingerprint database
www.tigr.org/tdb/at/atgenome/bac_end_search/bac_end_search.html	BAC end-sequence database
aims.cps.msu.edu/aims/	*Arabidopsis* Biological Resource Centre, DNA resources, including communal genomic libraries, and seed resources
genome-www.stanford.edu/Arabidopsis/	*Arabidopsis thaliana* database (AtDB), general source for all aspects of *Arabidopsis* information, including physical mapping and the genome sequencing project

Section 4.2, *Protocol 4*) for complementation of mutant phenotypes and gene identification. The recent report of biolistic transformation of plants using YACs retro-fitted with plant selectable markers (15) means that they may be useful directly for complementation analysis. For all of these applications, the ability to prepare high molecular weight DNA from YAC-containing yeast strains is essential. The method we routinely use is given in *Protocol 1*.

Protocol 1. Preparation of YAC chromosomal DNA

Equipment and reagents

- YEPD medium (per 600 ml): 6 g yeast extract, 12 g peptone, 12 g glucose; autoclave
- 1 M sorbitol
- SCEM: 1 M sorbitol, 100 mM sodium citrate pH 5.8, 10 mM EDTA, 30 mM 2-mercaptoethanol
- InCert agarose (FMC)
- 100 μl block formers (LKB/Pharmacia)
- 100 mg/ml Novozyme 234 in SCE (1 M sorbitol, 100 mM sodium citrate pH 5.8, 10 mM EDTA); filter sterilized
- ESP(1.0): 100 mM EDTA pH 8, 1% (w/v) Sarkosyl, 1 mg/ml proteinase K

Protocol 1. *Continued*

- ES: 100 mM EDTA pH 8, 1% (w/v) Sarkosyl
- ESP(0.5): 100 mM EDTA pH 8, 1% (w/v) Sarkosyl, 0.5 mg/ml proteinase K
- TE: 10 mM Tris–HCl pH 8.0, 1 mM EDTA
- 1 mM phenylmethylsulfonyl fluoride (PMSF) in 1 mM EDTA pH 8.0; freshly made up using a stock solution of 100 mM PMSF in isopropanol

Method

1. Grow culture to stationary phase (two days) at 30°C in 50 ml of YEPD medium.
2. Collect cells by centrifugation at 3000 r.p.m. for 10 min in a Sorvall RC3C.
3. Resuspend cells in 10 ml of 1 M sorbitol and transfer to a 15 ml screw-capped centrifuge tube.
4. Centrifuge at 3000 r.p.m. for 10 min in a MSE Centaur2 benchtop centrifuge.
5. Resuspend cells in 1 M sorbitol to a final volume of 0.6 ml.
6. Add 0.6 ml of InCert agarose made up in 1 M sorbitol and cooled to 50°C.
7. Mix in thoroughly and quickly dispense into 100 μl block formers (should make c. 12 × 100 μl blocks). Let set in ice for 10 min.
8. Press blocks out into 13 ml SCEM in a 15 ml screw-capped centrifuge tube. Add 0.12 ml 100 mg/ml Novozyme 234 solution. Screw on cap and place on cell mixer at room temperature for 4 h.
9. Transfer blocks to a 15 ml screw-capped centrifuge tube containing 7.5 ml ESP(1.0), fill tube with ES, and incubate at 50°C overnight.
10. Drain blocks, add fresh ESP(0.5), and incubate at 50°C for a further 24 h.
11. Transfer blocks to a 15 ml screw-capped centrifuge tube and wash 2 × 1 h with 1 mM PMSF, then 6 × 20 min with 1 mM EDTA pH 8, and finally 1 × 20 min with TE. Store at 4°C.

3. Genome mapping with BAC and P1 clones

3.1 Communal resources

Although YAC clones have the advantage for physical mapping of containing large inserts, they are not easy clones to handle, are difficult to use for moderate (10 kb) resolution mapping, difficult to confirm as collinear with genomic DNA, and unsuitable for efficient sequencing. In recent years, two clone types have been developed that combine the ability to propagate large inserts with the ease of handling clones propagated as supercoiled molecules in *E. coli*. These are bacterial artificial chromosomes (BACs) (16) and vectors

9: Physical mapping: YACs, BACs, cosmids, and nucleotide sequences

Table 2. Minimum set of YACs to represent the genome[a]

Chromosome 1

cM; Marker	YAC tiling path	cM; Marker	YAC tiling path
0; RS10	CIC10C4*	68.3; m402	CIC12F3
	YUP20D1		CIC6D4
	CIC3H3		CIC12A8
	CIC9A2		YUP18A5
	CIC7D5		YUP5F5
	YUP19H6**		YUP11F1
	YUP8H12		YUP10E6
	CIC9A9		YUP4H1
	YUP23H9	79.4; mi63	CIC6G1
	YUP6B2		YUP10G1
	YUP15G10		YUP17C8
	CIC5C1		YUP9E12
	CIC8C10		YUP16A9
	YUP19H2*		YUP5C9
14.2; nga63	YUP14G6		CIC5G9*
	YUP21H8	108.4; m213	CIC11F11
	CIC10E2		YUP3E4
	CIC9G11		YUP10C3
	CIC12A9		YUP15B2
	CIC3G6		CIC12H7
	CIC11C3		YUP4F10
23.7; ve007	CIC5D11		CIC6G10
	CIC4B9		CIC2E1
	CIC10B2		YUP12G3
	YUP9H2		CIC6H10
	YUP19H2*		YUP2D12
48.4; ve037	CIC5F4		YUP19H6**
	CIC1E7		YUP14G1
	YUP5A1		YUP21A4
	YUP13A3		CIC2H1
	YUP19E10		CIC12G10
	YUP1C1		CIC12G5*
	YUP12E5	123.2; mi353	YUP15C11
	YUP13C8		YUP17D11
51.7; mi265	CIC4F5		YUP24F6
	YUP21E5		YUP1H4
	CIC1D2		CIC9H12
	CIC11B7		CIC3F2
	CIC3E6		CIC9C4
	YUP7H5	149.2; nga11	YUP24G1
	YUP4E3		CIC4F1
	YUP6D9		CIC11F4
	CIC10B6		YUP11A8
	CIC9E5		YUP11B6
	YUP6F5		YUP3G5
	YUP15D4		YUP7C7
	CIC9C12		YUP17A11

Table 2. *Continued*

Chromosome 1

cM; Marker	YAC tiling path	cM; Marker	YAC tiling path
	CIC4B10		YUP6A6
	CIC12C12		YUP7D10
	CIC12A12		YUP3B7
			CIC1E4

Chromosome 2

cM; Marker	YAC tiling path	cM; Marker	YAC tiling path
	CIC7C11		
10.5; ve012	CIC11A4	48.4; g6842	CIC10F1
	CIC12G8		CIC7B5
	CIC2F1		CIC4B4
	CIC2G1		CIC12C3
	CIC6C8		CIC9H2
	CIC2G5		CIC5D4
	CIC4F6	74.4; ve017	CIC12G5*
	CIC9D11		CIC5G4
	CIC8G9		CIC10F12
22.4; g4532	CIC10F4		CIC2G9
	CIC9A4		CIC10A6
	CIC2A10		CIC11C8
	CIC8E10		CIC2E7
	CIC8B10		CIC8H2
	CIC6E8	88.3; mi79a	CIC6C3
38.7; mi148	CIC9G10		
	CIC11C11		
	CIC7E2		
	CIC6F9		
	CIC6C7		
	CIC10A9		
	CIC8H12		

Chromosome 3

cM; Marker	YAC tiling path	cM; Marker	YAC tiling path
	CIC11F3		
2.4; mi74b	CIC12H2	47.0; mi413	CIC10F3
	YUP1B6		CIC2B7
	CIC6D2		CIC11E8
	CIC11E4		CIC11H8
	CIC11D1		CIC11A9
	CIC1C12		CIC6C11
12.2; mi467	CIC7A12		CIC9H8
	CIC12H8		CIC6B1
	CIC3F3		CIC6F10
	CIC4H8		CIC3D5
	CIC11H12		CIC5B5
	CIC6H6		CIC3D2
	CIC12D7	81.8; nga6	CIC11G6
	CIC11H4		CIC4E6*

204

9: Physical mapping: YACs, BACs, cosmids, and nucleotide sequences

Table 2. *Continued*

Chromosome 3 *continued*

cM; Marker	YAC tiling path	cM; Marker	YAC tiling path
	CIC12C2		CIC9D1
19.9;m228	YUP4B5		CIC8E1
	YUP4B5	68.9; m457	CIC6F4
	YUP5A8		YUP3B5
	YUP19D2		CIC11G5
	YUP11G6		CIC5D3
	YUP23B9		CIC7F7
	YUP14E8		CIC8F3
	CIC6F6		CIC8D10
	CIC7B12		CIC11A1
	CIC12H5		YUP14A11
32.8; mi386	CIC4G6		CIC7D1
	CIC12H6		CIC9B7
	CIC10C1		CIC10C4*
	CIC11D3	81.8; nga6	CIC2A7
	CIC6B11		
	CIC1C10		
	CIC12D2		
	CIC10A8		

Chromosome 4

cM; Marker	YAC tiling path	cM; Marker	YAC tiling path
1.4; BIO217	CIC5B11	60.3; AG	CIC7D12
	CIC11A2		CIC2B3
	YUP4A6		YUP19G1
	CIC10H3		YUP1F6
	EG13E11		CIC5A4
	CIC4A7		CIC1A2
	CIC11B4		CIC1G9
	YUP8Eq		CIC3C11
19.3; g2616	CIC8B1		CIC12E11
	CIC7C3		CIC10E1
	CIC3F1		CIC5E5
	CIC3C8		CIC6A3
	CIC5H5		CIC11C6
	CIC9D3		CIC10E9
	CIC8B12		YUP5A6
	CIC4C6		CIC6A9
	CIC11E12		CIC4C4
	CIC3F10	80.6; g30	YUP23F8
	YUP17E10		CIC4E8
	CIC10F9		CIC7A10
	CIC9G5		YUP11E4
	CIC11G3		CIC12G2
40.7; g4108	CIC5E2		CIC6C9
	CIC12H12		CIC3H2
	YUP16B9		YUP1A3

Table 2. *Continued*

Chromosome 4 *continued*

cM; Marker	YAC tiling path	cM; Marker	YAC tiling path
	EW10D7	105.6;g3	EW11A7
	CIC1C2		
	EW14E4		
	EG9F3		
	EW16B10		
	EW9F3		
	EW2D3		
	YUP24F4		
	YUP16E6		
	EW2E9		
	EW5A7		

Chromosome 5

cM; Marker	YAC tiling path	cM; Marker	YAC tiling path
	YUP8A9	92.1; mi19	CIC6F12
	CIC2B9		CIC4E12
7.1; g3715	CIC5F11		CIC7G6
	CIC3A2		CIC7E12
	EW16C3		CIC11A11
	CIC10D2		YUP7F12
	YUP18F1		CIC10B7
	EG5H1		CIC11F10
	EG4F3		CIC8D5
	YUP14A12		EW3A5
	YUP19G12		EG6F5
	EG2E2		CIC8G12
	CIC6F3		YUP19G5
	CIC1B8		EW17B7
	YUP2E4	110.2; m4	CICB9
	CIC2F10		YUP5G3
	CIC6H5		CIC10H7
	CIC7A7		CICP2
	CIC7F6		YUP6B6
	EG1B4		YUP23H3
29.1; mi174	YUP19H12		YUP8C4
	CIC5H3		YUP13D4
	CIC9B8		CIC9C9
	YUP20B5		CIC9E2
	CIC3F11		YUP6G10
	CIC9F1		CIC4D8
	YUP17F7		CIC3A3
	CIC9A8		YUP5P19
	CIC2A4		CIC9H3
	YUP6D6		CIC9B5
	CIC5B2		YUP12E6
	CIC9B3		CIC5B3
	CIC7E11		CIC9G4

9: Physical mapping: YACs, BACs, cosmids, and nucleotide sequences

Table 2. *Continued*

Chromosome 5 *continued*

cM; Marker	YAC tiling path	cM; Marker	YAC tiling path
	CIC8B11		EW12E6
	CIC4D4		YUP4G7
	CIC12F2	132.6; mi3	YUP17F6
60.4; mi219	CIC2E3		
	CIC6D8		
	CIC3A1		
	CIC12F8		
	CIC5G9*		
	CIC4B3		
	CIC8C8		
	CIC7C1		
	CIC4E7		
	YUP19H6**		
	YUP11F9		
	CIC4E6*		
	YUP5F5		
	EG1D1		
	YUP10H3		

* = YAC mapped to two positions.
** = YAC mapped to three positions.
[a] The list of YACs was compiled from data shown on Internet sites listed in *Table 1*.

based on bacteriophage P1 (17). Both propagate as very low copy plasmids in *E. coli* and stably maintain their inserts. The BAC system has become more widely used, as clones can carry larger inserts (300 kb or more) than P1 clones (*c*. 85 kb). One P1 library and two BAC libraries are in widespread use: Mitsui P1 (18), TAMU BAC (19), and IGF BAC (20). All were constructed from genomic DNA of *Arabidopsis* ecotype Columbia and are available from ABRC as whole libraries, colony filter sets, and individual clones (see Chapter 2). The average insert sizes for the BAC libraries are about 100 kb, and for the P1 library about 85 kb. Taken together, the two BAC libraries have been shown to provide excellent overall coverage of the genome, though the redundancy of coverage varies widely (21).

The TAMU and IGF BAC libraries have been used to develop further communal resources. These include a genome-wide contig representation in IGF BACs, assembled using iterative hybridization of end probes to colony filters, a fingerprint database developed by agarose gel electrophoresis of endonuclease *Hin*dIII digests, and an end-sequence database (see *Table 1* for Internet sites of access). Physical maps displaying BAC clones used in the *Arabidopsis* genome sequencing project are shown on several Web sites, some of which display BAC-based maps integrated with other data such as genetic maps and nucleotide sequences (see *Table 1*). In practice, it is unlikely that chromosome walking at the level of YACs or BACs will be necessary in *Arabidopsis* ecotype Columbia, although confirmation of clone identity and local genome organization is still advisable.

3.2 Construction of BAC libraries

Although genomic clone libraries and physical maps are available 'off the shelf' for *Arabidopsis* ecotype Columbia, the same is not true of other ecotypes. The growing realization of the opportunities afforded by natural variation in *Arabidopsis*, as exemplified by the study of pathogen interactions (22), will result in the requirement to obtain genomic clones of alleles from a range of ecotypes. In some instances it may be adequate to identify a cDNA clone from the novel ecotype, by homology to its Columbia allele. Often, a new genomic library will have to be constructed. Genomic cosmid libraries are relatively easy to construct. However, cosmids do not provide complete representation of plant genomic DNA, and libraries approaching theoretical complete coverage are very large. A more common requirement is likely to be to make a BAC library, for example, for the cloning of disease resistance loci, which can be large and complex (23). It is also possible that some quantitative traits may be specified by multiple, clustered genes, again necessitating large insert clones if vectors are to be used that will allow the transfer of inserts into plant chromosomes. A BAC-based plant transformation vector, BIBAC (24) is available. In *Protocol 2* we provide a generalized method for the construction of new BAC libraries suitable for physical mapping.

Protocol 2. Construction of a BAC library

Equipment and reagents

- Pulsed-field gel electrophoresis equipment
- Nucleobond AX10 000 kit (Bio/Gene Ltd.)
- GeneCapsules (Geno Technology Inc.)
- 0.025 μM VS filters (Millipore)
- Cell-Porator with voltage booster (Gibco BRL)
- Lambda ladder markers
- 2 × CA buffer: 40 mM Tris pH 8, 14 mM MgCl$_2$, 200 mM KCl, 400 μg/ml BSA
- Calf intestinal alkaline phosphatase (CIAP) (New England Biolabs)
- TE, ESP(0.5), 1 mM PMSF (see *Protocol 1*)
- LB medium with selective antibiotic
- HindIII or BamHI or Sau3AI

- SeaKem GTG agarose (FMC)
- Agarose (electrophoresis grade)
- T4 DNA ligase: 2#000#000 U/ml (New England Biolabs)
- Electromax DH10B cells (Gibco BRL)
- Freezing broth (per 1 litre): 10 g tryptone, 5 g yeast extract, 6.3 g di-potassium hydrogen orthophosphate, 0.45 g trisodium citrate, 0.09 g magnesium sulfate, 0.9 g ammonium sulfate, 1.8 g potassium di-hydrogen orthophosphate, 44 g glycerol, adjust to pH 7.2; autoclave
- 10 × TBE: 108 g Tris base, 55 g boric acid, 9.3 g Na$_2$EDTA, make up to 1 litre with deionized water

A. Preparation of the vector

1. Grow 4 litres of bacterial culture harbouring the BAC vector plasmid at 37 °C overnight in LB medium plus appropriate antibiotic.

2. Prepare plasmid DNA using a Nucleobond AX10 000 kit as specified

by the manufacturers, except elute the DNA from the column with 4 × 25 ml aliquots of buffer N5 rather than 1 × 100 ml aliquot, and discard the first 10 ml of buffer off the column.

3. Linearize 30 µg of plasmid DNA with 100 U of the appropriate restriction endonuclease (e.g. HindIII for pBeloBAC11, BamHI for BIBAC2) in a total volume of 400 µl of 1 × CA buffer at 37 °C overnight.
4. Check that all the plasmid has been linearized by analysing a 3 µl aliquot by agarose gel electrophoresis.
5. To dephosphorylate, add 30 U CIAP to the linearized plasmid and incubate at 37 °C for 1 h.
6. Transfer to 65 °C for 20 min to inactivate the CIAP.
7. Extract once with phenol:chloroform, add 40 µl of 3 M sodium acetate, and 1 ml of ethanol. Place on ice for 30 min to precipitate.
8. Collect precipitate by centrifugation and dissolve vector DNA in 30 µl of 0.1 × TE.

B. *Preparation of the insert DNA*

1. Prepare plant cell nuclei and embed in InCert agarose block as described in the Texas A&M BAC Training Manual (http://http.tamu.edu:8000/~creel/TOC.html).
2. Add blocks to 13 ml ESP(0.5) in a 15 ml screw-capped centrifuge tube. Incubate at 50 °C overnight.
3. Exchange buffer for fresh ESP(0.5) and incubate at 50 °C for a further 24 h.
4. Transfer blocks to a fresh 15 ml screw-capped centrifuge tube and wash for 2 × 1 h with 1 mM PMSF in TE, then 6 × 20 min with TE. Store blocks at 4 °C in TE.

C. *Calibration of restriction endonuclease for partial digestion of the DNA preparation*

1. Transfer four blocks to a 15 ml screw-capped centrifuge tube containing sterile water, place tube on a cell mixer for 30 min.
2. Cut the blocks in half and place each in a microcentrifuge tube.
3. To each, add 50 µl of 2 × CA buffer and incubate at 37 °C for at least 2 h.
4. Replace buffer with fresh 1 × CA and incubate on ice for 20 min.
5. Add 5 µl of the appropriate restriction endonuclease diluted in 1 × CA buffer (e.g. HindIII for cloning into pBeloBAC11, Sau3AI for cloning into BIBAC2) and incubate on ice for 30 min. Use a twofold dilution series of enzyme in order to add 0, 0.125, 0.25, 0.5, 1.0, 2.0, 4.0, and 8.0 units per half block.

Protocol 2. *Continued*

6. Transfer tubes to a 37 °C water-bath for exactly 1 h.

7. Transfer tubes to ice and immediately replace buffer with 50 μl of ice-cold 50 mM EDTA pH 9.0 and incubate on ice for 1 h.

8. Resolve samples by pulsed-field electrophoresis (PFGE), using lambda ladder markers and appropriate pulse parameters, in order to identify the partial digests that produce the maximum amount of DNA in the 130–200 kb size range.

D. *Size fractionation and purification of the insert DNA*

1. Perform a large scale digest on 12–16 full blocks, using the procedure described above, but with one whole block per tube, double volumes, and with the optimum quantity of restriction endonuclease added (remembering to take account of the doubling of solution volumes and the quantity of DNA in each tube).

2. Prepare a 1% (w/v) SeaKem GTG agarose gel using 0.5 × TBE buffer. Cast without wells, reserving some molten agarose.

3. Cut a trough near one end of the gel and load the blocks of partially digested plant DNA. Load the blocks in two ranks of six or eight blocks, placed in contact with each other and the 'front' face of the trough. Place lambda ladder markers in the trough flanking the digested plant DNA. Fill the remainder of the trough with the reserved agarose to seal in the blocks, and let set.

4. Run PFGE for 18 h using pulse parameters selected so as to resolve to *c.* 250 kb.

5. Cut off the gel outer edges, containing the lambda markers and the edges of the digested plant DNA smear. Stain with ethidium bromide and visualize under UV illumination.

6. Measure the exact migration distance of DNA in the size range 130–200 kb.

7. Based on the figure for migration distance, excise the slice of gel containing 130–200 kb DNA from the unstained portion of gel.

8. Rotate the gel slice through 180° and load onto a fresh 1% (w/v) SeaKem GTG agarose gel, such that the largest size DNA is now furthest along the gel.

9. Run the second pulsed-field gel, using exactly the same parameters as used for the first size fractionation.

10. Cut off the gel outer edges containing the edges of the digested plant DNA. Stain with ethidium bromide and visualize under UV illumination.

9: Physical mapping: YACs, BACs, cosmids, and nucleotide sequences

11. Most of the DNA should be concentrated to a sharp band. Measure the exact migration distance of the band.
12. Guided by the migration distance, recover the DNA from the unstained portion of the gel by electroelution. We use GeneCapsules, following the protocol supplied by the manufacturer.
13. Pool eluted DNA and analyse a 10 µl sample by PFGE to check integrity and estimate yield.
14. Float a 0.025 µM VS filter on 50 ml of 0.1 × TE in a 50 ml screw-capped centrifuge tube. Gently pipette eluted DNA solution onto the membrane and place at 4°C overnight to dialyse. Store dialysed DNA at 4°C.

E. *Ligation, electroporation, and arraying the library*

1. Set up ligation reactions using 200 ng of insert DNA, a tenfold molar excess of vector DNA, 0.1 vol. of 10 × ligation buffer as supplied by the manufacturer, 2000 U (1 µl) of T4 DNA ligase in 100 µl final volume. Incubate at 4°C overnight. Multiple 100 µl ligations should be set up in parallel. As a control, set up one ligation lacking the insert DNA.
2. Float dialyse the ligations, as in part D, step 14. Store at 4°C.
3. Electroporate 1 µl aliquots of the ligation into 20 µl Electromax DH10B cells, using an electroporator with voltage booster and the protocol supplied by the manufacturer. Electroporation parameters: fast charge, low Ohms, 330 µF, voltage booster set to 4000 Ohms. Spread cells on appropriate selective plates (LB plus 40 µg/ml kanamycin and 5% (w/v) sucrose for BIBAC2, LB plus 12.5 µg/ml chloramphenicol for pBeloBAC11).
4. Check that the transformation efficiency is adequate for library construction and that the ligation mixes including insert DNA produce at least 50-fold more transformants than ligations without insert DNA. If the background of non-recombinants is too high (shown by a lower ratio), a new preparation of vector must be made.
5. Plates of transformed cells can be stored at 4°C for several days before picking. Constructs in pBeloBAC11 must be stored at 4°C for several days to allow the blue colour of non-recombinant clones to fully develop.
6. Use toothpicks to pick colonies and inoculate 384-well microtitre plates containing 50 µl per well freezing broth and appropriate antibiotic. Grow at 37°C for two days, stir, and freeze on dry ice. Store at –80°C.

3.3 Genome mapping with BACs

Two main approaches have been taken to whole genome mapping using clones such as BACs and P1s. One is to generate a fingerprint of all clones in the library, by one of a number of methods, and use automated image analysis and assembly of contigs based on matching bands in the fingerprints of individual clones (25). The other is to use iterative hybridization (or PCR amplifications) using clone end-specific probes (or primers designed to clone end-sequences for PCR approaches) to assemble contigs. Both are excellent approaches for the assembly of genome-wide contigs, though the former cannot assemble contigs through regions of the genome poorly represented in the clone libraries and the latter is prone to establishing false links caused by repetitive sequences and chimeric clones. Researchers are most likely to need to assemble contigs of clones from new genomic libraries in specific regions of the genome only. Appropriate techniques would therefore be ones which involve more directed approaches, as have been used for physical mapping of defined regions of the *Arabidopsis* genome in preparation for genome sequencing (21).

The *Arabidopsis* YAC maps provide an excellent resource to help construct BAC-based physical maps of defined regions of the genome. All of the YAC-based physical maps are well integrated with the genetic map of *Arabidopsis*, so YAC clones known to represent the region of the genome containing a locus of interest can be easily identified. The next step is to use these YACs to identify clones in the BAC library. This is most easily and economically achieved by gel purification of YAC DNA from the yeast strain harbouring the clone, and hybridization of that DNA to BAC colony filters. This method is described in *Protocol 3*. It is particularly important to ensure that the yeast chromosomes prepared are largely intact. If a strong background 'smear' is visible on analytical pulsed-field gels, the gel purified YAC DNA will be heavily contaminated with degradation products of the larger chromosomes. Probes so contaminated produce a high non-YAC-specific background on colony hybridizations, obscuring the true positives. Once clear YAC-specific hybridizing positive clones can be identified, DNA is prepared from them by standard alkali lysis methods (see Chapter 8, *Protocol 8*). Reliable clone contigs can then be assembled by restriction endonuclease digestion and Southern blotting of all clones, followed by iterative hybridization with clones from the set (as described in ref. 21). It should be noted that BACs can be scored as positive for hybridization with the YAC on colony filters, but fail to assemble into the final contigs. A preliminary hybridization of the BAC digest Southern blots with the original gel purified YAC DNA can be very valuable for the identification of both false positives (caused by particulate contaminants in the hybridization buffer, or the results of erroneous scoring of the autoradiograph) and BACs that contain some sequences homologous to the YAC, but do not represent the same genomic region. The latter are quite common and are usually identified by the characteristic that the YAC hybridizes to few restriction fragments of the BAC.

9: Physical mapping: YACs, BACs, cosmids, and nucleotide sequences

Protocol 3. Hybridization analysis of colony filters

Equipment and reagents
- Pulsed-field gel electrophoresis equipment
- 65°C incubator
- 2 × CA buffer (see *Protocol 2*)
- Beta-agarase I (New England Biolabs)
- 10 mg/ml yeast RNA
- Random prime labelling kit (Gibco BRL)
- 20 × SSPE: 3.6 M NaCl, 0.2 M NaH$_2$PO$_4$, 0.02 M Na$_2$EDTA pH 7.7
- 50 × Denhardt's solution: 10% (w/v) Ficoll 400, 10% (w/v) polyvinyl pyrrolidone, 10% (w/v) BSA; store at –20°C
- Hybridization buffer: 5 × SSPE, 5 × Denhardts, 0.5% (w/v) SDS, 5 µg/ml salmon sperm DNA
- Hybridization boxes (e.g. Bunzl 6064)
- 20 × SSC: 3 M NaCl, 0.3 M sodium citrate
- Hybridization wash buffer: 0.1 × SSC, 1% (w/v) SDS
- Membrane neutralization buffer: 0.1 × SSC, 0.1% (w/v) SDS, 0.2 M Tris pH 7.5
- Phenol (saturated)
- Phenol:chloroform (1:1)
- 3 M sodium acetate
- 100% ethanol
- 0.4 M sodium hydroxide
- Alpha *Taq*I (New England Biolabs)

A. Gel purification of YAC clones to be used as hybridization probes

1. Run yeast chromosomal preparations on a pulsed-field gel using pulse times selected to achieve maximum resolution of the YAC. Visualize under long-wave UV and excise minimum sized gel slice containing the YAC.

2. Wash the gel slice with at least ten volumes of water in appropriate sized tube on a cell mixer.

3. Transfer the gel slice to 1.5 ml centrifuge tube and place tube at 65°C for 10 min.

4. Add melted gel slice to approximately equal volume of 2 × CA buffer pre-warmed to 65°C and mix.

5. Add 10 U *Taq*I and incubate at 65°C for 30 min.

6. Add 0.05 vol. of 0.5 M EDTA, transfer the tube to a 37°C water-bath, and equilibrate for 10 min.

7. Add 1 U agarase per 150 µl of the original 1% (w/v) gel slice and incubate at 37°C overnight.

8. Centrifuge tube briefly and heat to 65°C for 20 min.

9. Equilibrate the tube at 37°C, add 0.5 U agarase per 150 µl of the original gel slice, and incubate at 37°C for 4 h.

10. Heat to 65°C and add to a centrifuge tube containing 0.1 vol. of 3 M sodium acetate, 2 µl yeast RNA carrier, and an equal volume of phenol. Shake to mix and resolve by centrifugation for 10 min.

11. Transfer aqueous phase to a fresh tube containing an equal volume of phenol:chloroform. Shake to mix and resolve by centrifugation for 2 min.

Protocol 3. *Continued*

12. Transfer the aqueous phase to a fresh microcentrifuge tube and precipitate the DNA with 2.5 vol. of ethanol at −20°C overnight.

B. *Hybridization of filters*

1. Label c. 50 ng of the gel purified YAC DNA using a random primers labelling kit. We usually use [α-^{32}P]dCTP, but other labels could be used. There is no need to remove unincorporated ^{32}P-labelled nucleotides before hybridization.

2. High density gridded colony filters of *Arabidopsis* genomic libraries can be obtained from the *Arabidopsis* Biological Resource Centre. Follow any specified pre-hybridization processing steps. The following protocol is designed for use with sets of 24 membranes of size 11.7 × 7.6 cm, and should be scaled up or down as appropriate.

3. Rinse filters in 2 × SSC then place 60 ml of hybridization buffer in a small hybridization box.

4. Add membranes, one-by-one, ensuring that each is fully submerged before adding the next.

5. Pre-hybridize at 65°C with gentle shaking for at least 4 h.

6. Boil the labelled probe and add to a fresh hybridization box containing 30 ml of hybridization buffer.

7. Add membranes, one-by-one, ensuring that each is fully submerged.

8. Hybridize at 65°C with gentle shaking overnight.

9. Transfer membranes, one-by-one, to a larger box containing 250 ml of hybridization wash buffer. Incubate at 65°C with gentle shaking for 20 min.

10. Pour off wash buffer and replace with fresh solution. Incubate at 65°C as before.

11. Repeat step 10 to give membranes a third wash.

12. Seal filters in plastic bags (12 per bag if 35 × 43 cm autoradiography cassettes are available) and expose to film overnight, with the film to the DNA side of the membranes.

13. Adjust length of autoradiography if necessary. Re-wash membranes two more times if membrane background is too high. An ideal autoradiograph should show most of the colonies on each filter as a just visible background, with the positive signals clearly stronger.

14. Filters can be stripped by incubation in 0.4 M sodium hydroxide for 2 × 20 min with gentle shaking at 45°C, followed by a 15 min wash in membrane neutralization buffer.

9: Physical mapping: YACs, BACs, cosmids, and nucleotide sequences

Figure 1. Autoradiograph of a high density colony filter of binary cosmid clones probed with a BAC. The clones were gridded in a 16 off-set array (four rows by four columns), in duplicate, from 384-well microtitre plates. The relative positions of the duplicate positives allows the determination as to which of the eight source microtitre plates contained the clone. The plate row and column coordinate is determined from the position of the square containing the positive relative to the remainder of the grid. In this example, there are three positive clones, each from different plates.

The protocol for the gel purification and hybridization of YACs to BAC filters can readily be adapted to allow the use of a variety of clone types. The most easily interpreted autoradiographs are usually from BAC clones used as hybridization probes to binary cosmid libraries (see *Figure 1* for an example). For both YAC and BAC probes it is particularly important to use high quality DNA (see Chapter 8, *Protocol 8*). Each preparation of BAC DNA to be used as a probe should be checked by digestion with *Not*I, to release the insert, and resolved on a pulsed-field gel. If more than a faint background smear of degraded *E. coli* DNA is visible, the clone insert fragment should be gel purified as described for YACs. When the complete genome sequence is available, it will become possible to identify *in silico* all cDNA clones (that contributed EST data to public databases) corresponding to strings of genes in any defined genomic region. These may also be used as hybridization probes, either pooled or singly. In the latter case the collinearity of the genome represented in the new library with that of Columbia ecotype can be assessed, and detailed BAC contigs can be assembled from the colony hybridization data. The use of PCR to amplify products from multi-dimensional pooled DNA is an alternative to using colony hybridization with cDNAs. However, it is more prone to false negatives (due to poor primer design or alterations in nucleotide sequence relative to Columbia ecotype) and does not permit the identification of homologous members of gene families.

4. High resolution mapping with cosmids
4.1 Approaches to mapping with cosmids
Cosmids were the first clone type to be used for a systematic attempt at contiguous representation of the *Arabidopsis* genome (26). A physical map consisting of 750, mostly unanchored, contigs was constructed. These were estimated to represent 90–95% of the genome, but contigs could be identified to represent less than half of a region of the genome analysed in detail (27). The future use of cosmids for genome mapping in *Arabidopsis* is likely to be focused on the representation of defined regions in transformation-competent vectors as a part of positional cloning experiments (see Chapter 8). 'Binary' cosmids are vectors capable of replication in both *E. coli* and *Agrobacterium* and by a process involving factors provided by other plasmids, transferred by *Agrobacterium* into plant genomes. One binary cosmid vector, pCLD04541, is available from ABRC and has been widely used for gene identification in *Arabidopsis* (27, 28).

Cosmid libraries can be screened by hybridization in the same way as BAC libraries. Contig assembly approaches are also similar. In the case of clones made using pCLD04541, the average insert size in libraries is around 17 kb, compared to a 29 kb vector. This allows the construction of high resolution maps, with each clone containing only three or four genes on average, but clones are relatively inefficient to sequence by standard shotgun methods. For contig assembly, approaches involving the digestion of clones with a combination of restriction endonucleases (e.g. *Hin*dIII and *Sal*I for clones with inserts in the *Cla*I cloning site) and matching of restriction fragment patterns can lead to the assembly of contigs and restriction maps, without the use of iterative hybridization (27).

4.2 Construction of cosmid libraries
Cosmid libraries can readily be constructed from a range of source DNA. A reliable method that is particularly suitable for the construction of libraries in the binary cosmid vector pCLD04541 involves the partial digestion of the source DNA with *Taq*I, size selection to 17 kb plus and ligation, at high concentration, into phosphatase-treated *Cla*I cut vector. Commercial packaging extracts are then used to package the ligated DNA as *c*. 48 kb 'head fulls' into lambda phage. These are used to infect *E. coli*, in which the DNA is circularized and propagated as a plasmid. A detailed method for the preparation of a library by subcloning a gel purified YAC is given in *Protocol 4*. This is the most demanding protocol for cosmid library construction, as the source DNA is available in limiting amounts. The protocol can be modified to utilize BAC or genomic DNA as the source for inserts by conducting a series of large scale (10 µg) partial digests and gel purification of insert DNA from the one yielding most fragments in the range 17–22 kb. Once a library has been prepared

9: Physical mapping: YACs, BACs, cosmids, and nucleotide sequences

and arrayed, colony filters for hybridization analysis can be prepared as described in *Protocol 5*. The same protocol can be used for preparing colony filters of BAC clones, and can be modified to allow the preparation of colony lift filters if the library is not to be arrayed.

Protocol 4. Construction of a binary cosmid library by subcloning a YAC

Equipment and reagents

- Pulsed-field gel electrophoresis equipment
- Corex tubes
- Cosmid vector pCLD04541 (obtainable for the *Arabidopsis* Biological Resource Centre)
- Beta-agarase I (New England Biolabs)
- 2 × CA buffer (see *Protocol 2*)
- Calf intestinal alkaline phosphatase (CIAP) (New England Biolabs)
- Low melting point (LMP) agarose (FMC)
- ESP(0.5) (see *Protocol 2*)
- 1 mM PMSF (see *Protocol 2*)
- Alpha *Taq*I (New England Biolabs)
- M. *Taq*I (New England Biolabs)
- T4 DNA ligase and ligation buffer (New England Biolabs)
- 20 mg/ml BSA
- Gigapack XL or Gold lambda packaging extracts (Stratagene)
- Freezing broth (see *Protocol 2*)
- 0.5 M EDTA pH 8.0
- TE (see *Protocol 1*)
- Phenol (saturated)
- Phenol:chloroform (1:1)
- 0.3 M sodium acetate
- 100%, 70% (v/v) ethanol
- 10 mg/ml yeast RNA carrier

A. Preparation of the vector

1. Digest 20 μg of pCLD04541 DNA to completion with *Cla*I in 1 × CA buffer.
2. Resolve the digested DNA on 0.8% (w/v) LMP agarose gel and excise linearized DNA.
3. Wash the gel slice and adjust to 1 × CA buffer at 37 °C.
4. Add 10 U CIAP and incubate at 37 °C for 30 min.
5. Add 0.5 M EDTA to 25 mM final concentration and incubate at 65 °C for 20 min.
6. Cool to 37 °C and recover DNA by agarase treatment, following the protocol supplied by the manufacturer.
7. Dissolve DNA in 20 μl water.

B. Size fractionation and purification of the insert

1. Prepare a large-format 1% (w/v) LMP agarose gel for pulsed-field gel electrophoresis without wells. Reserve some of the molten agarose.
2. Cut a trough near one end of the gel and load lambda ladder marker and 24 × 100 μl blocks of YAC chromosome prep. The YAC-containing blocks should be loaded in two ranks of 12 blocks, placed in contact with each other and the 'front' face of the trough. Fill the

Protocol 4. *Continued*

remainder of the trough with the reserved LMP agarose and allow to set.

3. Run the gel under appropriate PFGE conditions to achieve good resolution of the YAC DNA to be purified.
4. Stain the gel with ethidium bromide, visualize under long-wave UV, and excise the YAC gel slice, aiming for a total volume of 1.8–2.0 ml.
5. Cut the gel slice into six pieces, place in a 15 ml screw-capped centrifuge tube containing 12 ml ESP(0.5), and incubate at 50°C overnight.
6. Transfer the blocks to a 50 ml screw-capped centrifuge tube and wash for 2 × 1 h with 1 mM PMSF, then 6 × 20 min with 1 mM EDTA pH 8, and finally 1 × 20 min with TE. Complete the washes in one day and store the blocks at 4°C over one night only.
7. Equilibrate the blocks in water, transfer into two microcentrifuge tubes, and melt at 65°C for 20 min.
8. Ensure the volume is at least 1.8 ml then add aliquots of 300 μl to six tubes containing 300 μl of 2 × CA buffer, mix in well, and incubate at 65°C for a further 10 min.
9. Prepare dilutions of alpha *Taq*I in 1 × CA buffer: 0.128, 0.064, 0.016, 0.004, 0.001, and 0.00025 U/μl.
10. Prepare methylation buffer: 273 μl water, 33 μl NEB buffer 4 (supplied with M. *Taq*I), 3.3 μl 20 mg/ml BSA.

Perform steps 11 to 13 quickly.

11. Add 8.3 μl of 32 mM SAM (supplied with M. *Taq*I) and 13.2 μl (132 U) M. *Taq*I to the methylation buffer and dispense into 6 × 50 μl aliquots.
12. To each aliquot of methylation mix add 5 μl of an alpha *Taq*I dilution to give final enzyme quantities of: 0.64, 0.32, 0.08, 0.02, 0.005, and 0.00125 units.
13. Add the partial digestion mixes to aliquots of melted gel slice, mixing gently, but thoroughly, by stirring with pipette tip. Incubate at 65°C for 30 min.
14. Add 35 μl of 0.5 M EDTA pH 8 (pre-warmed to 65°C) to each and transfer to 37°C for 10 min.
15. Add 2 U beta-agarase I to each and incubate at 37°C overnight.
16. Centrifuge briefly, heat to 65°C for 20 min.
17. Equilibrate tube at 37°C, add 1 U beta-agarase I to each, and incubate at 37°C for 4 h.
18. Heat to 65°C then pool all six aliquots in a 15 ml screw-capped centrifuge tube containing 0.4 ml of 3 M sodium acetate and 4 ml phenol. Add 5 μl of 10 mg/ml yeast RNA carrier.

9: Physical mapping: YACs, BACs, cosmids, and nucleotide sequences

19. Shake the tube and centrifuge at 3500 r.p.m. for 10 min in a Sorvall Centaur2 centrifuge to resolve the phases.
20. Transfer the aqueous phase to a tube containing 4 ml of phenol: chloroform, shake, and resolve the phases by centrifugation as before.
21. Transfer the aqueous phase to a 30 ml Corex tube, add 2.5 vol. ethanol, and place at −20°C overnight.
22. Centrifuge tube at 10 000 r.p.m. for 10 min at 4°C using a HB4 rotor in a Sorvall RC5B centrifuge.
23. Pour off the supernatant and let drain for *c.* 10 min.
24. Dissolve the precipitate in 200 µl of TE, transfer to a 1.5 ml microcentrifuge tube, add 20 µl of 3 M sodium acetate, 550 µl ethanol, and place at −20°C overnight.
25. Collect the precipitate by centrifugation for 10 min, then dissolve in 10 µl of TE.
26. Add 2.5 µl of 5 × gel loading buffer and place on ice for 1 h.
27. Centrifuge the tube for 5 min then transfer the supernatant to a fresh tube.
28. Run DNA on a 0.8% (w/v) LMP agarose gel, including ethidium bromide. Run the gel slowly and for a short distance only (bromophenol blue to migrate *c.* 8 cm). Recover, by beta-agarase I treatment (using the protocol supplied by the manufacturer), DNA in the region of the gel from just in front of the wells to the region of peak fluorescence at the position corresponding to the compression zone for the markers. Ethanol precipitate the recovered DNA.

C. *Ligation and plating of the library*
1. Collect the precipitated DNA. Dissolve in 17 µl of TE.
2. Add 1 µl (1 µg) of dephosphorylated vector DNA and 2 µl of 3 M sodium acetate. Add 50 µl ethanol and place at −20°C for at least 1 h.
3. Collect the precipitate, rinse the pellet with 70% (v/v) ethanol, and air dry for *c.* 30 min.
4. Dissolve the pellet in 2 µl of water, add 1 µl of 3 × ligation buffer, and 0.2 µl T4 DNA ligase. Incubate at 4°C for three days.
5. Package using Gigapack XL or Gold lambda packaging extract, following the protocol supplied by the manufacturer.
6. Plate the whole mixture on *E. coli* Sure Tc^S strain (or similar), spreading cells over six 140 mm selective plates (LB plus 10 µg/ml tetracycline). (NB: spin down cells and resuspend in fresh LB before plating to avoid adding MgCl$_2$ to tetracycline plates.)

Protocol 4. *Continued*

7. Pick colonies into 384-well microtitre plates containing 50 µl freezing broth plus 10 µg/ml tetracycline per well. Grow at 37°C for two days. Freeze at −80°C.

Protocol 5. Preparation of colony filters for hybridization analysis

Equipment and reagents

- 384-prong replicator
- Hybond N$^+$ membrane (Amersham)
- 3MM paper (Whatman)
- Denaturation solution: 0.5 M NaOH, 1.5 M NaCl
- Neutralization buffer: 1 M Tris pH 7.5, 1.5 M NaCl
- 2 × SSC (prepare as 20 × stock, see *Protocol 3*)
- Scrape solution: 5 × SSC, 0.5% (w/v) SDS

Method

1. Cut individual membranes 11.7 × 7.6 cm from 20 cm wide rolls of Hybond N$^+$ membrane. Label using a pencil or fine, permanent marker pen and place on the surface of LB agar medium (containing 10 µg/ml tetracycline or other appropriate antibiotic) in 140 mm diameter Petri dishes.

2. Using a 384-prong replicator, spot the colonies of one 384-well microtitre plate onto each membrane. Prepare several replicates for simultaneous hybridization analysis.

3. Invert the plates and grow at 37°C overnight.

4. Lift the membrane from the Petri dishes and place on a sheet of Whatman 3MM paper soaked in denaturation solution. Leave for 3 min.

5. Transfer to a fresh sheet of 3MM paper soaked in denaturation solution. Leave for 3 min.

6. Transfer to a sheet of 3MM paper soaked in neutralization buffer. Leave for 6 min.

7. Rinse the membranes in 2 × SSC and air dry thoroughly, placing in sleeves of 3MM paper when they begin to curl.

8. Bake at 80°C for 2 h.

9. Soak the filters for at least 2 h at 42°C in scrape solution.

10. Scrape off all bacterial debris using a paper towel. Rinse membrane in 2 × SSC and start the hybridization procedure immediately, or store dry.

9: Physical mapping: YACs, BACs, cosmids, and nucleotide sequences

5. Nucleotide sequences — the ultimate mapping tool

5.1 The EST sequencing project

The most cost-effective approach to gaining sequence data of genes within a genome is to systematically sequence the transcripts of those genes as represented in cDNA libraries. Sequences so derived are termed expressed sequence tags (ESTs). EST data are usually acquired by single-pass sequencing of one or both ends of the cDNA inserts in plasmid vectors. Only genes represented in the mRNA pool used to construct the cDNA library can be represented in the EST dataset. The proportion of genes that can practicably be identified by this approach is the subject of debate. Analysis of present genomic sequence data suggests that a little over half of the *Arabidopsis* genes have been identified by an EST. A major source of error in this estimate is the relatively poor quality of EST sequence data. This results in the inability to differentiate between imperfect sequence matches being the result of sequencing errors in individual EST reads representing the same gene, and the origins of the ESTs being transcripts from different, related, genes.

The generation of more ESTs from *Arabidopsis* is unlikely to be cost-effective for gene identification, as new sequences will be rare (derived from genes transcribed at a low level or in specific spatial, temporal, or inducible patterns) amongst more abundantly represented transcripts, which have already been sequenced. Recent efforts have focused on maximizing the use of present data for the elucidation of gene structure and function. The existing EST sequences are being assembled into sequence contigs, providing greater sequence coverage of each transcript. This work may be extended to completely sequencing, to high accuracy, a non-redundant set of EST clones. This would contribute substantially to the determination of the exact structures of genes, as current methods for gene prediction and modelling using genomic DNA sequences are far from perfect. The inserts from *Arabidopsis* EST clones are being used for the analysis of gene expression, as an approach contributing to the understanding of gene function. The approach involves spotting DNA of the EST clone inserts, at high density, onto a membrane or glass support matrix, and differential hybridization with labelled cDNA from populations of mRNA derived from different plant tissues or treatments (29).

The EST sequence data are available to researchers via a number of databases, including AtDB, and most of the clones are available from ABRC (see Chapters 2 and 10).

5.2 The genome sequencing project

EST sequencing provides an excellent start to large scale gene identification projects. In organisms with large genomes, composed mostly of repetitive sequences, EST projects provide the only practicable approach. In species with smaller genomes, composed mostly of low copy, gene-containing sequences, such as *Arabidopsis*, sequencing of the whole genome is feasible. Although more costly than EST projects, genome sequencing provides more data. The identification of *all* genes is possible, regardless of transcription levels, the arrangement of the genes relative to each other, and structures contributing to chromatin structure and gene expression can be identified. There are presently significant limitations on our abilities to fully interpret the genome sequence data from *Arabidopsis*, particularly gene prediction and intron–exon structure modelling. However, as data accumulate and gene prediction programs are more thoroughly trained on *Arabidopsis* datasets, such abilities are steadily improving. Highly accurate modelling should eventually be possible, as the sequence data are being generated with the intention of being a definitive resource, with an accuracy of no more than one base error in 10 000.

A pilot project for large scale sequencing of the *Arabidopsis* genome commenced in the EU in 1993 aiming to sequence *c*. 2 Mb of the genome, mainly from two regions of chromosome 4. That early project initially attempted a cosmid-based sequencing strategy, using primarily the cosmid contigs that were available at the time (26). However, cosmids proved unsuitable for large scale sequencing of the *Arabidopsis* genome, and BACs were demonstrated to be preferable (23, 27). The Kazusa Institute in Chiba province of Japan started production scale sequencing of chromosome 5 in 1995, using the Mitsui P1 library. In 1996, production scale sequencing commenced in the US and EU, using the TAMU and IGF BAC libraries. By November 1998, 32 Mb of complete, annotated sequence had been released to GenBank. It is expected that the low copy portions of the genome, *c*. 100 Mb in all, will have been sequenced by the end of 2000, with annotation complete in 2001. There will probably be very few gaps in these sequences. At the time of writing, the EU project had successfully closed a 4.5 Mb sequence contig on the long arm of chromosome 4, which is expected to be extended to include the whole of the chromosome arm (excluding only the telomere and centromeric repeats) by mid 1999.

5.3 Sequence-based mapping

Many opportunities for the advancement of science will be opened up by the availability of the first complete genome sequence of a plant, i.e. that of *Arabidopsis*. The data are presently partial and tools for facile analysis and utilization are not fully developed. However, there are analyses that can be conducted, and productive approaches that will be possible upon completion of the sequence can be anticipated. The databases needed for the facile exploitation of the sequence data are being developed by AtDB and AGR (see *Table 1* for URLs).

For physical mapping, it will become possible to map, by sequence matches, segments of DNA to their position in the genome in a manner highly analogous to nucleic acid hybridization techniques involving probe DNA and immobilized target DNA. The 'probe' will be sequence data associated with the DNA to be positioned, and the 'target' will be the genome sequence. The 'hybridization' will be a sequence alignment using *BLAST* (30), or other alignment programs, conducted *in silico*. Mapping will be done by using sequence data generated, for example, as the sequences flanking transposon insertions in the genome, or an EST sequence. If a clone is to be aligned, end sequencing, even of large clones such as BACs, is relatively fast and inexpensive to conduct. The clone ends or tag sequences will be positioned very precisely, to the base pair, in the genome. Partial homologies will also be detectable, and will allow the identification of gene families. As well as being extremely rapid, *in silico* analysis is particularly sensitive for the detection of similarities, and can use indirect comparisons, for example of predicted amino acid sequences of hypothetical translation products, to identify more divergent homologues than is possible by hybridization.

A number of new mapping approaches for use in research projects involving *Arabidopsis* can be foreseen. These include:

(a) The identification of genomic integration sites, including the identification of genes disrupted, with T-DNA and transposons, using sequences immediately flanking the insertion site.

(b) Anchoring of transformation-competent individual clones and whole clone libraries, using clone end-sequences, to identify the sets of genes to be used in trait transfer or mutant complementation in *Arabidopsis*.

(c) Alignment of clones from new libraries of different ecotypes with that of Columbia ecotype, using end-sequences.

(d) Comparative mapping of regions of, or whole, genomes using the sequences of ordered EST clone sets or complete genome sequence data.

Many other applications of the genome sequence data are being developed, particularly for functional genomics and genetic marker development. It will provide a common reference resource and medium for pooling data from experiments involving *Arabidopsis* and many other plants.

References

1. Leutwiler, L. S., Hough-Evans, B. R., and Meyerowitz, E. M. (1984). *Mol. Gen. Genet.*, **194**, 15.
2. Pruitt, R. E. and Meyerozitz, E. M. (1986). *J. Mol. Biol.*, **187**, 169.
3. Goodman, H. M., Ecker, J. R., and Dean, C. (1995). *Proc. Natl. Acad. Sci. USA*, **92**, 10831.
4. Burke, D. T., Carle, G. F., and Olson, M. V. (1987). *Science*, **236**, 806.
5. McCormick, M. K., Shero, J. H., Cheung, M. C., Kan, Y. W., Hieter, P. A., and Antonarakis, S. E. (1989). *Proc. Natl. Acad. Sci. USA*, **86**, 9991.
6. Ecker, J. R. (1990). *Methods*, **1**, 186.

7. Ward, E. R. and Jen, J. C. (1991). *Plant Mol. Biol.*, **14**, 561.
8. Grill, E. and Somerville, C. (1991). *Mol. Gen. Genet.*, **226**, 484.
9. Creusot, F., Fouilloux, E., Dron, H., Lafleuriel, J., Picard, G., Billault, A., et al. (1995). *Plant J.*, **8**, 763.
10. Schmidt, R., West, J., Love, K., Lenehan, Z., Lister, C., Thompson, H. et al. (1995). *Science*, **270**, 480.
11. Zachgo, E. A., Wang, M.-L., Dewdney, J., Bouchez, D., Camilleri, C., Belmonte, S. et al. (1996). *Genome Res.*, **6**, 19.
12. Camilleri, C., Lafleuriel, J., Macadre, C., Varoquaux, F., Parmentier, Y., Picard, G. et al. (1998). *Plant J.*, **14**, 633.
13. Schmidt, R., Love, K., West, J., Lenehan, Z., and Dean, C. (1997). *Plant J.*, **11**, 563.
14. Bancroft, I., Westphal, L., Schmidt, R., and Dean, C. (1992). *Nucleic Acids Res.*, **20**, 6201.
15. Mullen, J., Adam, G., Blowers, A., and Earle, E. (1998). *Mol. Breed.*, **4**, 449.
16. Shizuya, H., Birren, B., Kim, U. J., Mancino, V., Slepak, T., Tachiiri, Y. et al. (1992). *Proc. Natl. Acad. Sci. USA*, **89**, 8794.
17. Oannou, P. A., Amemiya, C. T., Grames, J., Kroisel, P. M., Shizuya, H., Chen, C. et al. (1994). *Nature Genet.*, **6**, 84.
18. Liu, Y. G., Mitsukawa, N., Vazquez-Tello, A., and Whittier, R. F. (1995). *Plant J.*, **7**, 351.
19. Choi, S. D., Creelman, R., Mullet, J., and Wing, R. A. (1995). *Weeds World*, **2**, 17.
20. Mozo, T., Fischer, S., Schizuya, H., and Altmann, T. (1998). *Mol. Gen. Genet.*, **258**, 562.
21. Bent, E., Johnson, S., and Bancroft, I. (1998). *Plant J.*, **13**, 849.
22. Holub, E. B., Beynon, J. L., and Crute, I. R. (1994). *Mol. Plant Microbe Interact.*, **7**, 223.
23. Bevan, M., Bancroft, I., Bent, E., Love, K., Goodman, H., Dean, C. et al. (1998). *Nature*, **391**, 485.
24. Hamilton, C. M., Frary, A., Lewis, C., and Tanksley, S. D. (1995). *Proc. Natl. Acad. Sci. USA*, **93**, 9975.
25. Marra, M. A., Kucaba, T. A., Dietrich, N. L., Green, E. D., Brownstein, B., Wilson, R. K. et al. (1997). *Genome Res.*, **7**, 1072.
26. Hauge, B. M. and Goodman, H. M. (1992). In *Methods in Arabidopsis research* (ed. C. Koncz, N.-H. Chua, and J. Schell), pp. 191–223. World Scientific Publishing Co. Pty. Ltd., Singapore.
27. Bancroft, I., Love, K., Bent, E., Sherson, S., Lister, C., Cobbett, C., et al. (1997). *Weeds World*, **4**, 1.
28. Macknight, R., Bancroft, I., Page, T., Lister, C., Schmidt, R., Love, K., et al. (1997). *Cell*, **89**, 737.
29. Schena, M., Shalon, D., Davies, R. W., and Brown, P. O. (1995). *Science*, **270**, 467.
30. Altschul, S. F., Gish, W., Miller, W., Meyers, E. W., and Lipman, D. J. (1990). *J. Mol. Biol.*, **215**, 403.

10

Web-based bioinformatic tools for *Arabidopsis* researchers

SEUNG Y. RHEE and DAVID J. FLANDERS

1. Introduction

This chapter provides *Arabidopsis* researchers with an overview of bioinformatic tools and resources on the World Wide Web (Web).

The introduction briefly defines the term 'bioinformatics' and describes sources of bioinformatic data. The body of the chapter is divided into four sections, followed by a short glossary:

(a) How to get connected to the Internet and how to use Web browsers.
(b) A flowchart describing an experimental scenario that uses bioinformatic tools.
(c) Description of the tools and resources named in the flowchart, divided into four subsections: gene information resources, genetic and physical maps, sequencing, and sequence analysis tools. Each is followed by a table of URLs described or mentioned in that section.
(d) A discussion of current issues and future directions of bioinformatics.
(e) A glossary of bioinformatic and Internet terms. Words that appear in the glossary are bold-typed at their first use in the chapter.

1.1 What is bioinformatics?

The science of bioinformatics has been defined as 'a scientific discipline encompassing all aspects of biological information acquisition, processing, storage, distribution, analysis, and interpretation' that use computer science, engineering, and biological knowledge to extract biological significance from experimental and derived data (1). Understanding the biological significance of processes will enable researchers to predict phenotypes of organisms, mechanisms of evolution, and mechanisms of pathogenicity to name a few.

Bioinformatics has changed rapidly and today is understood to include: the study of information infrastructure construction; computational research to understand biological information (formally, computational biology); the study of an organism's DNA (genomics); and the study of an organism's proteins (proteomics) (1, 2).

1.2 Sources of *Arabidopsis* bioinformatic data

Biological information is being acquired systematically in ever-increasing amounts by laboratories in both the public and private sectors. Storage of these huge volumes of data, and rapid access to them, has been accommodated by genomic databases. These are the 'public window on the high-throughput genome projects' (2) that serve as the medium for storage and distribution of bioinformatic data. The rate and quantity of data accumulation will make databases the main publication source for scientific information in the near future (1, 3, 4).

Bioinformatics databases that are pertinent to this chapter can be divided into two main types:

(a) Molecular biological databases (e.g. GenBank, SWISS-PROT, PDB) and their associated retrieval systems (e.g. Entrez, SRS). These deal with particular sets of data, such as nucleic acid and protein sequences, and crystallographic structures.(Note: in this chapter, GenBank is used as a generic term for the three, international sequence databases: GenBank, EMBL, and DDBJ.)

(b) Organismal databases, e.g. AtDB (*Arabidopsis*), SGD (yeast), FlyBase (*Drosophila*), MaizeDB (corn), ACeDB (*C. elegans*), MGD (mouse). These are 'expert domain' databases that contain integrated information for a specific organism.

There are several databases that house the bioinformatic data of plants (5). In the public domain, at least, most of these data come from *Arabidopsis thaliana*, due to its popularity as an experimental organism and suitability for bioinformatic research. Major databases for *Arabidopsis* research include: the *Arabidopsis thaliana* database (AtDB); the *Arabidopsis* Genomic Resource (AGR) at the Nottingham *Arabidopsis* Stock Centre (NASC); the *Arabidopsis* Information Management System (AIMS) at the *Arabidopsis* Biological Resource Centre (ABRC); The Institute for Genome Research (TIGR); and the University of Minnesota (U.Minn).

Currently, bioinformatic tools and databases are accessible through networked servers (the Internet) that use Web browsers (**browsers**). The Internet, and improvements to the **Web servers** (servers) have revolutionized the extent to which bioinformatic tools can be used by the individual bench-scientist. No longer is it necessary to upgrade programs and update databases constantly on a stand-alone computer; researchers now have nearly instant access to the most current and extensive data via the Web. Improvements to interface features and design, and the databases they access, will enable researchers to obtain and analyse vast amounts of bioinformatic data more easily.

2. Basic tools for the Internet

The Internet provides a mode of communication among researchers through electronic mail (e-mail) and file transfer protocol (**FTP**). Software programs and databases are accessed through the browsers on the Internet. This section describes how to access and use the basic Internet tools and is divided into five subsections: Web basics; Getting onto the Web; Using a browser; Browser tips and errors; and Privacy issues. If you're 'online', you may want to skip to Section 3, perhaps casting your eyes over 'Browser tips and errors' (Section 2.4) on the way.

2.1 Web basics

To get onto the Web, you need a computer, a connection to the Internet, and a browser. The two most popular browsers are Netscape's Communicator and Microsoft's Internet Explorer, both of which operate in a similar way. Unless otherwise mentioned, all that follows refers to the use of Netscape version 4.0.7., although Internet Explorer should work the same.

2.1.1 URLs and Internet addresses

The key to finding anything on the Web is its location—its Uniform Resource Locator (**URL**)—e.g. http://genome-www.stanford.edu/Arabidopsis/search.html (AtDB's Search URL). The '**http**://' prefix, which stands for Hypertext Transfer Protocol, is the standardized method used to transfer information across the Internet and provides **hotlinks** (highlighted, active text, or images that take you to a new location) from one document to another. In this chapter, all URLs are given without the http:// prefix. You do not need to add the http:// because the browsers assume it is there.

Web pages are made available on the Internet from computers set up to act as servers. The name of the server (or host) comes immediately after the 'http://' prefix and ends at the first forward slash. In the URL given above, this is genome-www.stanford.edu/. The server name is the human readable form of what the computer uses, an Internet Protocol (**IP**) address. Translation of host names into IP addresses is performed on the Internet by servers called Domain Name Servers (**DNS**s). AltaVista provides useful definitions of these and other Internet-related terms. The part of the host name between the last dot and the first forward slash gives the domain that is the country's location (if not in the US), and other elements indicating the type of the organization, such as a company (e.g. '.com' in the US, '.co.uk' in the UK), an educational institution (e.g. '.edu' in the US, '.ac.jp' in Japan), a non-profit organization ('.org'), or an Internet-related organization ('.net').

2.1.2 Web file names and HTML

The path by which specific information is sent by the server appears after the host name (each directory is separated by a forward slash). (The path may be a directory on the server's disk.) In the URL given above, the path is `Arabidopsis/search.html`, where '`Arabidopsis`' is a directory and '`search.html`' is a file. The file's suffix tells the browser, and you, more about the file. For example, files ending in **.txt** are plain text files and those ending in **.gif** (Graphics Interchange Format) or **.jpeg** (Joint Photographic Experts Group) are images, but the vast majority of Web pages end in .html. This stands for **Hypertext Markup Language** (**HTML**) and is the standard language for Web presentation.

2.1.3 Making your own Web pages

Turning a document into HTML to put onto the Web is easy. Many programs now have a 'Save as... HTML' option. Furthermore, sophisticated Web pages can be created quickly by using the browser's Composer. In addition, there are many useful Internet guides to HTML, such 'The Bare Bones Guide to HTML' (`werbach.com/barebones`), which is available in several languages. If you are creating your own pages, particularly if they may be accessed by many people—it's generally best to keep things relatively simple. Flashing signs and animations using, for example, **Javascript** (a language that makes Web pages more dynamic) or **Java applets** (small programs that are downloaded by your browser and make pages more interactive) may look good, but at best they slow down the loading time for your users—as they take up more **bandwidth**—and at worst, they can cause your user's browser to crash.

2.2 Getting onto the Web

2.2.1 Getting a connection

If you are at an academic institution or a company, you are already likely to have a connection to the Internet. If your computer is at home, or elsewhere, you will probably need to get access to the Internet via an **Internet Service Provider** (**ISP**) and a means to access the ISP. In general, this means will be using a standard telephone-line and a '**modem**' connected to or built into the computer. The modem changes the digital signals of the computer into sounds that conventional telephone-lines can send. Other options include an **Integrated Services Digital Network** (**ISDN**) line, which is obtained from local telephone companies and uses standard telephone wires, but allows digital signals to be passed at fairly high speeds (up to about four times faster than the current 56 k modems). A quicker connection is also available through cable television systems. You can even access the Internet from your laptop (portable) computer by using a mobile (cellular) telephone or, in some areas, radio-based services.

2.2.2 Setting up your browser

Once you are connected to the Internet, you can customize your Web browser to work more efficiently. Suggestions for doing this are described in *Protocol 1*.

10: Web-based bioinformatic tools for Arabidopsis researchers

> **Protocol 1.** Setting up your Netscape browser
>
> A. *Recommended*
>
> 1. Go to the 'Edit:Preferences:Mail' and Groups menu. (The colon denotes a submenu.)
> 2. In 'Identity', type in your e-mail address, e.g. jsmith@weed.univ.edu.
>   - This allows you to send e-mail from your browser.
>   - You can add other information should you wish.
> 3. In 'Mail Server' field, type in your user name (which will usually be your login name, e.g. jsmith) and the name of the mail server, which is the computer that looks after your mail (weed-mail.univ.edu in the above example).
>   - This allows you to receive and read e-mail from your browser.
> 4. In 'Groups Server', type in the name of your news server[a] (a computer that sends and receives news messages over the Internet).
>   - This allows you to read messages sent to newsgroups.
>
> B. *Optional*
>
> 1. Use the 'Edit:Preferences' menu to customize other settings on the browser, such as:
>   - Type and size of font.
>   - Colour and style of hotlinks.
>   - The browser's home page (the one with which your browser opens).
> 2. The 'User Profile Manager' (see your browser's help) is a file on your computer, which allows you to store your settings in an individual profile. This allows several users to customize a given browser on machines with more than one user.
>
> [a] If you don't know the name of your news or mail server, ask your ISP help staff or local network administrator.

2.3 Using your browser

Once linked to the Internet, you can: send and receive e-mail or files; access newsgroups; and browse Websites. Browsers let you do all three in a near-seamless fashion.

2.3.1 E-mail and attachments

To understand what is possible when using your browser for e-mail, it helps to know that computer files are of two main types: text and **binary**. For many

229

years, e-mail has been used to send text messages. Text files comprise just text, the characters represented by the keys on a standard US keyboard. Binary files, such as word processed documents and images (e.g. .gif), need specific software to be interpreted by computers so they can be presented to humans.

Binary files can be 'attached' to e-mail messages easily by using browsers. The user simply clicks on 'Attach' and selects the file to include with the mail message. When attaching documents to e-mail messages, care has to be taken to ensure that the recipient has the correct software to read and edit the document on his or her computer. A document produced in the latest version of word processing software, for example, may not be readable by earlier versions of the same software. One solution is to save the file as an earlier version of the program and also send *another* version that can be read by many programs. For example, save: Word processor (such as Word) documents as text only or, ideally, Rich Text Format (**RTF**, which preserves most formatting); spreadsheet (Excel and the like) files as 'tab-delimited text'; graphics (PhotoShop, Corel Draw, etc.) either as .gif or .jpeg. Be aware, however, that some software features may be lost when converting to these common formats.

2.3.2 Sending files over the Internet: FTP

Large attachments can make an e-mail too big for some mail systems. To move around large datasets on the Internet, FTP is recommended. This is the quickest method to transfer files and can be performed using your browser. The prefix '`ftp://`' in the URL indicates that you are connected to an FTP site. At an FTP site, you can use your browser's 'File menu' to download (the 'Get' command) or upload ('Put') files. Most FTP sites require a username and password, in which case, enter 'anonymous' as the username, and then type your full e-mail address as the password. This is referred to as 'anonymous FTP'. Free, non-browser programs are available that allow you to do FTP easily, such as 'wsftp' for Windows and 'Fetch' for Macintosh. The use of Fetch is described in Appendix 1.

2.3.3 Newsgroups and mailing lists

Newsgroups (bulletin boards, discussion groups, Usenet) are read by newsreading software, which is generally an integral part of browsers. Newsgroups are accessed from the browser in the same way that your personal e-mail is (see *Protocol 1*), i.e. through the 'Communicator:Message Center' menu. The browser Help can be useful here as various options are available.

People post messages onto newsgroups and others read and reply. Some newsgroups are 'moderated'—meaning that undesirable messages are screened out by volunteers. This is the case with arab-gen (bionet.genome.arabidopsis), the electronic *Arabidopsis* newsgroup, which is a valuable source of information for the *Arabidopsis* researcher. Instructions on how to read and sub-

10: Web-based bioinformatic tools for Arabidopsis researchers

scribe to this newsgroup are given at: `genome-www.stanford.edu/Arabidopsis/newsgroup.html`. Instructions on how to access many newsgroups and to search old messages to them can be found at `www.dejanews.com`. Suggested rules of behaviour for using newsgroups and other Web manners—so-called Netiquette—are in Netscape's browser's 'Help' menu.

Some newsgroups, including arab-gen, also act as mailing lists, whereby messages sent to the list are then re-sent to all subscribers. Mailing lists alone (not newsgroups) are often used to keep relatively small groups of people in touch with each other. The usual way to get on a mailing list is to send a message with the word 'subscribe' and the name of the newsgroup to a specified e-mail address (which will be *different* from the address used to send messages). (Type 'unsubscribe' and the name of the newsgroup to get off the list.)

2.3.4 Searching the Web

An entire new industry that provides Web search services has recently developed (see *Table 1*). These use different engines, which search in slightly different ways and tend to give different results, so it is necessary to use the correct syntax and case. Capitalization of words will affect some searches—especially when using **boolean operators** such as AND, OR, NOT—as will quotation marks, which usually give only exact matches. AltaVista, for example, found no hits for the quoted term, 'Arabidopsis thaliana Database Home Page', while the same term without quotes produced hits, but different again from results obtained when the term was entered in all lowercase.

Search Spaniel provides a simultaneous search of the major engines. Specific sites, such as WhoWhere? and Four11, are available for finding someone's e-mail address or telephone number.

Web search engines collect their information through software termed **spiders**, which search the Web and FTP pages, extract words and put them into an index. This creates massive databases that are scanned when you search—you are not checking the whole of the Internet live! Therefore, a search will sometimes result in a 'URL not found' error (see Section 2.4) because the link is out of date on the results page.

In general, search engines will not find data displayed on the Web that originates from databases (as opposed to information on stand-alone Web

Table 1. Some Web search engines

Yahoo!	www.yahoo.com
eXcite	www.excite.com
Lycos	www.lycos.com
AltaVista	www.altavista.com
Search Spaniel	www.searchspaniel.com

pages). This is because database sites prevent spiders indexing all their potential Web pages, as the amount of indexing required would slow access considerably for users.

2.3.5 mailto and contacting Webmasters

Many Web pages have a 'mailto' hotlink, which enables you to send an e-mail message to someone connected with the information on that page. Technical problems with the site (e.g. a link within that site that consistently doesn't work) should be addressed to the site's 'Webmaster', while data-related problems are generally sent to a named individual or a curator. Always give the URL when reporting a problem.

2.4 Browser tips and errors

This section provides protocols for browser use and trouble-shooting. *Protocol 2* describes the causes of the four most frequent error messages. *Protocol 3* describes the remedies for further common problems, and *Protocol 4* gives tips to assist in smooth Internet access.

Protocol 2. Common browser-related error messages

A. *'The server does not have a DNS entry. Check the server name in the Location (URL) and try again'.*

1. This is usually caused by the server's host name in the URL being incorrectly typed—by you or the person who wrote the hotlink on the site you're viewing.
2. Carefully check the URL *up to* the first forward slash.
3. Sometimes this error occurs when the host name has just been changed and the DNS has not yet been updated.

B. *'404 Not Found' or similar error*

1. This is usually caused by a typing error in the filepath of the URL (the part after the host name).
2. Carefully check the URL *after* the first forward slash.
3. This error can also occur if something is 'down' (not working) at the site to which you are connecting.
4. If you are sure the URL is correct, try to link again in a few hours.

C. *'Broken pipe'*

1. This is usually the result of a bad cgi program or a severed connection somewhere along the way. It tends to occur more frequently when a modem is used.
2. Try reconnecting by clicking on the hotlink again.

10: Web-based bioinformatic tools for Arabidopsis researchers

D. 'No data'
1. The server took too long to reply. If the browser (or server) has not received anything in 90 sec, it will automatically sever the link, although you may not realize this. This can occur when you are attempting an extensive search of a database.

In addition to the error messages outlined above, there are several common problems that are encountered when using browsers. These and others are described in *Protocol 3*. In some cases, the answer to your problem may be on the site you're visiting. Browse around for a 'Help' or 'Frequently Asked Questions' (FAQ) page.

Protocol 3. Dealing with common problems when linking to Websites

Before giving up or complaining to a Website, check the following:

A. Is the problem only temporary?
1. If you get an error message, make sure it is not just a temporary problem; check again later.
2. A page may briefly be unavailable for several reasons (e.g. while being updated or a short-term network problem at the server).

B. Am I in the right place?
1. Look at the URL (optionally displayed in the browser's menu bar) to check that you are where you think you are.
   (a) Hotlinks can make it difficult to realize you've moved away from a site.
   (b) A good guide is to check that the server's host name (such as genome-www.stanford.edu), is the same or similar to where you started.
   (c) The messages displayed at the bottom of the browser can help in this regard, as it tells you the URL to which a hotlink will connect.
2. Check for a logo or site title on the page or frame.

C. Is it a problem at my end?
1. If possible, check that the problem you're having isn't related to your browser or the way you're accessing the Internet.
2. If your browser is old, try using a later version, a different browser, or another computer.

Protocol 3. *Continued*

3. **Firewalls**, computers set up to control Internet traffic into a site, can sometimes be a problem.
   (a) Check with the person responsible for your network if you're getting, for example, a blank page back from a query over the Web after you've hit a 'submit' button.
   (b) Contact a friend with an account on a different network and ask them to try. If it works for them, the problem is most likely to lie somewhere on your computer or network.

D. *Caches and unchanged pages*

1. If you expect a Web page to have changed and it hasn't (you may have altered something on your own pages or submitted data somewhere and expected to see it incorporated) and it doesn't appear to have done so, you may be looking at on old version in your **cache**. This is where the browser stores recently visited pages.
2. Try clicking the 'Reload' button while holding down the shift key. This forces the browser to go the server and get a fresh version of the Web page.
3. If this doesn't do the trick, go to the 'Preferences:Advanced:Cache' menu and clear the cache. For this reason, it's a good idea not to make your cache too large (5–10 MB is ample).

Protocol 4. Useful tips for smooth Internet access

A. *Opening another window*

Some hotlinks are designed to open another window.

1. To *force* a hotlink to open a new window, hold down the mouse button (middle on Unix, right on Windows) while clicking on the link.
2. This is sometimes a useful tactic when you want to compare two or more pages, or when the site uses frames (panels where only part of the window changes when you click on a hotlink).

B. *Saving information*

It is important to realize that the information displayed on your browser is now on your computer, which means you can usually save it in a form you can use later.

1. Go to the 'File:Save As...' menu.
   (a) Saving as 'text' will give you the information you're reading.
   (b) Saving as 'source' will, with .html files, save all information received by the browser. This allows you to save the HTML codes.

10: Web-based bioinformatic tools for Arabidopsis *researchers*

2. This is a good way of using other people's HTML for your own pages! Be aware, however, that the 'Save As...' source will generally not work if a program or database has produced the page. This is usually indicated by the directory path in the URL containing the term 'cgi' (see Section 2.1.2).

3. To save images, hold down the (right) mouse button over the images and then use the 'Save Image As...' option from the menu that opens.

C. *Slow connections*

When using a slow connection, such as a modem from home, you may want to modify the way you use your browser.

1. Go to the 'Preferences:Edit:Advanced' menu and turn off the automatic loading of images. This speeds things up considerably; you can always click the 'Images' button to load them should you wish. Even with a 'fast' connection to the Internet, some sites, particularly if trans-Atlantic, can be slow. This can be caused by:

    (a) Heavy traffic on the Internet.

    (b) The site having a slow connection to the Internet.

2. Connect when the Internet is less busy, e.g. outside US office hours, or when the site itself is quieter. Some sites have usage statistics that show the average number of connections ('hits') the site receives through the day. This can be used to pick a quieter time.

3. Be aware of the browser timing-out (see *Protocol 2*D). Keep an eye on the messages displayed at the bottom of the browser.

2.5 Privacy issues

You should be aware that browsing the Web is far from an anonymous activity. Your IP and, sometimes, e-mail address can be recorded and your movements logged when you visit a Website. **Cookies**, for example, are designed to pass information back to the Website and can be very useful. Once you've typed in your password to a site, the cookie will obviate the need to do this when you logon on again from the same machine. If you are worried about potentially less benign motives, cookies (and Java and Javascript) can be turned off in the Preferences:Advanced menu of your browser. For information on Web security issues, go to www.anonymizer.com.

3. Scenarios of bioinformatic use in *Arabidopsis* research

By this point in the chapter, the reader should be reasonably familiar with the Internet and browsers. The use of Web-based bioinformatic tools in *Arabidopsis*

From a Mutant Phenotype to a Gene Function

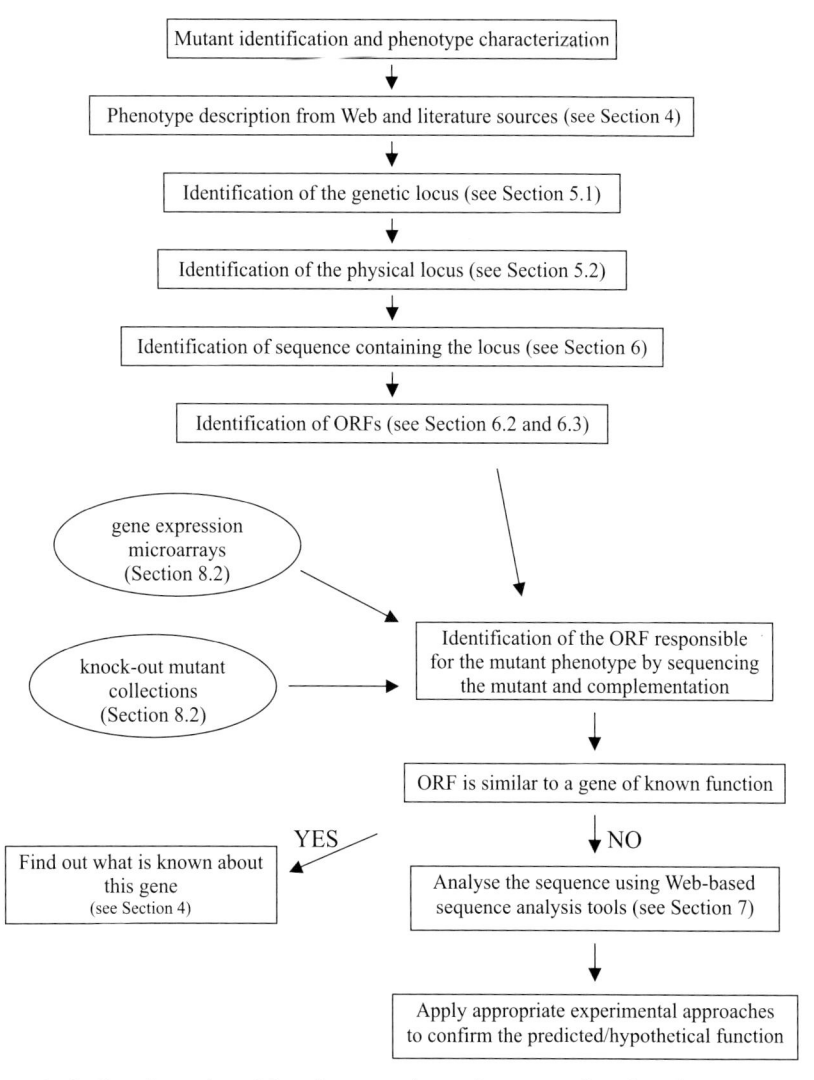

Figure 1. A flowchart describing how to determine gene function from a mutant phenotype for an *Arabidopsis* gene. Section numbers from this chapter are given where relevant bioinformatic tools for a given step are described.

research can now be addressed. To illustrate a typical use of these tools, a flowchart describing how to determine gene function from a mutant phenotype of an *Arabidopsis* gene is presented in *Figure 1*. This chart also points the reader to the relevant parts of this chapter.

4. Gene information resources
4.1 General gene information
Entrez, NCBI's search tool (*Table 2*) is an excellent place to find biological information of genes. Entrez contains structural data cross-linked to bibliographic information from PubMed, to the sequence databases, and to taxonomy. There is also a 3D-structure viewer, Cn3D, for easy interactive visualization of molecular structures from within Entrez. For literature searches, PubMed provides extensive services including links to many journals available on the Web.

If the gene of interest is, or is predicted to be, an enzyme, look for metabolic pathway maps in the Kyoto Encyclopaedia of Genes and Genomes or in EcoCyc, which contains all known *E. coli* metabolic pathways. To research the phylogenetic relationship of genes, start with the Tree of Life project, which contains information about the diversity of organisms on Earth, their history, and characteristics.

In some cases, information relevant to your gene of interest may be gleaned from other organisms. Currently, database interoperability (see Section 8.1) is under development, with a goal of common vocabulary so that information from different organism databases can be accessed easily. For now, gene information in a specific organism is best reached at organism-specific databases.

Table 2. General gene information resources

Entrez	www.ncbi.nlm.nih.gov/Entrez
PubMed	www.ncbi.nlm.nih.gov/PubMed/
Kyoto Encyclopedia of Genes and Genomes	www.genome.ad.jp/kegg/
EcoCyc	ecocyc.PangeaSystems.com/ecocyc/server.html
Online Journals	www.nih.gov/science/journals/
Tree of Life	phylogeny.arizona.edu/tree/phylogeny.html
SGD	genome-www.stanford.edu/Saccharomyces/
FlyBase	flybase.bio.indiana.edu/
MouseDB	www.jax.org/

4.2 *Arabidopsis* gene information
AtDB is the most comprehensive public database for bioinformatic data of *Arabidopsis* (*Table 3*). It also provides a useful launching pad to other sites that specialize in the types of information they contain, such as phenotypes of mutants (AIMS), *Arabidopsis* expressed sequence tags (ESTs) (TIGR, U.Minn), and genetic markers (NASC).

For *Arabidopsis* gene information, AtDB provides a specialized search engine, Arabidopsis Gene Hunter. This finds information about a given

Arabidopsis gene from a selectable list of *Arabidopsis*-related Websites and databases. The result is a concatenation of clickable Web pages from the sites searched. AtDB also provides information on *Arabidopsis* papers and a comprehensive list of *Arabidopsis* research laboratories. Information about individual researchers in the *Arabidopsis* community is provided at AtDB, NASC, and AIMS.

The *Arabidopsis* Genome Resource (AGR) is an additional information resource being developed at NASC (see also Chapter 1) as part of the UK CropNet comparative mapping project. The AGR aims to integrate sequence data with physical and genetic maps for *Arabidopsis* and with other crop species. This is an ongoing development which currently contains the physical data for chromosome 4 and 5, sequence, and new analysis of AGI sequence. This database serves as a resource for comparative mapping studies, assisting in gene function analysis and allowing crop orthologues to be identified.

Large numbers of mutants have been isolated and characterized in *Arabidopsis*. David Meinke's homepage provides links to detailed information on mutant gene symbols, rules of gene nomenclature, linkage data, genetic maps, and e-mail addresses of contributing laboratories (*Table 3*). The stock centres (NASC, AIMS, Lehle Seeds) provide a comprehensive collection of seeds of mutants, accessions, and ecotypes (see Chapter 1 and 2).

Functional genomic experiments, such as gene expression microarrays and identification of null alleles of all genes, are currently under development. In the foreseeable future, data on all *Arabidopsis* genes and their expression patterns, cellular roles, functions, and evolutionary relationships will probably be available online.

4.3 Plant gene information

For information on plant genes in general (*Table 4*), the Mendel database project (Mendel DB) aims to develop a common nomenclature, based on

Table 3. *Arabidopsis* gene information resources

AtDB	genome-www.stanford.edu/Arabidopsis/
AIMS database	aims.cps.msu.edu/aims/
TIGR	www.tigr.org/tdb/at/at.html
U.Minn	www.cbc.umn.edu/ResearchProjects/Arabidopsis/index.html
NASC	nasc.life.nott.ac.uk/home.html
Lehle Seeds	www.arabidopsis.com/
Arabidopsis Gene Hunter	genome-www.stanford.edu/cgi-bin/AtDB/geneform
Arabidopsis Labs	genome-www.stanford.edu/Arabidopsis/labs.html
Meinke's gene registry	mutant.lse.okstate.edu/
Arabidopsis Genome Resource (AGR)	synteny.nott.ac.uk/agr/agr.html

10: Web-based bioinformatic tools for Arabidopsis *researchers*

Table 4. Plant gene information resources

Mendel DB	jiio6.jic.bbsrc.ac.uk/index.html
Mendel DB (Stanford mirror)	genome-www.stanford.edu/Mendel/
USDA list of plant genome databases	probe.nalusda.gov:8300/plant/index.html
UK CropNet	synteny.nott.ac.uk/
Internet Directory of Botany	herb.biol.uregina.ca/liu/bio/botany.html
Rice Genome Research project	www.staff.or.jp/
Cyanobase	www.kazusa.or.jp/cyano/cyano.html
Snapdragon database	www.mpiz-koeln.mpg.de/~stueber/snapdragon/snapdragon.html

gene families, for sequenced genes from all photosynthetic organisms, the organelle genomes of both photosynthetic and non-photosynthetic organisms (fungi, algae, and protozoa), and plant viruses. Mendel DB contains information on: gene name, gene synonym, accession number, gene product name and synonym, expression (in a few cases at present), coding sequence coordinates, and SWISS-PROT or PID numbers. For faster access from the US, Mendel DB is mirrored at AtDB.

The USDA list of plant genome databases at the National Agriculture Library provides an interface to many plant databases that use ACEDB software. UK CropNet uses ACEDB software to manage information for comparative mapping and genome research on *Arabidopsis*, barley, Brassicas, forage grasses, and millet.

The Internet Directory of Botany provides an extensive, searchable list of botany-related Websites. Useful plant organismal databases include: the Rice Genome Research project; Cyanobacteria database; and the Snapdragon database.

5. Maps

In addition to knowing what genes do, it is also important to know where they lie in relation to one another, both genetically and physically. Genetic and physical maps that have been constructed for *Arabidopsis* are described in this section.

5.1 Genetic maps

There are several types of genetic maps for *Arabidopsis*. AtDB has the latest versions of three of these in its database: Lister and Dean recombinant inbred (RI) (6), classical genetic (7), and mi-RFLP (8). There are other types of genetic map that are not currently in the database, such as the Simple Sequence Length Polymorphism (SSLP) (9). These can be found in the Web pages under Maps section at AtDB. A list of URLs for genetic maps is provided in *Table 5*.

5.1.1 Lister and Dean RI map

The Lister and Dean (RI) map comprises molecular markers that are polymorphic between the ecotypes (accessions) Columbia and Landsberg *erecta* and is currently the most extensive genetic map available (7). Each marker has been mapped using the RI lines and its ecotype-specific pattern is used to map new loci relative to the markers. The RI map is prepared by Sean May and co-workers at NASC (see Chapter 3). This map is regularly updated and currently comprises 953 markers that include restriction fragment length polymorphisms (RFLPs), simple sequence length polymorphisms (SSLPs), cleaved amplified polymorphic sequences (CAPS) (10), cloned genes, ESTs, and the ends of bacterial (BAC) and yeast (YAC) artificial chromosomes.

5.1.2 Classical genetic map

The classical genetic map is maintained by David Meinke and colleagues at Oklahoma State University (7). This map is produced by analysing segregating phenotypes in the F2 generation following self-pollination of F1 plants. It currently contains over 450 mutant genes.

5.1.3 Other genetic information

There are two types of molecular genetic markers based on PCR that are worth mentioning here: CAPS and SSLP markers.

(a) CAPS are co-dominant, ecotype-specific, PCR-based markers (see Chapter 3); new markers continue to be developed by members of the *Arabidopsis* research community. Eliana Drenkard and Fred Ausubel (Massachusetts General Hospital) maintain the current list of markers and provide it in tabular form for each of the five chromosomes. Many CAPS markers are mapped on to the Lister and Dean RI map. The latest list of CAPS markers maintained by Drenkard and Ausubel (and by other groups) can be found at AtDB from its Maps page (see *Table 5*).

(b) An SSLP primer pair amplifies a unique segment of genomic DNA containing a simple tandem repeat. The PCR products differ in length between Columbia and Landsberg *erecta*, and often among other ecotypes. Information on SSLP markers can be found on the *Arabidopsis thaliana* Genome Center (ATGC)'s Website (see *Table 5*). The primers themselves are available from Research Genetics Inc.

5.2 Physical maps

At the time of writing all five chromosomes of *Arabidopsis* are completely covered by overlapping clones except at the centromeric and nucleolar organ-

Table 5. *Arabidopsis* physical and genetic maps

Genetic maps

AtDB genetic map displays	genome-www.stanford.edu/Arabidopsis/maps.html
Lister and Dean RI maps	nasc.nott.ac.uk/new_ri_map.html
Classical genetic maps	mutant.lse.okstate.edu/classical_map_dir.html
CAPS markers	genome-www.stanford.edu/Arabidopsis/aboutcaps.html
SSLP maps	genome.bio.upenn.edu/SSLP_info/SSLP.html

Physical maps

Unified physical map display	genome-www3.stanford.edu/cgi-bin/AtDB/Pchrom
Other physical maps	genome-www.stanford.edu/Arabidopsis/maps.html

izing (NOR) regions (11, 12). AtDB has assembled currently available physical map data into a single map for each chromosome (*Figure 2*). This display does not attempt to resolve conflicts or ambiguities, but instead presents the user with the most consistent interpretation of the comprehensive clone and probe information. It provides links between genetic and physical maps in a comprehensive and unambiguous manner. This aspect of the unified display can be useful for researchers wishing to position new genetic loci on the physical map.

The unified display of the physical map consists of three graphics, drawn directly from information in the database. The first graphic displays the five *Arabidopsis* chromosomes with landmark probes. Bars lie below each chromosome. These represent contigs, coloured according to the method of the mapping. All the contigs and the chromosomes are hotlinks and will take the user to the next graphic. The second graphic presents all the clones and probes in the particular region that is clicked. In this graphic, the most comprehensive contig is used as the framework. Other contigs are placed relative to the framework contig, using probes that were used in both contigs. Each clone or probe in each contig, when clicked, spawns a new page containing details of hybridizing clones and probes and the source and type of the hybridization information, clone library source, and probe source and type. If the probe is also a genetic marker, comprehensive marker information is provided. If a clone is undergoing AGI sequencing, its status is provided with links to: AtDB's AGI graphical display, the relevant annotation page of the sequencing groups, and GenBank. Each clone presented in this page has a link back to the appropriate physical map page.

In addition to the unified display of the physical maps at the AtDB, physical maps of different regions of the genome can be found on the mapping groups' Websites from AtDB (see *Table 5*, and Chapter 9).

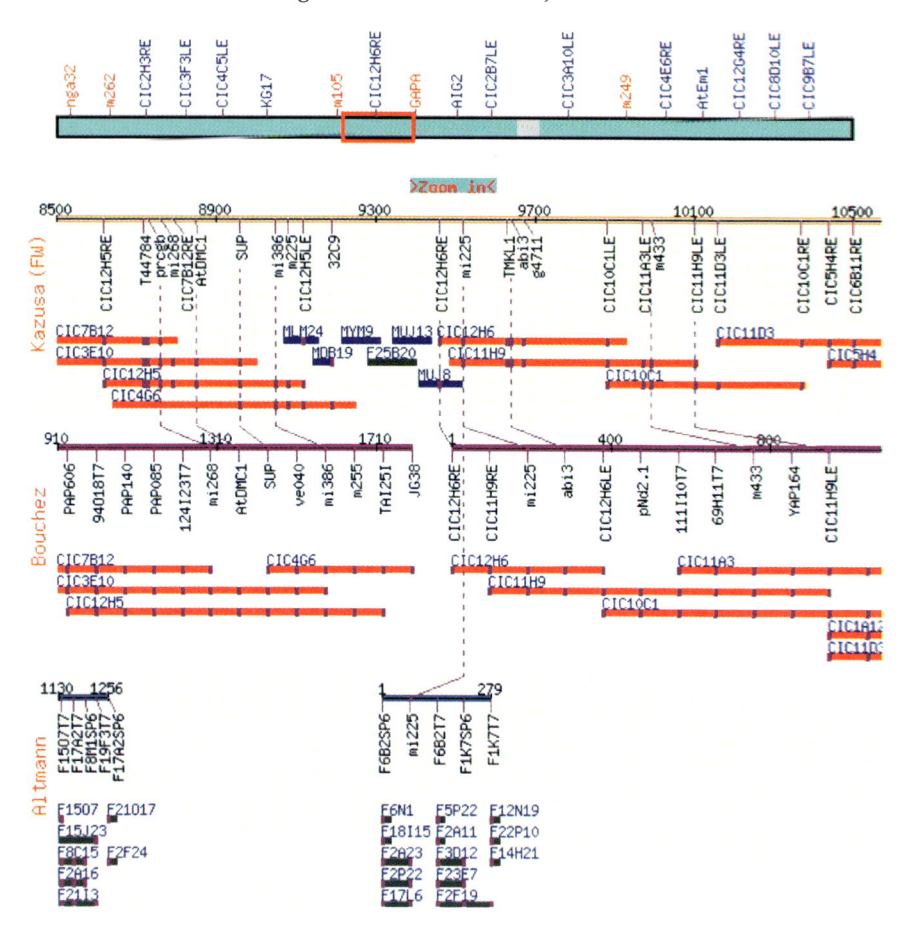

Figure 2. Physical map display from AtDB showing a region of chromosome 3 between 8500 and 10500 kb. Different types of contigs from various research groups covering the region (named at left) are shown. Each contig type is presented as a narrow bar with kb coordinates and probes along it. Below the contigs are clones of differing type. Hybridizations are indicated on the clones by tick marks. Dashed lines show probes common to more than one contig. Details on clones and probes are obtained by clicking on them. (URL: genome-www3.stanford.edu/cgi-bin/AtDB/Pmap?chr=3&beg=8500&end=10500)

6. Sequencing

Having the sequence of a whole genome allows bioinformatics to play a crucial role in uncovering all the genes and gene complements of an organism. *Arabidopsis* has one of the smallest genomes of any flowering plant (~ 130 Mb) and is forecast to be sequenced by the end of the year 2000.

10: Web-based bioinformatic tools for Arabidopsis *researchers*

6.1 The *Arabidopsis* genome initiative (AGI)

AGI is the international consortium to sequence the entire genome of *Arabidopsis*. In November 1998, 39 Mb of AGI sequence is in GenBank, approximately 34% of the expected total of 130 Mb (see *Table 6*). In addition, there are approximately another 8 Mb of annotated *Arabidopsis* sequence in GenBank that originate from non-systematic sequences.

AtDB provides the central station for information about AGI progress by collating information through a combination of direct submission by AGI participants and programmatically scanning new or changed *Arabidopsis* sequences from GenBank. These data are presented in graphical displays, which are updated automatically from the database and include a 'progress meter' of AGI sequence in GenBank, and a table listing sequencing status. For further information on AGI sequencing, see Chapter 9.

6.2 Annotation of sequences by AGI

Annotation (definition of features of the DNA sequence, including ORF and gene-function designation) is an essential and time-consuming aspect of genome sequencing. Annotation of AGI-sequenced clones is provided on AGI participants' Websites. Information about annotation methods used by different AGI groups can be found from the Websites (see *Table 6*). Most of the AGI groups' annotation methods involve using sequence homology searches such as *BLAST* (see Section 7.1) and *FASTA* (see Section 7.2), followed by the use of exon and splice site identification programs such as *NetPlantGene*, *GRAIL*, and *GeneFinder* (see Section 7.4). Some groups also subject their sequences to further analyses using motif search programs (see Section 7.6). Annotation currently requires manual curation to make a decision on the identity of open reading frames (ORFs). This task is assisted by the extensive collection of ESTs, which, together with other cDNA data, help identify real ORFs and splice sites (see Section 7.3).

Results of these endeavours place genes into one of five main categories. Those that:

(a) Exactly match existing known genes, often where the sequencers have re-sequenced well studied *Arabidopsis* genes.
(b) Show similarity to known genes from *Arabidopsis* or other organisms. These are usually described as 'similar to' or 'putative'.
(c) Show similarity to unknown proteins.
(d) Show no similarity to defined proteins, but have EST matches.
(e) Are predicted to be transcribed, but with no known similarity. These are described as 'hypothetical proteins'.

The lower down the list a 'gene' is, the more cautious you should be when using the data. For these less certain annotations, further sequence analysis (see Section 7) should be carried out, preferably in conjunction with biological evidence.

Knowledge-based, expert systems of annotation, in which programs will be able to make the 'best judgement' on the identity of ORFs and gene function, are currently under development. This, along with more biological data such as gene expression microarrays and knock-out phenotypes, will facilitate the speed and accuracy of annotation.

6.3 Caveats in annotation

Much of the annotation relies on information from public sequence databases such as GenBank. The great advantage of GenBank is that anyone can deposit sequences and their annotations. However, errors are easily introduced and propagated. The types of errors you should look out for include:

(a) Redundant gene names. The same gene name may have been used for two different proteins. An example is the *Arabidopsis* gene name AAT1, which has been used for an aspartate aminotransferase (ATAAT1) and a cationic amino acid transporter (ATLBAATI, ATAAT1G).

(b) Misplacement of gene families: annotation errors. Using the search term 'Arabidopsis GLP3' at Entrez brings up nine separate entries for the GLP3 gene. At first sight, these may appear to be part of an extensive gene family. Closer inspection of the entries, however, reveals that eight of them have very minor differences in sequence and probably represent the same locus. One entry, ATGLP3, is, however, clearly different from all the other GLP3s and is therefore a different gene. In fact, it appears to be a GLP2, a different member of the family.

(c) Clustering of genes: very similar sequences. Occasionally, members of a closely related gene family will have very similar sequences. An example of this is the vacuolar H^+-pumping ATPase family from *Arabidopsis*. In this case, the amino acid sequences of the genes ava-p and ava-p1 appear to be identical, and both of these are, in turn, very similar to ava-p2e. However, a nucleotide-based *BLAST* search reveals that there are a number of differences between these loci, suggesting that, unlike eight of the GLP3 entries mentioned above, these do represent different loci. Thus the amino acid sequence does not provide enough detail to reveal the differences between these genes.

(d) Compounded errors. An ever-growing problem is that of error propagation. This is particularly so with the increasing amount of '(semi-)automatically' annotated genomic sequences, where annotation is based upon prediction and homology to existing entries, without experimental corroboration. Examples exist (28) whereby gene1 that has domains A and B has been ascribed function and name according to domain A only. A second entry containing domains B and C, without careful checking, will be described incorrectly as gene1-like, owing to the presence of B. If a third entry is entered with domains C and D, it, through homology to C in the second entry, will be described as gene1, without having any connection to it.

6.4 Sequence contigs from AtDB

AtDB is currently constructing sequence contigs to anchor non-redundant, contiguous sequences to the unified physical map and to provide graphical annotation. In addition to the different annotation methods provided by individual AGI participants, GenBank entries from non-AGI sequences often have annotations without information on which programs were used. The sequence contigs from AtDB will provide a consistent presentation of annotation methods and provide links to the genetic markers from the physical map.

Table 6. AGI information

Sequencing summaries

Sequencing progress meter	genome-www.stanford.edu/Arabidopsis/agi.html
AGI Labs	genome-www.stanford.edu/Arabidopsis/AGI/AGI_links.html

Sequencing annotations

ATGC	genome.bio.upenn.edu/annotation/annotation.html
CSHL	nucleus.cshl.org/protarab/AnnotationExternalWeb.html
KAZUSA	www.kazusa.or.jp/arabi/about.html
MIPS	pedant.mips.biochem.mpg.de/index.html
PGEC	pgec-genome.pw.usda.gov/annotation.html
Stanford U.	www-sequence.stanford.edu/ara/ann_methods_new.html
TIGR	www.tigr.org/tdb/at/atgenome/annotation/annotation.html

7. Sequence analysis tools

Once genomic sequence is available, the next task is to identify gene functions. The quality and quantity of sequence analysis tools are increasing. The purpose of this section is to introduce the reader to the rationale and limitations of some of the key sequence analysis tools and databases. To facilitate systematic identification of genes, several computer algorithms have been developed. Functional assignment to new sequences can be divided into two parts: comparative and compositional analyses.

(a) Comparative methods of sequence analysis include identification of sequence similarity by searching databases for pairwise or multiple alignments of: total sequence, short sequence patterns (motifs), or protein structure patterns.

(b) Compositional methods include gene modelling from exon prediction and splice site prediction programs. There are several Websites that provide further information and links to sequence analysis tools (see Section 7.8).

An important issue when considering sequence analysis tools is that of specificity versus sensitivity. Specificity is defined as the mode of identifying significant homologies to structure (primary, secondary, or tertiary) and patterns of structure that results in the least number of 'false positives'. Sensitivity is the mode of identifying the highest number of possible homologies to structure and patterns of structure. There is a trade-off between sensitivity and specificity in any sequence analysis tool, depending on its format and parameters.

7.1 *BLAST*

The most common and fastest analysis to find the identity of a gene is *BLAST* (Basic Local Alignment Search Tool) (13), which searches nucleotide or protein datasets for overall sequence similarity by a pairwise alignment.

7.1.1 AtDB's Web *BLAST* and datasets

AtDB provides a *BLAST* Web service (*Table 7*), which searches against all publicly available *Arabidopsis* sequences (see *Protocol 5*). In addition to nucleotide sequences, a non-redundant *Arabidopsis* protein sequence dataset is created by merging all entries with identical sequences to form a unique set of sequences from GenPept, SWISS-PROT, and PIR. AtDB's *BLAST* datasets include the Mendel database-curated set of higher plant proteins, and all higher plant DNA sequences from GenBank. Subsets of the *Arabidopsis* datasets may be searched, such as GenBank sequences new or changed in the last month, or BAC end-sequences. *BLAST* provides facilities to search translations from nucleotide to protein of both the query and the dataset. Note that the *BLAST* server at Stanford is a single machine, users are currently restricted to one search at a time.

AtDB's datasets are updated three times a week; a weekly summary of *Arabidopsis* accessions that are new or have changed is automatically generated and posted to the *Arabidopsis* electronic newsgroup (see Section 2.3.3).

Protocol 5. Using AtDB's *BLAST*

First, go to AtDB's *BLAST* form from AtDB's home page (genome-www.stanford.edu/Arabidopsis).

A. *Basic use*

1. Type or paste in your sequence. It must be longer than 15 characters and contain no comments. Numbers are ignored and so can be included.

10: Web-based bioinformatic tools for Arabidopsis researchers

2. (Optional) Type a comment into the box.
3. From the pop-up menu, select the *BLAST* program you wish to run:
   (a) blastn = nucleotide query to nucleotide dataset.
   (b) blastp = protein query to protein dataset.
   (c) blastx = translated (six frames) nucleotide query to protein dataset.
   (d) tblastn = protein query to translated (six frames) nucleotide dataset.
   (e) tblastx = translated (six frame) nucleotide query to translated (six frame) nucleotide dataset.
4. Choose one of the sequence datasets:
   (a) GenBank = all *Arabidopsis* GenBank DNA, including EST and BAC ends.
   (b) Recent = *Arabidopsis* GenBank sequences new/changed in the last month.
   (c) NRAT = non-redundant *Arabidopsis* protein.
   (d) BACEND = BAC end-sequences for *Arabidopsis*.
   (e) EST = EST sequences for *Arabidopsis*.
   (f) Annotated = *Arabidopsis* GenBank entries minus EST and BAC ends.
   (g) CSHLPrel = CSHL/WashU. Preliminary *Arabidopsis* genomic sequences.
   (h) AtRepbase = *Arabidopsis* DNA repeat database from CSHL.
   (i) Mendel = Mendel plant gene nomenclature database (higher plant proteins).
   (j) PlantDNA = all higher plant sequences from GenBank.
5. Press the submit button. If you are in a queue, a moving graphic shows how your search is progressing (and keeps the browser connected; see *Protocol 4*).
6. A results page will appear, an example of which is shown in *Figure 3*.

B. *Advanced use*
1. Many parameters can be changed by the user when doing a *BLAST* search with AtDB's Web form. Some of these options are described in Section 7.1.2.

7.1.2 *BLAST* search parameters

The user should understand the action of gapping and filtering. Other parameters to be aware of include expect threshold and comparison matrices.

(a) Gapping. AtDB uses the *BLAST2* program from Washington University (Warren Gish, unpublished), which allows gaps (deletions and insertions) in the alignment. This is the AtDB default setting and can be switched by the user to a non-gapped alignment.

(b) Filtering. The user can also define sequence filtering. If used, this masks out regions with low compositional complexity (such as internal repeats or poly(A) sequences) from the query sequence. Filtering can eliminate statistically significant, but biologically uninteresting, reports from the *BLAST* output. A low complexity region found by a filter program is substituted in the search program by the letter 'N' in nucleotide sequence (e.g. 'NNNNN') and the letter 'X' in protein sequences (e.g. 'XXXXX').

(c) Expect threshold. The expect threshold ('E') reflects the number of matches expected to be found by chance, based upon the size of the dataset. Decreasing the E threshold will increase the stringency of a search and so fewer matches will be reported. The level of E should be increased when searching with short sequences, as these are more likely to occur by chance in the dataset.

(d) Comparison matrices. Key elements in evaluating the quality of a pairwise sequence alignment are the comparison (scoring) matrices (e.g. BLOSUM62, PAM250), which are used to align and identify regions of sequence similarity. These can be changed to affect the results. The default BLOSUM-62 matrix is good for detecting most weak protein similarities and is probably suitable for most users.

7.1.3 AtDB's *BLAST* results page: general

Figure 3 shows the results of a typical *BLAST* search. At the top of the *BLAST* result pages is a graphical summary of the result, which shows the query sequence and the position and size of the best matches to it. Shown below is information about the type of search performed and a series of one-line descriptions of all the matching sequences. Next is a set of the actual alignments of the query sequence with database sequences, and the last section lists the parameters used and the statistics generated during the search.

7.1.4 AtDB's *BLAST* results page: HSPs

The one-line descriptions give information about dataset sequences that form a high scoring segment pair (HSP) with the query sequence. *BLAST* uses HSPs to identify sequences of similarity. An HSP is created when two sequence fragments (one from the query sequence and the other from the dataset sequence) show a locally maximal alignment for which the alignment exceeds a pre-defined cut-off score, which can be changed by the user.

Each one line description has a hotlink to the GenBank entry for the matching sequence and two important HSP-related statistics. The first is the high score, which gives the score value of the highest scoring (and therefore

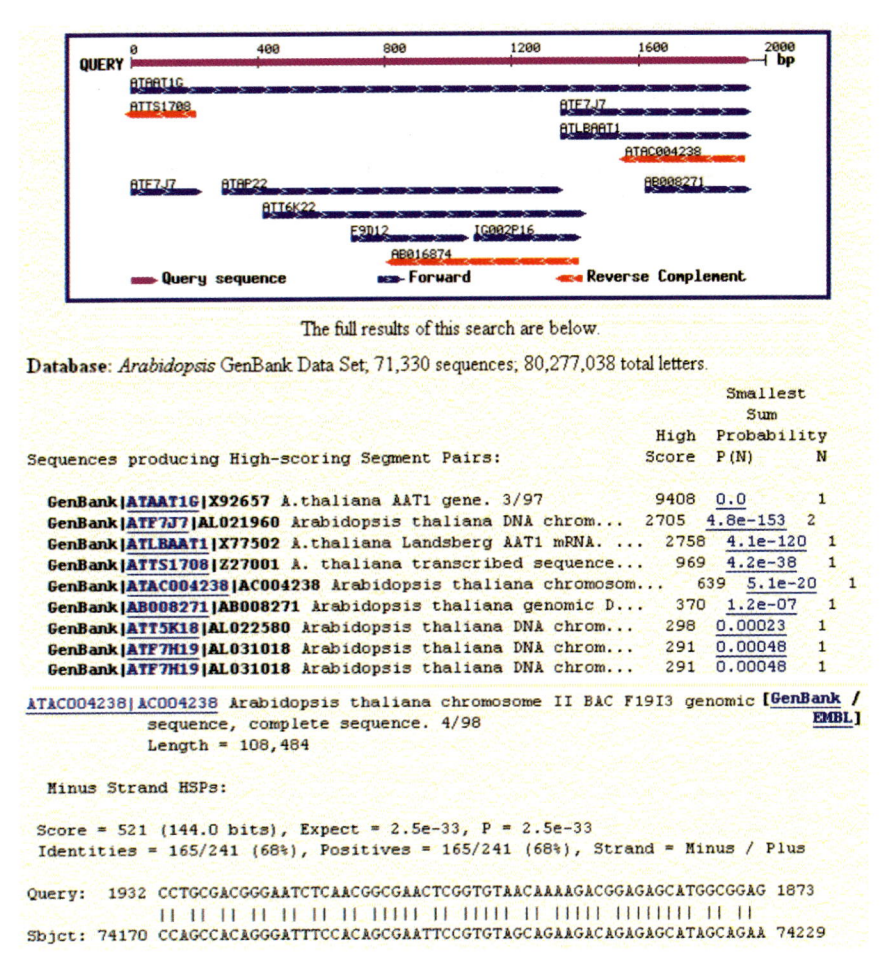

10: Web-based bioinformatic tools for Arabidopsis researchers

The full results of this search are below.
Database: *Arabidopsis* GenBank Data Set; 71,330 sequences; 80,277,038 total letters.

```
                                                            Smallest
                                                              Sum
                                                 High     Probability
Sequences producing High-scoring Segment Pairs:  Score      P(N)       N

GenBank|ATAAT1G|X92657 A.thaliana AAT1 gene. 3/97          9408   0.0         1
GenBank|ATF7J7|AL021960 Arabidopsis thaliana DNA chrom...  2705   4.8e-153    2
GenBank|ATLBAAT1|X77502 A.thaliana Landsberg AAT1 mRNA. ... 2758  4.1e-120    1
GenBank|ATTS1708|Z27001 A. thaliana transcribed sequence... 969  4.2e-38     1
GenBank|ATAC004238|AC004238 Arabidopsis thaliana chromosom... 639 5.1e-20    1
GenBank|AB008271|AB008271 Arabidopsis thaliana genomic D... 370  1.2e-07     1
GenBank|ATT5K18|AL022580 Arabidopsis thaliana DNA chrom... 298   0.00023     1
GenBank|ATF7H19|AL031018 Arabidopsis thaliana DNA chrom... 291   0.00048     1
GenBank|ATF7H19|AL031018 Arabidopsis thaliana DNA chrom... 291   0.00048     1

ATAC004238|AC004238 Arabidopsis thaliana chromosome II BAC F19I3 genomic [GenBank /
            sequence, complete sequence. 4/98                              EMBL]
            Length = 108,484

Minus Strand HSPs:

 Score = 521 (144.0 bits), Expect = 2.5e-33, P = 2.5e-33
 Identities = 165/241 (68%), Positives = 165/241 (68%), Strand = Minus / Plus

Query:   1932 CCTGCGACGGGAATCTCAACGGCGAACTCGGTGTAACAAAAGACGGAGAGCATGGCGGAG 1873
              || || || || || || || ||||| || ||||| || ||||| |||||||| || ||
Sbjct:  74170 CCAGCCACAGGGATTTCCACAGCGAATTCCGTGTAGCAGAAGACAGAGAGCATAGCAGAA 74229
```

Figure 3. Result page of an AtDB *BLAST* search. At the top is a graphical summary showing the query sequence and the position and size of the best matches to it. Shown below is a series of one-line descriptions of all the matching sequences (not all shown). Next is an example of part of an HSP (high scoring pair) alignment of the query sequence to a database sequence. In the actual page, all pertinent alignments are shown.

best) HSP for alignments of the query to that database sequence. There will often be more than one HSP for each hit sequence. The second is the smallest sum probability, or P-value, which is the probability of the HSP of that score or higher occurring in the dataset by chance and is dependent upon the size and composition of the dataset and of the query. The lowest (and best) P-value is given. The *larger* the *high score* and the *smaller* the *P-value*, the more likely the match is to be significant statistically. Note, however, that this may or may not relate to biological significance. When interpreting *BLAST* results, the overall sequence alignment and any information about the query or matching sequence should be considered.

7.1.5 AtDB's *BLAST* results page: sequence alignment

The sequence alignments of every HSP (above the cut-off score) for each database match to the query sequence are shown with the query sequence at the top and the aligned database sequence at the bottom. Dashes indicate where gaps were introduced to achieve the alignment. Conservative amino acid changes are indicated by a plus sign between the aligned residues. The start and end points of the areas of similarity are shown at the left and right of the aligned sequences. It is important to note the extent of the alignment, as a good match may only occur over a small part of either the query or matching sequence. This can be significant, for example, when annotating an unknown sequence with an annotated match from the dataset, as other domains may be present in either the query or database sequence, apart from the HSP.

7.1.6 Other *Arabidopsis BLASTs*

Other *Arabidopsis*-specific Web *BLASTs* are available at: Washington University, for their AGI sequence (see Section 6.1); Cold Spring Harbor Laboratory, for their database of repeated sequences in *Arabidopsis*; TIGR for *Arabidopsis* BAC end-sequences; and MIPS, for ESSA sequences. All of the sequences provided by these sites are also incorporated into AtDB's *BLAST* datasets.

7.1.7 Other *BLAST* programs

BLAST searches can be performed over the Web at NCBI to screen for matches against sequences from all organisms in GenBank. This is useful if you're looking for non-plant similarities. NCBI also provides variations of the *BLAST* program for downloading by FTP (see Section 2.3.2), some of which can be used over the Web. These include:

(a) *PSI-BLAST*, which can be more sensitive than gapped *BLAST* for weak, yet biologically relevant similarities. It starts with a single sequence and combines multiple alignment with similarity searching by taking the best hits from a conventional *BLAST* search and using the regions of alignment in subsequent iterative *BLAST* searches.

(b) Pattern Hit Initiated (*PHI-*)*BLAST*, which combines matching of regular expressions (see below) with local alignments surrounding the match. This allows the user to find other protein sequences that contain a given pattern and are homologous to the query sequence in the vicinity of the pattern.

7.2 *FASTA*

An alternative to *BLAST* for comparing a DNA or protein sequence with a dataset of sequences is *FASTA* (14). *FASTA* initially identifies sequences

10: Web-based bioinformatic tools for Arabidopsis researchers

with a degree of similarity to the query sequence (but in a different way from *BLAST*; see Section 7.2.1 and 7.2.2) and then conducts a second comparison on the selected sequences. *FASTA* is slower, may give different results, and is better for motif (see also Section 7.6) searching than *BLAST*. As with the *BLAST2* used by AtDB, *FASTA* tolerates gaps in the aligned sequences.

7.2.1 AtDB's Web *FASTA* and datasets

AtDB provides a *FASTA* Web service (*Table 7*), which is used in a similar way to the *BLAST* form (Section 7.1) and searches the same *Arabidopsis* and plant datasets. Note that a major difference is that *FASTA* only searches one DNA strand at a time. An option is provided to search the reverse complement and is strongly recommended after each sense-strand search. In addition *FASTA* searches can also be conducted from other websites (see *Table 7*).

7.2.2 *FASTA* search parameters

If the FASTA search results in no, or very few, matches, then you can alter the search parameters to increase the sensitivity. There are fewer variables than with *BLAST*, but they include comparison matrices (as for *BLAST*), the expected number of matches (increase for short queries), and optimization threshold (a parameter that controls the trade-off between speed and sensitivity of the search).

7.2.3 AtDB's *FASTA* results page: general

The results page of a *FASTA* search is presented in sections. First is a brief description of the type of search performed, followed by a histogram showing the number of sequences in the dataset with different degrees of identity to

Table 7. BLAST and FASTA URLs

Arabidopsis BLAST services

AtDB BLAST	genome-www2.stanford.edu/cgi-bin/AtDB/nph-blast2atdb
AtDB FASTA	genome-www2.stanford.edu/cgi-bin/AtDB/nph-fastaatdb
CSHL	nucleus.cshl.org/protarab/AtRepBase.htm
MIPS	muntjac.mips.biochem.mpg.de/arabi/blast_arabi.html
TIGR	www.tigr.org/tdb/at/atgenome/bac_end_search/bac_end_search.html
Washington U., St. Louis	genome.wustl.edu/gsc/arab/arabidopsis.html

Other *BLAST* services

NCBI	www.ncbi.nlm.nih.gov/BLAST/
BLAST2 (Wash. U.)	blast.wustl.edu/blast/README.html
BLAST help	www.ncbi.nlm.nih.gov/BLAST/blast_help.html

Databases

GenPept (GenBank)	www.ncbi.nlm.nih.gov/
SWISS-PROT	www.expasy.ch/sprot/
PIR	www-nbrf.georgetown.edu/pir/

the query sequence. Below this is a series of one line descriptions of the matching sequences. The next section shows the actual alignments generated between the query sequence and the dataset sequences.

The one-line descriptions give information about sequences that show statistically significant similarity to the query sequence and include several numbers. The most important of these is the Z-score, which is hotlinked to the appropriate alignment. The higher this is, the more similarity the query sequence shows to the subject sequence.

Overall similarity searches such as *BLAST* and *FASTA* can find closely related genes. However, distantly related genes may not be found from these searches. To analyse the sequence further, either to find other matching genes or to perform experimental research, it is helpful to define the ORFs of the sequence. Searching datasets of ESTs and using gene prediction programmes can be used to predict ORFs, which should be confirmed with experiments before analysing the gene further. These algorithms and datasets are described below.

7.3 EST databases

ESTs were developed to allow rapid identification of expressed genes by sequence analysis. All ESTs in GenBank (dbEST) can be searched using NCBI's *BLAST* and *FASTA*. ESTs are partial, single-pass sequences from either end of a cDNA clone, which have become valuable aids for annotation of genomic sequence (see also Section 6.2). In particular, they play an important role in confirming ORFs that have been predicted computationally. EST matches to a putative exon make that prediction much more likely to be correct. In addition, ESTs are assembled into virtual—often full-length—transcripts, termed tentative consensus (TC) sequences by TIGR (and clusters by others). *Arabidopsis* gene index (*Table 8*) at TIGR allows searches of TCs and ESTs by: nucleotide or protein sequence; the tissue used to construct the cDNA library; the name or identifier of the library; gene product name; or the database identifier (including GenBank). In addition, U.Minn provides a suite of software tools for the analysis of ESTs including:

(a) Search reports of EST similarity analysis. This allows a text search, such as 'kinase', of sequence analysis reports for ESTs from plant genome datasets for rice and loblolly pine, as well as *Arabidopsis*.

Table 8. EST databases

DbEST (GenBank)	www.ncbi.nlm.nih.gov/dbEST/index.html
Arabidopsis Gene Index	www.tigr.org/tdb/agi/index.html
EST similarity search	www.cbc.umn.edu/ResearchProjects/Search/index.html
Plant EST databases	www.cbc.umn.edu/cgi-bin/blasts/public_blasts/blastn.cgi

10: *Web-based bioinformatic tools for* Arabidopsis *researchers*

(b) Search plant EST databases. This allows you to search your DNA sequence for similarity against ESTs from a variety of plants, including *Arabidopsis* and maize.

7.4 Gene identification programs

If a new sequence matches strongly to ESTs, they and TCs (if available) can be used to predict ORFs. A series of gene prediction programs can be used to refine the designation of an ORF from the EST data, or to find ORFs if there were no matches to any ESTs. Generally, a suite of programs is employed to model a new sequence, where each program is used for a different aspect of gene prediction. For example, one could use *NetPlantGene* to identify splice sites, *GENSCAN* to find exons, and *GRAIL* for terminal exons (*Table 9*). A useful analysis of the efficacy of several gene identification programs for *Arabidopsis* has been conducted (15, and manuscript in preparation). This work is based on the analysis of 104 genes representing 731 exons represented in cDNAs aligned to genomic DNA sequence. In summary, the authors found that a combination of programs gives the best results. When a single program was used to predict the perfect gene model (defined as one without falsely predicted exons and with successful prediction of all splice sites, ATG, and terminal codon) they determined success rate to be: *GENSCAN* ~ 25%, *GeneFinder* ~ 20%, and *GRAIL* ~ 13% (Larry Parnell, personal communication).

7.4.1 *NetPlantGene*

NetPlantGene (NPG) (16) uses artificial **neural networks** combined with a rule-based system to predict intron splice sites in *Arabidopsis*. This approach increases specificity and so reduces the number of false positives. Users simply enter their sequence of between 200 and 80 000 bases (larger sequences can be processed via e-mail). The results are returned in a Web page and give lists of the position of each predicted splice site, its strand (direct or complement), the level of confidence, and 20 bases of sequence around the predicted site. These are given in turn for donor and acceptor sites for each strand, and contain predictions made by two detection levels for true sites: one level where approximately 50% of the true sites are detected with very few false positives; and another level where nearly all true sites are found, but with more false positives. The predictions are also displayed in graphical format, which can be downloaded to the user's computer.

A newer version, *NetGene2*, predicts splice sites in human and *C. elegans*, as well as *Arabidopsis* sequences. This works in a similar way to NPG, except that users can input a *FASTA* file directly from their computer. *NetGene2* has a better performance for predicting donor and acceptor sites as judged by plotting false positives against sensitivity.

7.4.2 GENSCAN

GENSCAN (17) uses a probabilistic model of gene structure and compositional properties to predict complete gene structures, including exons, introns, promoter, and poly(A) signals, in genomic sequences. *GENSCAN* also handles partial genes. Sequences up to 200 kilobases (kb) in length may be pasted into the form, and a list of predicted genes, together with the corresponding peptide sequences, is returned. A graphical version of the predicted exons is also available.

7.4.3 GRAIL

GRAIL (18) uses a neural network that combines a series of coding prediction algorithms to provide analysis of the protein coding potential of a DNA sequence. There are several flavours of *GRAIL*. Version 2 incorporates predictions 'trained' on *Arabidopsis* sequences and uses genomic context information (splice junctions, translation starts, non-coding scores of 60 base regions on either side of a putative exon) to enhance the exon recognition process. This, however, makes it inappropriate for sequences where the regions adjacent to an exon are not present. In addition to exons, *GRAIL* can search for poly(A) sites, CpG islands, repetitive DNA, and frameshift errors. *GRAIL* usually predicts the terminal exon accurately and so is much less likely to extend falsely a gene model into the next gene.

7.4.4 GeneFinder

GeneFinder (P. Green and D. Hillier, unpublished) systematically uses statistical criteria—mainly log likelihood ratios (LLRs)—to identify genes in plants, yeast, and others. Candidate genes are evaluated on the basis of 'scores' that reflect their splice site, translation start site, coding potential LLRs, and intron sizes. *GeneFinder* can be used via e-mail and is also usable via the Web. The user enters a sequence and chooses one of the following plant sequence analysis programs: *FGENEA*, which constructs a gene model by exon assembling; *FEXA*, which searches for potential 5', internal, and 3' coding exons; or *ASPL*, which searches for potential splice sites. The results can be returned via e-mail or in a Web page.

Table 9. Gene identification programs

NetPlantGene	www.cbs.dtu.dk/netpgene/cbsnetpgene.html
NetGene2	www.cbs.dtu.dk/services/NetGene2/
GRAIL	compbio.ornl.gov/tools/index.shtml
GeneFinder	dot.imgen.bcm.tmc.edu:9331/gene-finder/gf.html
GenScan	CCR-081.mit.edu/GENSCAN.html

10: Web-based bioinformatic tools for Arabidopsis *researchers*

7.5 Gene family analyses

Once the ORF has been designated, its protein sequence can be subjected to further analysis to predict functional identification. Overall sequence alignment tools such as *BLAST* and *FASTA* can easily detect closely related genes and proteins, but they may miss distantly related genes or proteins. However, these can be identified by alignments to clusters of sequence patterns (motifs), or domains. Motifs are discovered by multiple sequence alignments and motif analyses. Some of the tools used to do this are described in this and Section 7.6.

7.5.1 Multiple sequence alignments

The increasing amount of redundancy and number of sequences in databases make it more time-consuming to find and analyse sequence similarities. Identification of gene and protein families not only reduces the size of the datasets, but also increases the sensitivity of gene function prediction by looking for changes in conserved regions. Multiple sequence alignments are used to identify sequence similarity characteristics of a family and to estimate phylogenetic relationships of protein families. *ClustalW* is the most popular method of multiple sequence alignment that compares overall sequence similarity (*Table 10*). It is designed to increase sensitivity of aligning divergent protein sequences (19). There are several other multiple sequence alignment tools geared towards finding motifs, such as *MEME* (29) and *BlockMaker* (30).

Results of multiple sequence alignments can be edited and displayed using *BOXSHADE*, *Logos* (31), and *CINEMA* (32). *BOXSHADE* highlights conserved residues; *Logos* represents the degree of conservation by stacking conserved residues in a column; and *CINEMA* is an editing tool that allows the user to manipulate the alignments.

7.6 Motif analyses

Searching a database of motifs allows information from multiply aligned sequences to be reduced in noise. This increases sensitivity to distant relationships. Similar to searching against gene family databases such as Pfam (*Table*

Table 10. Multiple sequence alignment tools

ClustalW	www-igbmc.u-strasbg.fr/BioInfo/ClustalW/clustalw.html
MEME	www.sdsc.edu/MEME/meme/website/meme.html
BlockMaker	www.blocks.fhcrc.org/blockmkr/make_blocks.html
BOXSHADE	ulrec3.unil.ch/software/BOX_form.html
CINEMA	www.biochem.ucl.ac.uk/bsm/dbbrowser/CINEMA2.02/index2.html
Logos	www.blocks.fhcrc.org/about_logos.html

11), motif information is represented in a position-specific scoring table (profile), in which each column of the alignment is converted to a column of a table representing the frequency of occurrence of each of the 20 amino acids. A motif can be used to assign a newly sequenced protein to a specific family of proteins using the profiles and thus to formulate hypotheses about its function. There are several databases that have generated motifs from multiple sequence alignment of known and unknown proteins. Some of the most extensive and their associated search tools are briefly described in this section. For more details, the reader is encouraged to visit the individual Websites.

7.6.1 PROSITE

PROSITE is a curated database of protein domains and families. In addition to a few motifs found in the literature, most of the motifs in PROSITE are identified using multiple sequence alignments of protein families using parameters of biological significance such as catalytic sites, binding sites, or prosthetic group attachment sites (20). In addition to motifs, PROSITE contains scoring tables of position-specific amino acid weights and gap costs (profiles) that may be useful in identifying new members of families that exhibit extreme sequence divergence. PROSITE currently contains motifs specific to approximately a thousand protein families or domains. Each of these comes with documentation providing background information on the structure and function of the proteins. New sequences can be searched for motifs in PROSITE using programs such as *Block Searcher* and *Blimps*.

7.6.2 BLOCKS

Blocks are multiply aligned, ungapped segments corresponding to the most highly conserved regions of proteins. *Block Searcher*, *Get Blocks*, and *Block Maker* are programs used to detect and verify protein sequence homology (21). They compare a protein or DNA sequence, to a database of protein blocks, retrieve blocks, and create new blocks, respectively. The 2000 blocks in the BLOCKS database are made automatically by looking for the most highly conserved regions in groups of proteins represented in the PROSITE database and calibrating them against the SWISS-PROT database to obtain a measure of the chance distribution of matches.

The most powerful inference for finding a new family member in BLOCKS comes from a new sequence that contains more than one block, which strengthens the match and is represented by a P-value. P-values cannot be assigned to matches that only contain one matched block in the new sequence and interpretation of such matches should be made with caution. Another caveat with searching the BLOCKS database lies in interpreting repeats. Some repeats are reported patterns in PROSITE. Therefore, they will sometimes cause an alignment; although the repeat is real, the homology is not. In addition, compositional bias of the query can inflate the alignment score of a block that has a similar compositional bias. This will often appear as a repeat, with the block aligned at multiple positions along the compositionally biased segment.

7.6.3 PRINTS

PRINTS is a database of protein fingerprints. A fingerprint is a group of conserved motifs used to characterize a protein family. It is defined and refined by iterative scanning of sequences in the OWL database using multiple sequence alignment tools. The OWL database is a non-redundant composite of four publicly available protein databases. Although strict redundancy criteria are applied to the amalgamation of the primary sources, error checking of the sources themselves is not undertaken. Fingerprints can encode protein folds and functionalities more flexibly and powerfully than can single motifs. PRINTS thus provides a useful adjunct to PROSITE. Usually the motifs do not overlap, but are separated along a sequence, though they may be contiguous in 3D space. Currently PRINTS contains 990 entries, encoding 5701 individual motifs.

PRINTS can be browsed, or searched, with a new sequence using the programs *BLAST*, *FingerPRINTScan*, or *Blimps*.

7.6.4 EMOTIFs

The motifs in the databases described above are manually compiled using a given set of parameters that can limit the range of specificity and sensitivity of motif matches to a new sequence. *EMOTIFs* generates a set of motifs with a wide range of specificity and sensitivity by using an algorithm that incorporates biochemically-constrained amino acid positioning in enumerating the possible motifs (22). This program was used to generate 50 000 motifs from 7000 protein alignments in the BLOCKS and PRINTS databases. These motifs can be accessed from the IDENTIFY database.

7.6.5 *Arabidopsis* motif searching

AtDB's *Pattern Matching* searches the same datasets used by AtDB's sequence alignment programs, *BLAST* and *FASTA*, and is an alternative method for identifying motifs. These are identified using simple terms or regular expressions, which allow for ambiguity and for up to three mismatches, insertions, or deletions, as well as for direct matches. For DNA searches, either strand or both strands together can be searched. An example of a regular expression search is 'L{3,5}X{2}[AVLI] X>', which will find protein subunits that have leucine (L) three to five times({3,5}), followed by two ({2}) of any amino acid (X), then an aliphatic amino acid ([AVLI]), and then any amino acid (X) at the carboxyl terminal (>). Examples of hits from this search include LLLLAPLM and LLLEKAK.

U.Minn's *Motif Explorer* performs a similar function to *Pattern Matching*

Table 11. Motif analysis tools

Pfam	www.sanger.ac.uk/Software/Pfam/
PROSITE	expasy.hcuge.ch/sprot/prosite.html
BLOCKS	www.blocks.fhcrc.org/
PRINTS	www.biochem.ucl.ac.uk/bsm/dbbrowser/PRINTS/PRINTS.html
OWL db	www.biochem.ucl.ac.uk/bsm/dbbrowser/OWL/OWL.html
EMOTIFs	dna.stanford.edu/emotif/
IDENTIFY	dna.stanford.edu/identify/
AtDB's *Pattern Match*	genome-www2.stanford.edu/cgi-bin/AtDB/PATMATCH/nph-patmatch
U.Minn's *Motif Explorer*	lenti.med.umn.edu/gst/MotifExplorer.html

program, but searches U.Minn's EST *Arabidopsis* dataset or International Protein Sequence Database (PIR).

7.7 Protein structures

If any of the above sequence similarity search tools fails to result in useful matches, searching for protein structures may be an alternative. There are many programs that predict structural domains such as coiled-coil regions, transmembrane helices, solvent accessibility, and secondary structures (helix, strand, and loop) (*Table 12*). One drawback of using structural prediction programs is that each program has a different level of specificity, depending on the amount of experimental data and the parameters used. For example, a transmembrane helix can be predicted with higher than 95% accuracy, whereas predicting a secondary structure is about 70% accurate (23, 24). The PredictProtein Server is a useful structural prediction platform that allows all these programs to be used at one sitting; it also provides the observed frequencies of false positives for all of these programs. A list of useful URLs for structural analysis is provided by Sacch3D at SGD.

Most of the information in the structural prediction programs comes from the Protein Data Bank (PDB), the most comprehensive protein structure database. PDB currently contains over 6000, 3D coordinates. These data are organized into hierarchical classification schemes in derived databases, such as Structural Classification of Proteins (SCOP) and CATH. SCOP contains hierarchical classifications based on near and far evolutionary relationships (family and superfamily), as well as on geometrical relationships (folding patterns). In SCOP, new sequences can be searched for sequence similarity using *BLAST*, classes can be searched by keyword, and trees can be browsed.

If a new protein has 3D coordinates available, SCOP and CATH can be searched for structural similarity. In addition, the Vector Alignment Search Tool (VAST) algorithm can be used to compare related structures directly. VAST allows display and manipulation of 3D structural superpositions.

10: Web-based bioinformatic tools for Arabidopsis researchers

Table 12. Protein structure databases and search tools

Paircoil	nightingale.lcs.mit.edu/cgi-bin/score
Repeat Finder	www.proweb.org/proweb/Tools/selfblast.html
PredictProtein Server	www.embl-heidelberg.de/predictprotein/
Sacch3D	genome-www.stanford.edu/Sacch3D/urls.html
PDB	www.pdb.bnl.gov/
FSSP	www.embl-heidelberg.de/dali/fssp/
SCOP	scop.mrc-lmb.cam.ac.uk/scop/
Dali Server	croma.ebi.ac.uk/dali/
CATH	www.biochem.ucl.ac.uk/bsm/cath/
VAST	www.ncbi.nlm.nih.gov/Structure/vast.html

7.8 Comprehensive sequence analysis tools

The increasing number of sequence analysis tools is overwhelming even for the most sophisticated bioinformatics specialist. Thankfully, there are some Websites that provide multiple sequence analysis tools which allow you to visit several sites at one sitting (see *Table 13*). BCM Launcher at the Human Genome Center and GeneQuiz provide a list of sequence analysis tools in a logical manner starting with pairwise similarity search and ending with secondary structure predictions. SRS (Sequence Retrieval System) cross-references a wide range of molecular biology databases at many sites. For those interested in setting up their own set of sequence analysis tools, Genetic Data Environment (GDE) allows integration of several programs into a single analysis environment.

Cold Spring Harbor Laboratory provides a well-organized list of sequence analysis tools and resources. A less logical, but comprehensive set of tools can be found in Pedro's Molecular Tools.

Another useful service is the Sequence Alerting System, which performs daily searches of new protein sequence databank entries and alerts you to new

Table 13. Comprehensive sequence analysis tools and resources

BCM Launcher	kiwi.imgen.bcm.tmc.edu:8088/search-launcher/launcher.html
GeneQuiz	columba.ebi.ac.uk:8765/ext-genequiz/
SRS	expasy.hcuge.ch/srs5/
GDE	www.stanford.edu/~jeisen/GDE/GDE.html
CSHL Sequence Analysis page	formaggio.cshl.org/talks/BioWWW/handout.html
Pedro's Molecular Tools	www.public.iastate.edu/~pedro/research_tools.html
ExPASy	expasy.hcuge.ch/
Cross-reference Alerting System	www.ncbi.nlm.nih.gov/XREFdb/
Sequence Alerting System	www.bork.embl-heidelberg.de/Alerting/

homologues. This is currently provided by EMBL, MIPS, and SWISS-PROT. XREFdb additionally informs you of new mapping positions and phenotype descriptions, for human, mouse, fruit fly, worm, and *E. coli*.

8. Current issues and future directions in bioinformatics

8.1 Some important bioinformatic issues

Three important issues in bioinformatics that are relevant to bench researchers today are:

(a) *The need for experimentation* to base and confirm bioinformatic predictions and theories. Bioinformatics is not just computational modelling of biological information. Rather, it is about understanding the processes of an organism from the enormous amount of data resulting from recent technological developments. As such, the function of each gene needs to be defined by experimental data from, e.g. functional genomics studies. Computer programs and databases will make these processes more efficient, but will not replace experimentation.

(b) *Database interoperability*—the requirement for databases to talk to one another—both at the level of software development and controlled biological vocabulary. This will allow biologists to access and interpret the information from various databases in the most meaningful way (3, 4, 25). Organism-specific databases may not be easily understood by people not working on that organism. A step towards universal understanding is to ensure that the names and functions of genes conform to an agreed form.

(c) *Data accessibility*. The Web allows for easy display of data, but much useful information is not freely available as it remains behind company firewalls or is restricted to those with the means to pay for it. A related issue is that of online publication. Many journals publish electronic versions of their articles and may soon move to complete online publishing (see *Table 2*). There are many advantages of online publishing for biology, such as the ability to show rotating 3D structures moving images readily, to annotate reviews, and to cross-reference articles both forwards and backwards in time. The scientific community, however, may need to take steps towards preserving the integrity and quality of publication.

8.2 Bioinformatic tools currently under development for *Arabidopsis* research

Several aspects of bioinformatics are currently under development and will soon be available to the researcher over the Web. In computational biology, the ability to predict 3D protein structure from primary sequence is improving, but is still some time in the future. In genomics, many exciting projects are under development for *Arabidopsis* research. These include gene expression microarrays that depict the expression of all genes of an organism under various conditions (26). Similar to microarray technology, DNA chip technology from Affymetrix, which synthesizes oligonucleotides onto silicon to hybridize either cDNA or genomic DNA, will enable definitive assignment of ORFs and development of high resolution genetic markers, such as single nucleotide polymorphisms (SNPs). Currently, this technology is available for *Arabidopsis* and oligonucleotide-based arrays of approximately 400 SNPs have been developed (Ron Davis, Michael Mindrinos, and colleagues, unpublished data). A high resolution map will be essential for studying complex traits that arise from mutations at multiple loci. In addition, comprehensive collections of insertional mutants and identification of null alleles of all *Arabidopsis* genes are currently under development. In proteomics, protein expression arrays are being developed with 2D SDS–PAGE, combined with tandem mass spectroscopy (27). This technology will not only enable the designation of all proteins expressed under given conditions, but will also be useful for identifying proteins in protein–protein interaction screenings and protein complexes. In bioinformatics, the main challenge is to develop databases capable of housing comprehensive data that can be cross-referenced for all genes, this is currently under development in a few model organism databases such as SGD, FlyBase, and MGD (see *Table 2*).

9. Conclusion

Biological research is undergoing a paradigm shift in which the mode of data collection is changing from 'hunting and gathering' to 'global industrialization'. Individual researchers can now retrieve and analyse massive amounts of the information that is being produced at a tremendous rate. The change in data collection and organization should equip biologists with knowledge and tools to regard complex problems in biology, such as developmental, behavioural, and evolutionary biology, in a different light.

Acknowledgements

The authors thank Mike Cherry, Head, Genome Databases Group, Department Genetics, Stanford University, for support of this work and for valuable suggestions. They thank Larry Parnell for supplying unpublished information, and Laura Donohue, Margarita García-Hernández, Gail Juvik, Mark Schroeder, Gavin Sherlock, and Shauna Somerville for helpful comments on the manuscript.

References

1. Benton, D. (1996). *Trends Biotechnol.*, **14**, 261.
2. Safer, Jr. M. H. (1998). *Plant Physiol.*, **117**, 1129.

3. Sansom, C. (1997). *Nature Biotech.*, **15**, 1253.
4. Karp, P. D. (1998). *Trends Biochem. Sci.*, **23**, 114.
5. Somerville, C., Flanders, D., and Cherry, J. M. (1997). *Plant Physiol.*, **113**, 1015.
6. Lister, C. and Dean, C. (1993). *Plant J.*, **4**, 745.
7. Meinke, D. W., Cherry, J. M., Dean, C., Rounsley, S. D., and Koornneef, M. (1998). *Science*, **282**, 662.
8. Liu., Y. G., Mitsukawa, N., Lister, C., Dean, C., and Whittier, R. F. (1996). *Plant J.*, **10**, 733.
9. Bell, C. J. and Ecker, J. R. (1994). *Genomics*, **19**, 137.
10. Konieczny, A. and Ausubel, F. M. (1993). *Plant J.*, **4**, 403.
11. Rhee, S. Y., Weng, S., Bongard-Pierce, D. K., García-Hernández, M., Malekian, A., Flanders, D. J., et al. (1999). *Nucleic Acids Res.*, **27**, 79.
12. Mozo, T., Fischer, S., Meier-Ewert, S., Lehrach, H., and Altmann, T. (1998). *Plant J.*, **16**, 377.
13. Altschul, S. F., Gish, W., Miller, W., Myers, E. W., and Lipman, D. J. (1990). *J. Mol. Biol.*, **215**, 403.
14. Pearson, W. R. and Lipman, D. J. (1988). *Proc. Natl. Acad. Sci. USA*, **85**, 2444.
15. Schueller, C., Parnell, L., and Mayer, K. (1998). *9th Int. Conf. Arabidopsis Res. Abstract 355* (genome-www.stanford.edu/Arabidopsis/madison98/abshtml/335.html).
16. Hebsgaard, E., Korning, S. M., Tolstrup, P. G., Engelbrecht, N., Rouze, J., and Brunak, P. (1996). *Nucleic Acids Res.*, **24**, 3439.
17. Burge, C. and Karlin, S. (1997). *J. Mol. Biol.*, **268**, 78.
18. Uberbacher, E. C. and Mural, R. J. (1991). *Proc. Natl. Acad. Sci. USA*, **88**, 11261.
19. Thompson, J. D., Higginsl, D. G., and Gibson, T. J. (1994). *Nucleic Acids Res.*, **22**, 4673.
20. Bairoch, A., Bucher, P., and Hofmann, K. (1997). *Nucleic Acids Res.*, **24**, 217.
21. Henikoff, S. and Henikoff, J. G. (1994). *Genomics*, **19**, 97.
22. Nevill-Manning, C. G., Wu, T. D., and Brutlag, D. L. (1998). *Proc. Natl. Acad. Sci. USA*, **95**, 5865.
23. Rost, B. and Sander, C. (1993). *J. Mol. Biol.*, **232**, 584.
24. Rost, B., Casadio, R., Fariselli, P., and Sander, C. (1995). *Protein Sci.*, **4**, 521.
25. Ouzounis, C., Casari, G., Sander, C., Tamames, J., and Valencia, A. (1996). *Trends Biotechnol.*, **14**, 280.
26. Schena, M., Shalon, D., Davis, R. W., and Brown, P. O. (1995). *Science*, **270**, 467.
27. Clauser, K. R., Hall, S. C., Smith, D. M., Webb, J. W., Andrews, L. E., Tran, H. M., et al. (1995). *Proc. Natl. Acad. Sci. USA*, **92**, 5072.
28. Smith, T. F. and Zhang, X. (1997). *Nature Biotech.*, **15**, 1222.
29. Grundy, W. N., Bailey, T. L., and Elkan, C. P. (1996). *CABIOS*, **12**, 303.
30. Henikoff, S., Henikoff, J. G., Alford, W. J., and Pietrokovski, S. (1995). *Gene*, **163**, 17.
31. Schneider, T. D. and Stephens, R. M. (1990). *Nucleic Acids Res.*, **18**, 6097.
32. Parry-Smith, D. J., Payne, A. W., Michie, A. D. and Attwood, T. K. (1998). *Gene*, **221**, GC57.

10: Web-based bioinformatic tools for Arabidopsis *researchers*

Appendix 1. Using Fetch for Macintosh

A simple way to transfer files by FTP. Fetch is available at www.dartmouth.edu/pages/softdev/fetch.html

A. *Connecting to an anonymous FTP site*
1. Start Fetch.
2. Go to the 'File:New Connection' menu.
3. In 'Host', type the host name of the FTP site you're going to, e.g. genome-ftp.stanford.edu.
4. In 'User ID', type anonymous.
5. In 'Password', type your e-mail address (e.g. jsmith@weed.univ.edu).
6. Press the 'OK' button.

B. *Getting (downloading) files*
1. Connect to the site.
2. Select where you want to download files using the 'Customize:Preferences:Downloads' menu.
3. (a) Either 'Browse' to find the files or folders you want.
   (b) Or, use the 'Remote:Get Folders and Files...' menu and type in the file you want.
4. If you know whether the file is text or binary, click the appropriate button in the Fetch window. Otherwise click the 'Automatic' button. Fetch transfers the file by its extension (the suffix of the filename after the dot, e.g. .txt, .jpeg), so be careful if the file lacks an extension.
5. Press the 'Get File...' button.
6. Use 'Close' or 'Quit' to disconnect from the site.

C. *Putting (uploading) files*
1. Connect to the site.
2. Press the 'Put File' button and browse to select the file you wish to send. If you have several files or folders to send, use the 'Remote:Put Folders and Files...' menu.
3. From the Format section of the 'Put File...' window, select the correct format to send the file. Send:
   (a) Text only (.txt), rich text format (.rtf), tab-delimited text (.tab), or HTML (.html) files as 'text'.
   (b) Program (binary) files, such as from Word (.doc), or graphic files

Appendix 1. *Continued*

(.gif, etc.) as 'Raw Data'.

BinHex puts a Macintosh (binary) file in text form that can be sent through most electronic mail systems. MacBinary is used for storing a Macintosh file on a non-Macintosh system for later retrieval. It cannot go through most electronic mail systems.

4. Press the OK button.
5. Use 'Close' or 'Quit' to disconnect.

Glossary

Bandwidth	An Internet connection's capacity to transfer data (e.g. 56 kbytes/sec).
Binary files	Files that need specific software to be interpreted by computers and read by humans, including most word processed files and graphics files (see Text only).
Boolean operators	The use of AND, OR, or NOT to refine search terms.
Cache	Temporary store of Web pages on your computer.
Cookies	Small pieces of information that track where and what you've done on a particular Web site (see www.colorado.edu/infs/jcb/sinewave/technology/cookies/definition.html).
Domain name server (DNS)	A server that translates an IP number into a host name.
File transfer protocol (FTP)	A method to allow file transfer between computers on the Internet.
Firewalls	Devices used to control Internet traffic into a site.
Graphics interchange format (.gif)	One of the standard formats for images on the Web.
Hotlinks (hypertext)	Highlighted, active text, or images that point to another Web page.
Hypertext markup language (HTML)	The language used to write a Web page.
Hypertext transfer protocol (HTTP)	Standard protocol that defines how the information is transferred from server to client.
Integrated services digital network (ISDN)	A relatively high speed telephone connection to the Internet (~ 128 kbytes/sec).
Internet service provider (ISP)	A company that provides a service for a fee to connect you to the Internet.
Internet protocol (IP) number	The unique number that identifies a computer on Internet.
Javascript	A software language that make Web pages more dynamic.
Java applets	Programs written in Java language, which are downloaded by your browser. These make Web pages more interactive by allowing a program to run within your browser.
Joint photographic experts group (.jpeg)	One of the standard formats for images on the Web.
Modem (MOdulator, DEModulator)	A device that connects your computer to a telephone-line, and allows it to talk to other computers.
Neural network	An algorithm made up of simple learning processes that are fed large amounts of data from which it is able to learn and infer rules governing the data.

10: Web-based bioinformatic tools for Arabidopsis *researchers*

Glossary *Continued*

RTF (rich text format)	A general text format for saving word processed files that saves most formatting.
Spider	A software robot that explores the net collecting web page addresses and page contents.
Text only (.txt)	Text files (also called ASCII files) with a restricted set of characters that are readable by humans and can be understood by most computers.
Uniform resource locator (URL)	An unique address for information on the Web.
Web browser (browser)	Software that enables you to interact with Web pages.
Web servers (Servers)	A computer that provides Web services and pages to Internet users.
World Wide Web (Web)	An Internet protocol that makes use of HTML and HTTP to provide pages with links to other pages.

A1

List of suppliers

Agar, 66a Cambridge Road, Stansted, Essex, CM24 8DA, UK.
Affymetrix, 3380 Central Expressway, Santa Clara, CA 95051, USA.
Amersham
Amersham International plc, Amersham Place, Little Chalfont, Buckinghamshire HP7 9NA, UK.
Amersham International plc, Lincoln Place, Green End, Aylesbury, Buckinghamshire HP20 2TP, UK.
Amersham Corporation, 2636 South Clearbrook Drive, Arlington Heights, IL 60005, USA.
Anachem Ltd., 20 Charles Street, Luton, Bedfordshire LU2 0EB, UK.
Anderman
Anderman and Co. Ltd., 145 London Road, Kingston-Upon-Thames, Surrey KT17 7NH, UK.
***Arabidopsis* Biological Resource Center (ABRC)**, Ohio State University, 1735 Neil Avenue, Columbus, Ohio, OH 43210, USA.
Aracon Beta Developments, G. Callierlaan 72B-9000, Gent, Belgium.
Bando Chemical Industry Ltd., Kobe, Japan.
BDH Laboratory Supplies, Poole, Dorset BH15 1TD, UK.
Beckman Instruments
Beckman Instruments UK Ltd., Progress Road, Sands Industrial Estate, High Wycombe, Buckinghamshire HP12 4JL, UK.
Beckman Instruments Inc., PO Box 3100, 2500 Harbor Boulevard, Fullerton, CA 92634, USA.
Becton Dickinson
Becton Dickinson and Co., Between Towns Road, Cowley, Oxford OX4 3LY, UK.
Becton Dickinson and Co., 2 Bridgewater Lane, Lincoln Park, NJ 07035, USA.
Berthold Instruments
Bibby Sterilin Ltd., Stone, Staffordshire ST15 0SA, UK.
Bio
Bio 101 Inc., c/o Stratech Scientific Ltd., 61–63 Dudley Street, Luton, Bedfordshire LU2 0HP, UK.
Bio 101 Inc., PO Box 2284, La Jolla, CA 92038–2284, USA.

List of suppliers

Bio, Pan Britannica Industries Ltd., Britannica House, Waltham Cross, Hertfordshire EN8 7DY, UK.

Bio/Gene Ltd., 6, The Business Centre, Harvard Way, Kimbolton, Cambridgeshire PE18 0NJ, UK.

Bio-Rad Laboratories

Bio-Rad Laboratories Ltd., Bio-Rad House, Maylands Avenue, Hemel Hempstead HP2 7TD, UK.

Bio-Rad Laboratories, Division Headquarters, 3300 Regatta Boulevard, Richmond, CA 94804, USA.

Boehringer Mannheim

Boehringer Mannheim UK (Diagnostics and Biochemicals) Ltd., Bell Lane, Lewes, East Sussex BN17 1LG, UK.

Boehringer Mannheim Corporation, Biochemical Products, 9115 Hague Road, PO Box 504, Indianapolis, IN 46250–0414, USA.

Boehringer Mannheim Biochemica, GmbH, Sandhofer Str. 116, Postfach 310120, D-6800 Ma 31, Germany.

British Drug Houses (BDH) Ltd., Poole, Dorset, UK.

Calbiochem

Calbiochem, PO Box 12087, San Diego, CA 92119–4180, USA.

Calbiochem-Novabiochem (UK) Ltd., 3 Heathcoat Building, Highfields Science Park, University Boulevard, Nottingham NG7 2QJ, UK.

Clontech Laboratories UK Ltd., Unit 2, Intec 2, Wade Road, Basingstoke, Hampshire RG24 8NE, UK.

Courtaulds, Avonmouth Way, Avonmouth, Bristol BS11 9DZ, UK.

CPRO-DLO, BMF/Marketing, PO Box 16, 6700 AA Wageningen, The Netherlands.

Difco Laboratories

Difco Laboratories Ltd., PO Box 14B, Central Avenue, West Molesey, Surrey KT8 2SE, UK.

Difco Laboratories, PO Box 331058, Detroit, MI 48232–7058, USA.

DiverseyLever, Weston Favell Centre, Northampton NN3 8PD, UK.

Dow Elanco Ltd., Latchmore Court, Brand Street, Hitchin, Hertfordshire SG5 1HZ, UK.

Du Pont

Dupont (UK) Ltd. (Industrial Products Division), Wedgwood Way, Stevenage, Hertfordshire SG1 4Q, UK.

Du Pont Co. (Biotechnology Systems Division), PO Box 80024, Wilmington, DE 19880–002, USA.

Endecotts Ltd., Lombard Road, London SW19 3TZ, UK.

European Collection of Animal Cell Culture, Division of Biologics, PHLS Centre for Applied Microbiology and Research, Porton Down, Salisbury, Wiltshire SP4 0JG, UK.

Falcon (Falcon is a registered trademark of Becton Dickinson and Co.)

List of suppliers

Fargro Ltd., Toddington Lane, Littlehampton, West Sussex BN17 7PP, UK.
Fisher Scientific Co., 711 Forbest Avenue, Pittsburgh, PA 15219–4785, USA.
Flow Laboratories, Woodcock Hill, Harefield Road, Rickmansworth, Hertfordshire WD3 1PQ, UK.
Fluka
Fluka-Chemie AG, CH-9470, Buchs, Switzerland.
Fluka Chemicals Ltd., The Old Brickyard, New Road, Gillingham, Dorset SP8 4JL, UK.
FMC, Flowgen, Lynn Lane, Shenstone, Lichfield, Staffordshire WS14 0EE, UK.
GenLab Ltd., Tan House Lane, Widnes, Cheshire WA8 0SR, UK.
GenoTechnology, 3047 Bartold Avenue, Maplewood, MO 63143, USA.
Gerhardt Pharmaceuticals Ltd., Thornton House, Surbiton KT6 5AR, UK.
Gibco BRL
Gibco BRL (Life Technologies Ltd.), 3 Fountain Drive, Inchinnan Park, Paisley, PA4 9RF, UK.
Gibco BRL (Life Technologies Ltd.), Trident House, Renfrew Road, Paisley PA3 4EF, UK.
Gibco BRL (Life Technologies Inc.), 3175 Staler Road, Grand Island, NY 14072–0068, USA.
Arnold R. Horwell, 73 Maygrove Road, West Hampstead, London NW6 2BP, UK.
H. Smith Plastics Ltd., The Mayphil, Battlesbridge, Wickford, Essex SS11 7RJ, UK.
Hybaid
Hybaid Ltd., 111–113 Waldegrave Road, Teddington, Middlesex TW11 8LL, UK.
Hybaid, National Labnet Corporation, PO Box 841, Woodbridge, NJ 07095, USA.
HyClone Laboratories, 1725 South HyClone Road, Logan, UT 84321, USA.
Imperial, Imperial Laboratories (Europe) Ltd., West Parkway, Andover, Hampshire SP10 3LF, UK.
International Biotechnologies Inc., 25 Science Park, New Haven, Connecticut 06535, USA.
Invitrogen Corporation
Invitrogen Corporation, 3985 B Sorrenton Valley Building, San Diego, CA 92121, USA.
Invitrogen Corporation, c/o British Biotechnology Products Ltd., 4–10 The Quadrant, Barton Lane, Abingdon, Oxon OX14 3YS, UK.
J. Arthur Bower's, Firth Road, Lincoln LN6 7AH, UK.
Joseph Bentley, Barrow on Humber, South Humberside DN19 7AQ, UK.
Kenro Ltd., The Oppenheimer Centre, Greenbridge Road, Swindon SN3 4LH, UK.
Kodak: Eastman Fine Chemicals, 343 State Street, Rochester, NY, USA.

List of suppliers

Koppert UK Ltd., 1 Wadhurst Business Park, Faircrouch Lane, Wadhurst, East Sussex TN5 6PT, UK.
Lantor (UK) Ltd., St. Helens Road, Bolton BL3 3PR, UK.
Lehle Seeds, 1102 South Industrial Blvd., Suite D, Round Rock, TX 78681, USA.
Levington Horticulture Ltd., Paper Mill Lane, Bramford, Ipswich, Suffolk IP8 4BZ, UK.
Life Technologies Inc., 8451 Helgerman Court, Gaithersburg, MN 20877, USA.
LIP Equipment and Services Ltd., Dockfield Road, Shipley, West Yorkshire BD17 7SJ, UK.
Merck
Merck Industries Inc., 5 Skyline Drive, Nawthorne, NY 10532, USA.
Merck, Frankfurter Strasse, 250, Postfach 4119, D-64293, Germany.
Merck Ltd., Merck House, Poole, Dorset, BH15 1TD, UK.
Millipore
Millipore (UK) Ltd., The Boulevard, Blackmoor Lane, Watford, Hertfordshire WD1 8YW, UK.
Millipore Corp./Biosearch, PO Box 255, 80 Ashby Road, Bedford, MA 01730, USA.
Nalgene, Rochester, NY 14602–0365, USA.
New England Biolabs (NBL)
New England Biolabs (NBL), 32 Tozer Road, Beverley, MA 01915–5510, USA.
New England Biolabs (NBL), c/o CP Labs Ltd., PO Box 22, Bishops Stortford, Hertfordshire CM23 3DH, UK.
Nikon Corporation, Fuji Building, 2-3 Marunouchi 3-chome, Chiyoda-ku, Tokyo, Japan.
Northern Media, Phillip Harris Scientific, Unit 39, Nottingham South Industrial Estate, Wilford, Nottingham NG11 7EP, UK.
Nottingham *Arabidopsis* Stock Centre (NASC), Plant Science Division, School of Biological Sciences, The University of Nottingham, University Park, Nottingham NG7 2RD, UK.
Novo Enzyme products, 4 St. Georges Yard, Castle Street, Farnham, Surrey GU9 7LL, UK.
Perkin-Elmer
Perkin-Elmer Ltd., Maxwell Road, Beaconsfield, Buckinghamshire HP9 1QA, UK.
Perkin Elmer Ltd., Post Office Lane, Beaconsfield, Buckinghamshire HP9 1QA, UK.
Perkin Elmer-Cetus (The Perkin-Elmer Corporation), 761 Main Avenue, Norwalk, CT 0689, USA.
Perkin Elmer Applied Biosystems, Kelvin Close, Birchwood Science Park North, Warrington, Cheshire WA3 7PB, UK.

List of suppliers

Pharmacia Biotech Europe, Procordia EuroCentre, Rue de la Fuse-e 62, B-1130 Brussels, Belgium.
Pharmacia Biosystems
Pharmacia Biosystems Ltd. (Biotechnology Division), Davy Avenue, Knowlhill, Milton Keynes MK5 8PH, UK.
Pharmacia LKB Biotechnology AB, Björngatan 30, S-75182 Uppsala, Sweden.
Promega
Promega Ltd., Delta House, Enterprise Road, Chilworth Research Centre, Southampton, UK.
Promega Corporation, 2800 Woods Hollow Road, Madison, WI 53711–5399, USA.
Qiagen
Qiagen Inc., c/o Hybaid, 111–113 Waldegrave Road, Teddington, Middlesex TW11 8LL, UK.
Qiagen Inc., 9259 Eton Avenue, Chatsworth, CA 91311, USA.
Research Genetics Inc., 2130 Memorial Parkway, Huntsville, Alabama 35801, USA.
Roche, Boehringer Mannheim UK, (Roche Diagnostics Ltd.), Bell Lane, Lewes, East Sussex BN7, UK.
Sarstedt, Sarstedt Ltd., 68 Boston Road, Beaumont Leys, Leicester LE4 1AW, UK.
Sanyo Gallenkamp plc, Park House, Meridian East, Meridian Business Park, Leicester LE3 2UZ, UK.
Schleicher and Schuell
Schleicher and Schuell Inc., Keene, NH 03431A, USA.
Schleicher and Schuell Inc., D-3354 Dassel, Germany.
Schleicher and Schuell Inc., c/o Andermann and Co. Ltd.
Seikagaku Corporation, 1-5 Nihonbashi-honco 2-chome, Chuo-ku, Tokoyo 103, Japan.
Shandon Scientific Ltd., Chadwick Road, Astmoor, Runcorn, Cheshire WA7 1PR, UK.
Sigma Chemical Company
Sigma Chemical Company (UK), Fancy Road, Poole, Dorset BH17 7NH, UK.
Sigma Chemical Company, 3050 Spruce Street, PO Box 14508, St. Louis, MO 63178–9916, USA.
Silvaperl, William Sinclair Horticulture Ltd., Firth Road, Lincoln LN6 7AH, UK.
Sorvall DuPont Company, Biotechnology Division, PO Box 80022, Wilmington, DE 19880–0022, USA.
Stratagene
Stratagene Ltd., Unit 140, Cambridge Innovation Centre, Milton Road, Cambridge CB4 4FG, UK.
Stratagene Inc., 11011 North Torrey Pines Road, La Jolla, CA 92037, USA.

List of suppliers

Swiss Silk Blotting Mfg Co Ltd.,
United States Biochemical, PO Box 22400, Cleveland, OH 44122, USA.
Vector Laboratories, 30 Ingold Road, Burlingame, CA 94010, USA.
Wellcome Reagents, Langley Court, Beckenham, Kent BR3 3BS, UK.
Whatman, Whatman International Ltd., Maidstone, Kent, UK.

Index

ABRC 4,29
Activator (Ac) element 9, 144–147
Agrobacterium tumefaciens 8, 127–131, 158, 189–190, 193, 223
Agrobacterium tumefaciens library 49
alkaline hydrolysis 98
alkaline lysis 40
amplified fragment length polymorphism (AFLP) 62, 68, 82, 176–177
ARMS 62, 81, 174
autogamous 1

bacterial artificial chromosome (BAC) 30, 202, 207
BAC DNA isolation 190–191
BAC filters 39
BAC fingerprinting 211–212
BAC library 38, 207–211
BAC screening 213–215
BAC subcloning 191–193
bioinformatics 225–6, 260
bioinformatics glossary 264–265
biochemical markers 68
biochemical mutants 7
β-glucuronidase 134, 137, 147
BLAST 246–250, 255
BLOCKS 256
Bradford reagent 136
browser tips 232–5
browser errors 232–233

calcofluor 102
callose staining 86
cDNA 30, 42
cell ablation 126
chromosome landing 177
chromosome walking 171
cleaved amplified polymorphic DNAs (CAPS) 62, 67, 174, 240
clone integrity 43
clone verification 43
complementation analysis 81
colour mutants 7
cookies 235
cosmids 32, 64
cosmid library 216–220
cosmid screening 220
crossing 18, 79
CTAB DNA isolation 45, 57, 58
cytogenetics 105

DAPI staining 106, 120
databases 226
daylength 16
DEPC treatment 89
developmental stages 2
DIG-UTP 97, 100
diseases 19, 20
Dissociation (DS) element 9, 144–147
DNA isolation 45, 54, 56–58, 173
dTph1 transposon 157

ecotypes 6, 8
electron microscopy 87
email 229, 230
emasculation 79
emotifs 257
EMS mutagenesis 12, 13
Enhancer (EN) element 9, 144, 148, 150, 157
enhancer trap 9, 156, 160
environment 82
esterase activity 84
expressed sequence tag (EST) 30, 31, 62, 221, 223, 237, 243, 252, 253

FASTA 250–252, 255
Fetch 263
fixation 82–85, 87, 93
flowering conditions 16
flower mutants 7
fluorochrome staining 111
fluoroscein diacetate (FDA) 84
form mutants 7
formamide 99
forward genetics 12
FTP 230, 263

gene clustering 244
GeneFinder 254
gene identification programs 254
gene information 237–239, 251
gene names 244
genetic mapping 51, 240
genetic maps 239
genome size 3
genomic DNA 45
genomic sequencing 221–3
Genscan 254
germination 84
GFP 9
Grail 254

Index

growth characteristics 1
growth conditions 15–17, 23, 55
growth regulators 24
guanidinium isothiocyanate 90
GUS 9
gusA 134–6

habitat 1
haematoxylin-iron alum 118
hormone mutants 8
hybridization 61, 62, 97, 99, 213, 214, 220
8-hydroxyquinoline 108, 109
hygromycin phosphotransferase 147

infection 15
Inhibitor (I) element 9, 48, 150, 157
Intergrated Services Digital Network (ISDN) 228
Internet Protocol (IP) 227
IPCR 153–156, 165, 186–8
IPTG 40
in situ hybridization 92, 112, 114

Joinmap 69, 70, 74, 81

Kosambi mapping function 80

lacz promoter 38
labelling 97
lambda markers 63
licensing 15
lifecycle 1, 3
light microscopy 82, 86
linkage 80, 81, 172
lipid staining 53
luciferase 134, 138

M1 generation 14
map-based cloning 171
map position 69, 78, 175, 222, 223
Mapmaker 69, 70, 172
mapping programmes 69, 81
meiosis analysis 86, 117, 118, 121
metaphase analysis 113
methylation 127
microsatellites 68
minipreps 173
mitotic analysis 108, 109
molecular complementation 189
molecular databases 226
Motif analysis 255, 257
multimarker lines 10, 78

multiple sequence alignments 256
mutagenesis 4, 11–14, 77

NASC 4, 5, 70
NASC RI mapping 71
NetPlantGene 253
netscape browser 229
newsgroup 230
nuclear DNA staining 83
northern analysis 91
nucleolus organizing regions (NORs) 105, 240

pP1 vectors 30, 38
P1 DNA isolation 40
PCR 14, 34, 44, 46, 66
PCR screening 37
pests 19, 20
phage libraries 33
photography 117
physical map 241, 242
physical mapping tools 201
plasmid rescue 163, 164, 184–186
pollen viability 93
polyethylene glycol 132
poly-L-lysine 95
pooled DNA 44, 47
PredictProtein server 258
promoter trapping 156, 160
pronase E 95, 96
PRINTS 257
Prosite 256
proteases 95
protein data bank 258
protein structure 258, 259
proteinase K 95
protoplasts 131, 133
pulsed field gel electrophoresis 182, 183

random prime labelling 115
RAPDs 62, 66, 176
reporter gene 9, 10, 125, 134
restriction endonuclease digestion 60
reverse genetics 14
RFLP markers 30, 51, 54, 60, 62–64, 81, 174, 176, 239, 240
RI lines 11, 52–54, 62, 67, 68, 239
RI maps 72, 73
RNase A 99
RNA blotting 91
RNA electrophoresis 91
RNA isolation 89, 90
RNA staining 87
root growth 55

scanning electron microscopy 88

274

Index

scarification 18
secondary metabolites 69
seed 1, 4
seed harvesting 18
seed moisture 22, 23
seed storage 21
self-fertile 1, 78
sequence analysis tools 245, 246, 257, 259
sequencing 242–244
sequencing contigs 245
slide preparation 95, 122
Southern blotting 61
sterile culture 24, 25
stock centres 238
stock data 48
SSLPs 68, 174, 239, 240
Structural Classification of Proteins 258
superpools 44
surface spreading 122
Supressor-Mutator (Spm) 144
Swisprot 246
synaptonemal complex 107, 108, 122

TAC library 49
tagging 143–145, 148, 149, 151–152
targeted tagging 149, 151–152
T-DNA 8, 12, 14, 44, 62, 143, 157–159, 161, 162, 166, 223
tissue culture 125

transformation 126
transient gene expression 131
transmission electron microscopy 87
transposon 9, 143, 151
trisomics 11
two-hybrid system 30, 42

vacuole staining 83
vector alignment search tool 258
vernalisation 18
vesicle staining 83

web searching 231
www 227, 228

YAC 30, 33, 178, 199, 200
YAC chromosomal DNA isolation (plugs) 34–36, 182, 183, 201, 202
Yeast DNA miniprep isolation 36, 37, 183, 184
YAC contigs 203–207
YAC endprobes 184, 185, 186–188
YAC libraries 38, 178, 200
YAC library maintenance 34, 179
YAC pools 37–38
YAC screening 179–181
YAC storage 34
YAC subcloning 217–220